SPITTLEBUGS

SPITTLEBUGS

Biology, Ecology, and Economic Impact

Vinton Thompson

COMSTOCK PUBLISHING ASSOCIATES
AN IMPRINT OF CORNELL UNIVERSITY PRESS ITHACA AND LONDON

Visit our website at cornellpress.cornell.edu.

First published 2026 by Cornell University Press

Librarians: A CIP catalog record for this book is available from the Library of Congress.

ISBN 9781501786099 (hardcover)
ISBN 9781501786105 (pdf)
ISBN 9781501786112 (epub)

GPSR EU contact: Sam Thornton, Mare Nostrum Group B.V., Doelen 72, 4831 GR Breda, NL, gpsr@mare-nostrum.co.uk.

To my entomological colleagues—your friendship, help, and support have made my work with spittlebugs and this book possible.

Contents

Preface

In January 1953, when I was five years old, my family relocated temporarily to Chicago from our home in rural New Jersey. My father, a cranberry grower, had to attend the Army Quartermaster Subsistence Technology School to continue his active service in the Army Reserve, which provided significant cash income. We lived for six months at 9604 Hoxie Street, on the city's far South Side. One day that spring, my mother walked me down the street, or I walked her, to the nearby East 97th Place cul-de-sac, where weedy plants had sprung up adjacent to the curb. I noticed some wet white stuff on one of the plants, and, probing it with my fingers, found, to my surprise, a little green bug inside. Bless my mother for encouraging, or at least not discouraging, my curiosity. At this remove, it is a small miracle I remember the incident at all, and some details are lost. Nevertheless, that improbable urban encounter was my introduction to spittlebugs.

Fast forward to fall 1969. After a sixteen-year absence, I was back in town to study for a PhD in Evolutionary Biology at the University of Chicago, just 15 km from that curb on East 97th Place. During the intervening years I had shown no special interest in insects and have no memory of having noticed or thought about spittlebugs at all. But I had developed a passion for biology, fueled and guided by my undergraduate mentors at Harvard (E. O. Wilson, Stephen Jay Gould, and Ernst Mayr), who all directed me to Chicago, where Richard Lewontin, Richard Levins, and Daniel Janzen, among others, had created a vibrant program in population biology.

I arrived in Chicago intending to work in population genetics. My advisor, Leigh Van Valen, handed me a stack of offprints of a hundred or so journal articles (this was in the days before PDFs). He had a knack for spotting interesting, offbeat pieces of research. While dutifully working through the pile, I found two papers that were of special interest. Olli Halkka (Plate 16a), a Finnish biologist, had devoted years of study to color polymorphism in the meadow spittlebug, *Philaenus spumarius*. Better yet, he had worked out the simple Mendelian inheritance of multiple color forms. This was an organism amenable to population genetics study. And, even better, it clearly occurred in North America as well as Europe, making it economically accessible to casual study by an impoverished graduate student. I asked Leigh to buy me an insect net. He did, and I commenced studies on *P. spumarius* color polymorphism in June 1970.

Working out of a Volkswagen Beetle, Ruth Moscovitch and I sampled *P. spumarius* populations across the United States and Canada. I compiled the data and, pre-internet, air-mailed a hard copy to Halkka, suggesting that he might find the color form distributions of interest. A few months later he mailed back a draft of a joint manuscript on color polymorphism in North America, with me as first author. In the meantime, stimulated by an article by Karl Lindstrom in the journal *Nature*, I concluded that one of the color forms I was looking at in *P. spumarius* represented warning coloration based on escape ability. I submitted the evidence for this hypothesis in a letter to *Nature*, and, most improbably, they published it in March 1973 (just shy of 20 years after Watson and Crick published a short paper on DNA in the same venue). The paper with Halkka appeared shortly afterwards. This was heady stuff for a fourth-year graduate student. I was hooked on spittlebugs, or at least on *P. spumarius*.

I finished my PhD thesis on fruit fly population genetics, left professional science for five years, and then returned to teach at Roosevelt University, still in Chicago. In the meantime, I had continued to conduct *P. spumarius* population surveys in my after-hours as a travelling factory inspector. At Roosevelt, I started to expand my interests to spittlebugs beyond *P. spumarius*. In the Field Museum library, I found a set of a rare British compendium, the *Biologia Centrali Americana*, which includes a volume devoted to spittlebugs and related insects (Fowler 1997). One full page of hand-colored images of Central American spittlebugs was revelatory—as spectacular an advertisement for warning coloration as one could imagine.

I had to see these colorful spittlebugs for myself. In 1987, Ruth and I booked a flight to Costa Rica and hitched a late-night bus ride to a *pensión* adjacent to the famous Monteverde Cloud Forest. In the morning our host, Bob Law, accustomed to accommodating itinerant biologists, politely inquired as to my specialty. When I said "spittlebugs," he said, "For years farmers here have been asking why we don't get scientists who study something really important, like spittlebugs." It turned out that grass-feeding spittlebugs had almost destroyed the local dairy industry some years earlier (see Chapter 7). Hooked again!

So was set my trajectory to broader work on spittlebugs. The last piece came from teaching. In the early 1990s, Roosevelt assigned me to teach an undergraduate ecology course, a bit outside my core fields of genetics and evolution. This forced me to study up on many things I did not know. One of those topics was nitrogen cycling in ecosystems. The textbook I had chosen referred to non-nodular ("associative") nitrogen fixation in the very sorts of grasses the Monteverde spittlebugs had attacked. This complemented observations I had already made of associations between spittlebugs like *P. spumarius* and nodular nitrogen-fixing plants like alfalfa. Taken together, these observations stimulated a series of publications on spittlebug preferences for nitrogen-fixing plants (see Chapter 3). My lifelong engagement with spittlebugs was fully off and running.

Note on Content and Structure

This book synthesizes my work and that of many others on a biologically fascinating, economically important, but little recognized group of insects. The goal is to present an overview of the biology, ecology, evolution, and agronomics of spittlebugs and provide a gateway to the vast literature on the group, much of it in obscure agricultural venues (the literature review encompasses works published through 2024). I have tried to craft an approachable narrative, accessible to entomologists, ecologists, and agronomists, as well as to nonspecialists who are simply looking for a good read on interesting bugs. Like all overviews, it reflects the interests and experience of the author. My work with spittlebugs has focused on color polymorphism, warning coloration, and the ecology and evolution of host plant associations. Not surprisingly, these subjects are covered fairly thoroughly, as is the role of spittlebugs as major pests. I have consciously downplayed taxonomy and morphology, including just enough of each to provide a framework for understanding the broader biology and evolution of spittlebugs. Others can take up overviews of those topics where I have left off.

I have tried to provide enough information to give readers an in-depth perspective but not so much as to preclude reading for pleasure. If I have succeeded, you will have a deeper understanding of spittlebugs and their role in the world. I hope you will come to look upon them fondly (perhaps excepting the pests) and spend a bit of time hereafter keeping an eye out for their characteristic "spittles." Probing spittles to reveal the surprise nymphs inside is a great way to introduce children (and others with curiosity) to one small, resolvable mystery of nature.

The structure of the book is simple. Chapter 1 gives an overview of spittlebug biology, how they live, where they are found, and how they fit into the diversity of life. Chapter 2 explores how they got that way—when they first evolved, the lineages we know from the fossil record, and the fit of modern forms into DNA-based trees of life. Chapters 3 and 4 take up interrelated topics central to the evolution of spittlebugs: xylem sap feeding and the construction and biological function of spittle masses. Chapter 5 examines adult defense mechanisms, including jumping, reflex bleeding, warning coloration, and camouflage. Chapters 6 and 7 take a close look at spittlebugs as the most important pests of sugarcane and pasture grasses in tropical and subtropical parts of the

Americas. Chapter 8 looks at spittlebug pests of rice, sorghum, maize, and millet in Africa and China, as well as spittlebug pests of legume forage crops in North America and a variety of trees worldwide. Chapter 9 recounts the still-unfolding drama of a spittlebug-borne bacterium that has destroyed the olive industry in the boot heel of Italy and threatens other areas and crops. Chapter 10 looks at spittlebugs in the context of climate change—where they are vulnerable, where they might spread, and challenges and opportunities we might anticipate in the study of spittlebugs.

Anyone intrepid enough to make it through all ten chapters will have a thorough overview of this important but neglected group. Many will choose, according to their individual interests, to sample a subset of chapters. Toward that end, I have tried to make each chapter as self-contained as possible, with cross-references to direct readers to related materials in other chapters. If it accomplishes nothing else, I hope this book will give a broad audience a deeper understanding of spittlebugs and stimulate future study of their fascinating biology.

SPITTLEBUGS

1

WHAT IS A SPITTLEBUG?

This chapter presents an overview of spittlebugs, which includes their life cycle, the names by which they are known, their basic structure, their primary functions of eating, survival, and reproduction, and their distribution around the world. It also introduces the three characteristics that make spittlebugs particularly unique and fascinating: spittle formation, xylem sap feeding, and propulsive jumping. These will be discussed in detail in later chapters, but as you read, you may note that the importance of xylem feeding and its consequences, from spittle formation to host selection, form a recurring theme from this chapter through the last.

A Young Linnaeus Chances Upon Spittlebugs in Lapland

In 1732, at age thirty-four, years before he achieved world fame as the founder of modern biological classification, Carl Linnaeus undertook a botanical excursion to Swedish Lapland. His diary of the journey records that one day,

> even though I went about busily, there was nothing in particular to see apart from a sort of spittle on the grass everywhere. The common people call it frog-spit, and, when you wipe it off, a small insect is left behind; flesh-colored, like a small grasshopper though still immature. (Svanberg 2016)

Likewise, if you have taken nature walks in late spring or early summer in temperate parts of the world, you have likely come upon plants with little white frothy masses, looking like somebody spit on the plant (Plate 1). These spittle masses, often conspicuous and abundant on meadow plants or trees and shrubs, are the product of immature sapsucking insects called, appropriately enough, spittlebugs. Venture a bit further, wiping away the froth, and you can find one or more of these little bugs inside each mass (Plates 1a,b,d, 2d,e, 3a, 7e, 8b). Linnaeus himself did this, and upon coming back an hour later found the spittle mass reconstituted. He speculated that the foam protects the delicate bug from the heat of the sun, a hypothesis that has been recently validated (Chapter 4). He also left us a charming, simple sketch of a spittle and a baby bug revealed in a spittle (Linné 1913) (Fig. 1.1).

17. Fast än man flitigt gick omkring, war ingenting serdeles att observera, mehr än dett, att allestädz på gräset satt liksom spott, hkn gemene man kallar grodspot, hkn då dett strykes af, sitter allenast ett litet insect, liffärgat, likt en liten gräshoppa,[3] doch ännu omogen, till ett klart tekn, att doch af samma en sådan blifwer, fast hon ännu eij lärt giöra caprioler. Jag strök af samma wätska och lät creaturet sittia. En tima där äffter, då jag kom dit, satt åter en lika upblåst fradga, till tekn att den blifwer af samma fradga, hkn insectet blåser upp, och altså conserverar sig ifrån starka solens wärme, hwar af älliest hennes fina hull snart skulle uptorkas.

FIGURE 1.1. Excerpt from a 1913 printed edition of Linnaeus's handwritten 1732 Lapland expedition field notes, with reproductions of sketches embedded in the notes (Linné 1913). At the lower right are his crude renditions of a spittle (left) and a spittlebug nymph revealed within a spittle (right). The accompanying paragraph is his description in Swedish of the spittlebug he encountered, identified in a numbered footnote as "*Cicada spumaria*," the meadow spittlebug, now *Philaenus spumarius*. See text for an English translation of this description.

The resemblance of spittle masses to spit is superficial. They are the product of feeding by the immature spittlebugs, called nymphs, which create the foamy white masses from excreted plant juices. Unlike spit, the froth comes from the rear, not from the mouth. While the foam renders the bugs themselves nearly invisible, spittles of late-stage nymphs are usually conspicuous and easy to see. In north temperate climates they are most often encountered in May or June. In tropical and subtropical areas they may occur year-round but are often associated with the rainy season or the period shortly before the commencement of the rainy season. These spittles, which are unique in the insect world, are discussed in detail in Chapter 4. Chapter 1 introduces spittlebugs, with background on their basic biology and their broader place in nature. Subsequent chapters cover spittlebug evolution, ecology, and their importance as pests.

The Spittlebug Life Cycle: Egg to Nymph to Nymph to Nymph to Nymph to Nymph to Adult

Perhaps the earliest scientific reference to the relationship between spittles and nymphs came from the British naturalist and physician Thomas Moffet. In the late 1500s he observed that from

> a whitish worm in frothy dew that in May sticks to plants, a certain winged green creature is bred, in form like to the smallest kinde of Catterpillar, first it leaps, and afterwards it flies, and therefore I think fit to call it Locustella, a little Locust. The English call that frothy matter Woodsear, as if you would say the putrefaction of the wood. The Germans call it Cuckow-spittle. (Hardy 1862; Weaver and King 1954)

Due to problems finding a publisher, Moffet's work appeared decades posthumously. He got the taxonomy wrong. Spittlebugs are neither worms nor caterpillars (the larvae of moths and butterflies), nor locusts (by which he meant grasshoppers), but he was clearly describing spittlebugs.

The first detailed, scientific description of spittlebug natural history and behavior came from Linnaeus's countryman Charles De Geer (1720–1778), an entomologist, baron, parliamentarian, and, at the time, the wealthiest person in Sweden (Egerton 2008). De Geer grew up in Holland and arrived in Sweden at age nineteen to take possession of extensive inherited properties. Despite an absence of scientific publications, the Swedish Academy of Sciences soon elected him a member, evidently in the hope of generous donations, which did not materialize. Nevertheless, with the help of the most expensive English microscope on the market, he embarked on a very productive scientific career (Landin 2023). Then, as now, wealth had its privileges; De Geer put those privileges to good use. He produced a masterful eight-volume study of insects in French—Volume 3 of which includes seventeen full pages plus twenty-one numbered figures on the biology of *Philaenus spumarius* (Fig. 1.2), the meadow spittlebug, including elucidation of its life cycle (De Geer 1773).

Spittle-making nymphs are just one part of the spittlebug life cycle. Spittlebugs start life as eggs, which, depending on the species, may be deposited in soil; on dead leaf blades and stems; or in or on the living tissue of herbs, shrubs,

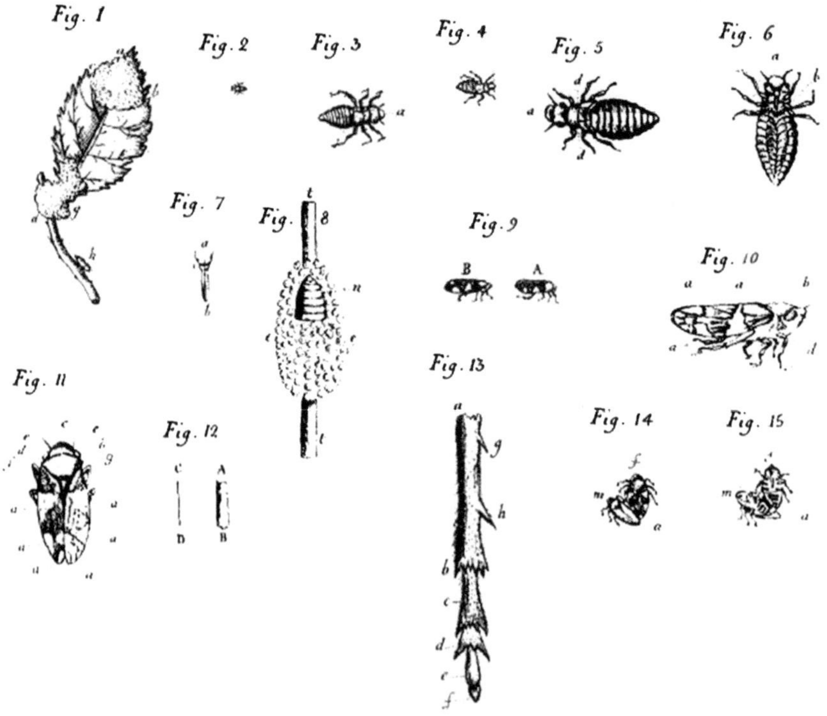

FIGURE 1.2. Excerpt from Plate 11 of Volume 3 of Charles De Geer's 1773 treatise on insects illustrating the life history and anatomy of *Philaenus spumarius*. In De Geer's numbering, Figure 1 illustrates two spittles on a leaf, one on the petiole and one at the tip, with a small nymph descending the stem below. Figures 2 and 3 show a young nymph, probably first instar, in natural size and magnified. Figures 4 and 5 show a fully mature nymph (fifth instar with wing pads) in natural size and magnified. Figure 6 illustrates the same fifth instar nymph in ventral perspective. Figure 7 illustrates the beak (rostrum) of the same fifth instar nymph at higher magnification. Figure 8 shows the abdomen of a nymph revealed in a spittle. Figure 9 illustrates adults of different color patterns in lateral perspective. Figures 10 and 11 show an adult in lateral and dorsal perspectives. Figure 12 shows the width of the thin antenna (left) compared to the width of a hair (right). Figure 13 illustrates the detailed anatomy of the tibial and tarsal portions of the hind leg. Figures 14 and 15 illustrate a copulating pair in dorsal and ventral perspective.

and trees (Fig. 1.3). The eggs hatch as tiny, first stage nymphs, typically about 1 mm long (Plate 2a, 2b). To live, they must almost immediately find a feeding spot on a plant, tap into one of the plant's xylem vessels with their beak, and begin pumping sap through their digestive system. All spittlebugs live on xylem sap, the part of the plant's plumbing that brings water from the roots to the

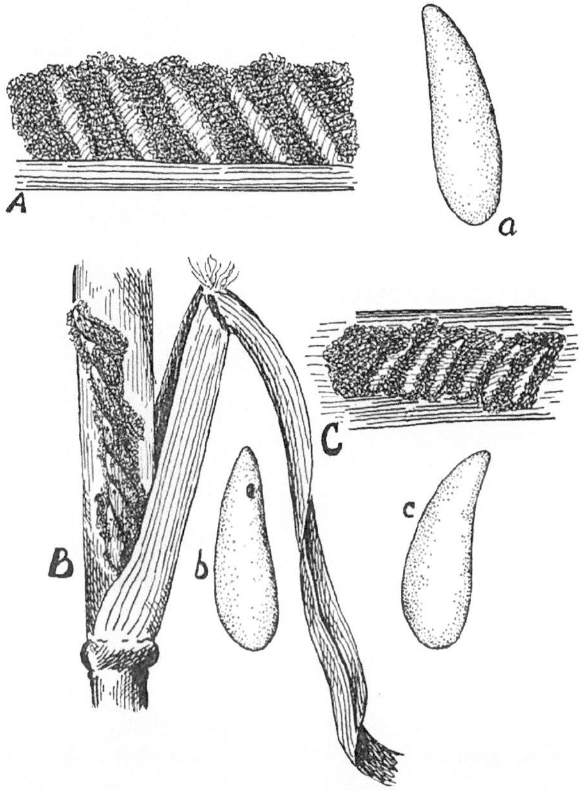

FIGURE 1.3. Deposited egg masses of three eastern North American spittlebugs. A) *Philaenarcys* sp., B) *Neophilaenus lineatus* (Plate 11d), C) *Philaenus spumarius* (Plate 10a–c), and their respective eggs (a, b, c). The eggs were deposited between stems and leaf sheaths of the grass *Cenchrus americanus*, side by side in groups of two to twenty-four, embedded in a white frothy material that hardened to be tough and inelastic (from Barber and Ellis 1922). Also see *P. spumarius* eggs in Fig. 8.2.

leaves. Xylem feeding has profound consequences for spittlebug biology and is covered in detail in Chapter 3. For the moment, note that xylem sap is extremely dilute and spittlebugs, even tiny ones, have to process a lot to make a living. The resulting voluminous excretion provides the starting point for the spittle. One bubble at a time, a new nymph builds its first tiny spittle, always inconspicuous, usually hidden in leaf axils or in the basal parts of plants near the soil, sometimes on roots at the soil surface or underground (Plate 1b).

Insects are divided into groups, such as bees, beetles, and butterflies, that form pupae and undergo a complete transformation between the larval and adult stage (*holometabolous* insects), and groups that transition more gradually (*hemimetabolous* insects). Spittlebugs are hemimetabolous. Growing nymphs undergo five molts, shedding their skins (exoskeletons) between stages. The first four molts each produce a larger and more developed nymph, leading, in all, to five nymphal stages (*instars*). Nymphal coloration varies from species to species and often from one instar to another (Plates 1d, 2, 4e, 7e, 8b–d). The nymphal period generally lasts four to eight weeks, varying by species and ambient conditions (Williams 1921; Garman 1923; Weaver and King 1954; Peck 2002; Schöbel and Carvalho 2021; Espitia B. et al. 2022). At least three temperate climate species (*Prosapia ignipectus*, *Cercopis sanguinolenta*, and *Makiptyelus dimorphus*) overwinter in the nymphal stage, extending the nymphal period by months (Xie 1994; Biedermann 2002; Thompson and Carvalho 2016). Interestingly, while the now-established pattern of five molts and five nymphal instars had been accurately described by the German entomologist August Johann Rösel von Rosenhof as early as 1749 (cited in Heidemann 1901), debate over the number of nymphal stages in some species lingered well into the twentieth century (Williams 1921; Weaver and King 1954). Later instars produce more spittle and are often more conspicuous, partly because nymphs tend to move to more prominently visible parts of plants as they develop. When you notice spittles, you are most often looking at fourth or fifth instar nymphs.

Fifth instar nymphs molt into adults, which differ from nymphs in three major ways: they do not make spittle, they have wings, and they are agile jumpers. Wings are the most conspicuous feature that distinguishes adults from nymphs. Depending on the group, the transformation to adulthood may take place inside or outside the spittle. If outside, the fifth instar nymph leaves its last spittle, walks a short distance down a twig or leaf, and then, while grasping the substrate, exits the shell of its last nymphal skin through a split in the dorsal side. This mirrors the emergence of adult cicadas. The exoskeletons of newly emerged (*teneral*) adults take a few hours to harden and assume final coloration. After that, adult morphology and coloration are fixed for life. In species that molt to adults inside the spittle, the late fifth instar nymph creates a drier final spittle with an interior air pocket and molts inside the pocket (Plate 1a, also see Chapter 4). Teneral adults remain in the air pocket until they harden and darken to final adult form, sometimes giving the false impression that adults live in spittle. Molting outside the spittle is characteristic of the genera *Clastoptera*, *Aphrophora*, and *Ptyelus* (Ball 1901; Heidemann 1901; Decary 1928; Doering 1931). Other spittlebugs of the families Cercopidae and Aphrophoridae seem to molt inside the spittle, but fifth instar molting behavior has been observed only in a small subset of species. Fifth instar molting in the Machaerotidae, a group without conventional spittle, is described in Chapter 4.

Adult spittlebugs, like nymphs, feed on the xylem sap of plants. Unlike nymphs, which are relatively sedentary and usually develop on just one or a few individual plants, adults crawl actively on and among plants, fly, and may shift hosts as the day or season progresses. In temperate climates, most spittlebugs have a single generation per year. At least two North American spittlebugs, the alder spittlebug *Clastoptera obtusa* and the pecan spittlebug *Clastoptera achatina* (Plate 5d), have two yearly generations in the southern parts of their range (Garman 1923; Tedders 1995). Tropical and subtropical spittlebugs often have multiple generations per year, limited only by the advent of local dry seasons, during which many spittlebugs do not reproduce. However, this is not universal. In China, the tropical/subtropical rice pest *Callitettix versicolor* (Fig. 8.3b, Plate 12a) has a single generation per year (Li et al. 2001).

Most North American spittlebugs lay eggs during the summer or fall and overwinter in the egg stage, but two species of the genus *Lepyronia* overwinter as adults and, as noted previously, *P. ignipectus* overwinters as a nymph. In tropical areas with intense dry seasons, many spittlebugs pass the dry period as dormant eggs. Oviposition sites vary from soil to dead plant debris to the tissue of living plants. The meadow spittlebug typically deposits eggs in clutches on plant litter, often on grass stubble (Fig. 1.3c). Many *Clastoptera* lay eggs in slits in the living twigs of shrubs and trees (Osborn 1916). Tropical pests of pasture grasses and sugarcane, whose nymphs usually live on plant roots, frequently deposit eggs directly into soil (Chapters 6 and 7). *Aphrophora permutata* and *Aphrophora saratogensis* nymphs that hatch in overstory pines fall into the surrounding understory—analogous to the life cycle of many cicadas—but instead of burrowing into the ground like cicadas, the nymphs develop on understory plants (Kelson 1964; Wilson 1987). In one apparently unique case, the Jamaican spittlebug *Leocomia cinchonae* lays eggs on the underside of leaves inside spittle masses created by its own nymphs (Pickles 1930).

What's in a Name?

Formally, spittlebugs are insects of the superfamily Cercopoidea in the order Hemiptera, which is divided into five families. Informally, they have a host of common names, with "spittlebug" (sometimes spelled "spittle bug") being the most frequently used in English, which is followed by "froghopper," a term based on a fancied resemblance to miniature frogs and the fact that they can jump. Occasionally, spittlebug has been used to refer solely to nymphs and froghopper to adults. This usage confuses the distinction between informal names for a taxonomic category and names for life stages. Note that the production of spittle is a unique feature of the group, whereas many other insects hop. The term spittlebug is a well-established usage for all life stages, with froghopper as a common alternative (Hamilton 1982; Dietrich 2002; Carvalho and Webb 2005; Soulier 2024). Both names are evocative and memorable.

In England, spittle masses are referred to as "cuckoo spittle" or "cuckoo spit," based on the false notion that they are the product of cuckoos, a usage echoed in the French *crachat de coucou* and the German *Kuckucksspucke*. Historically, by virtue of association with cuckoos, spittles were tarred as symbols of infidelity. Lavender, it was said, was "full of cuckow-spittles, to show that women's light thoughts make their husbands heavy heads" (Hardy 1862), which is a sly reference to the fact that cuckoos lay their eggs promiscuously in the nests of other birds.

In German, the common name for spittlebug is foam cicada (*Schaumzikada*), likewise in Chinese (*mò chán*), which is an appropriate term given that cicadas appear to be spittlebugs' closest relatives (Chapter 2). Foam appears in the Honduran name *espuma de sapo* (toad foam). Spit shows up in the common Latin American name *baba de culebra* (snake spit), as well as in the English names snakes' spit, frog spit, toad spit, and witches' spit. Spit is also mirrored in the German *Hexenspeichel. Salivaso* (or *salivazo*) and *salivita*, both meaning spittle, are widely used for spittlebugs in the Spanish-speaking New World, while *mion* (bedwetter) is common in Colombia and *mosca pinta* (spotted fly) in Mexico. In Venezuela, spittlebugs are called *candelilla*, a reference to the symptoms of their attacks on sugarcane (Chapter 6) and pasture grasses (Chapter 7). In Brazilian Portuguese, spittlebugs are *cigarrinhas*, a term that lumps them with related xylem-feeding leafhoppers and refers to the putative "little-cigar shape" of the adults. In Italy, spittlebugs take the delightful name *sputacchina*, the diminutive of *sputacchio* (spit), meaning a tiny spit (Chapter 9). Svanberg (2016), Weaver and King (1954), and Hardy (1862) provide useful overviews of common names, and Bentley and Rodríguez (2001) provide a list for Honduras.

While spittlebugs do not loom large in folk culture, there are some interesting folk beliefs and misconceptions surrounding them. Some of these are evident in the common names listed above, which often attribute spittle masses to spit from animals or supernatural beings. Spittles have also been attributed to the stars, to evaporation from the earth, and to perspiration by plants (Rösel von Rosenhof, cited in Heidemann 1901). In nineteenth-century Louisiana, spittlebug nymphs were believed to be the progenitors of greenhead horse flies (*Tabanus nigrovittatus*), a noxious biting pest (Webster 1892). Elder men of the indigenous Nuxalk (Bellacoola) people in Canada believed that spittle was good for sores (Smith 1925). In Kenya, smallholder farmers on the slopes of Mount Kilimanjaro see the appearance of spittles on trees as a predictor of an impending rainy season (Chengula and Nyambo 2016).

Spittlebug Body Parts: A Primer

Spittlebug nymphs and adults have three major body sections: the head, thorax, and abdomen (Fig. 1.4). The head has the feeding apparatus and a concentration of sensory organs. Most spittlebugs have a distinctive, swollen sucking pump on the face between the eyes (Figs. 1.5, 3.2, Plates 2c, 2f, 2g, 3b, 3d, 12b). This houses the muscles necessary to pump large amounts of xylem sap through the extended beak, or rostrum, as will be described in detail in Chapter 3. The head has a pair of eyes plus two mini-eyes (*ocelli*) that lie between the eyes on the top of the head, which is called the vertex. Cicadas and most dinosaur-age spittlebugs have three ocelli, which is one of the big anatomical differences between them and living spittlebugs (Chapter 2). There is a unique report based on a single female specimen of an African spittlebug genus with no ocelli (Linnavuori 1969).

The thorax holds the appendages responsible for movement, three pairs of legs and two pairs of wings, as well as the muscles required for their function. The legs

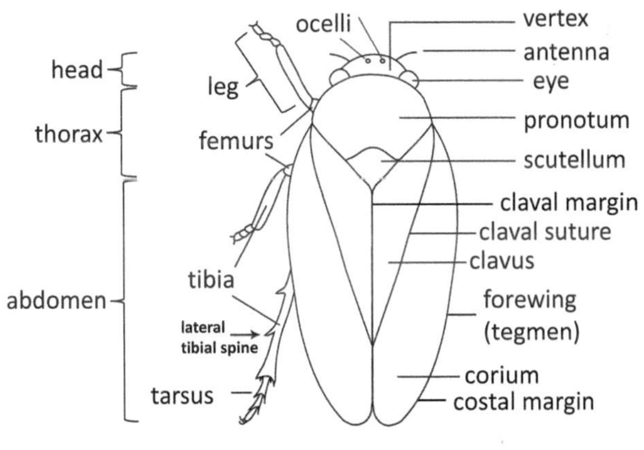

FIGURE 1.4. Simple schematic diagram of adult spittlebug body parts from a dorsal perspective. The model insect was a European species of the genus *Cercopis* (Cercopidae). Elke Freese (modified by Giancarlo Dessì), public domain, via Wikimedia Commons, https://commons.wikimedia.org/wiki/File:Morphology_of_froghoppers.svg. Modified to add body parts annotations.

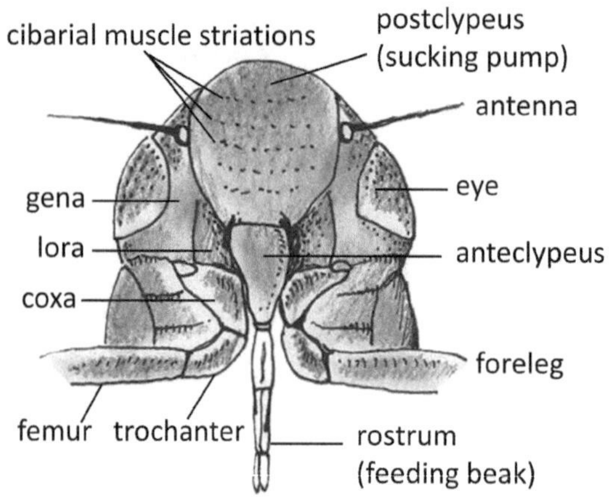

FIGURE 1.5. Details of spittlebug ventral head anatomy, the spittlebug's "face." Drawing by Halvard from Norway, public domain, via Wikimedia Commons, https://commons.wikimedia.org/wiki/File:Homo-frogghopper-sucing-mou.gif. Modified to add body part annotations.

allow both nymphal and adult spittlebugs to walk and adults to jump. For jumping, adult hind legs are specially equipped with characteristic circular rings of spines on the tibia and tarsi (Fig. 1.7). The hindwings (Figs. 1.9b,1.9d, 1.9f, 1.9j, 1.9n, 8.3b, 8.3c, 8.3f, 8.3h, 8.3j), used for flying, are hidden under the forewings when the bugs are not in flight. The forewings, called *tegmina* (sing. tegmen) or *elytra* (sing. elytron), serve as protective wing covers (Fig. 1.4). Along with the *pronotum* and *scutellum*, equivalent to the spittlebug's back, the tegmina form the visually prominent dorsal display side of spittlebugs, often marked by characteristic patterns and colors, which, depending on species, contribute to camouflage or warning displays (Chapter 5). The forewings also have a set of reinforcing veins that provide structural rigidity. Detailed forewing vein patterns serve as an important feature for distinguishing taxa within and among related groups. Nymphs are wingless, but from the third instar on have developing wing pads, the precursors to wings, which are most conspicuous in the fifth instar (Plate 2g).

The abdomen, which forms the largest part of the body, houses much of the digestive tract and organs called bacteriomes, which play host to symbiotic bacteria that have a critical role in nutrition (Chapter 3). In nymphs, the sides of the underbelly of the abdomen fold together to form a breathing tube that permits them to live in the semiaquatic environment of spittle masses (Chapter 4). In adults, the abdomen houses the reproductive organs, both internal and external. The characteristic shape of the tip of the female abdomen, usually elongated and pointed with a prominent median ovipositor, contrasts sharply with the tip of

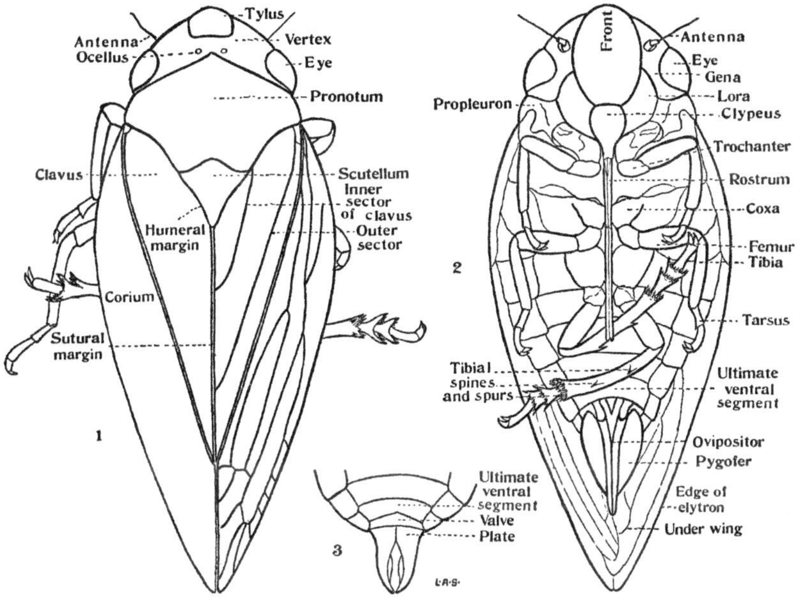

FIGURE 1.6. A more detailed schematic diagram of spittlebug anatomy, including the ventral portion and male and female external genitalia for an unidentified species of the family Aphrophoridae. Left: Dorsal view. Right: Ventral view of female. Center: Ventral view of tip of male abdomen. From Stearns (1923). Note that there are some inconsistencies in terminology with Figures 1.4 and 1.5, a reflection of both the early date of this illustration and lack of consensus terminology.

the male abdomen, which is usually shorter and blunter, often with a pair of prominent subgenital plates (Fig. 1.6). The plates, which are part of the male mating apparatus, vary immensely in size and shape, from short and rudimentary in the Family Clastopteridae to often prominent and sometimes elaborately sculpted in the Families Aphrophoridae and Cercopidae, where their shape and form is taxonomically important. The elongated, pointed female abdomen with ovipositor versus the short, blunt male abdomen, often with visible plates, is the simplest way to distinguish males and females (Fig. 1.6).

This section covers only the highlights of spittlebug adult anatomy. Readers looking for more depth and detail can consult Lallemand 1949, Kramer 1950, Evans 1966, Paladini et al. 2015, and Le Cesne et al. 2021. Be forewarned that there is no consensus on anatomical terms, particularly for details of the head, wing veins, and male genitalia, though for male genitalia, Le Cesne et al. (2021) have proposed a consensus terminology for the Family Cercopidae.

Where Do Spittlebugs Fit in the Diversity of Life?

Spittlebugs are sucking insects of the order Hemiptera. Here is how they fall in contemporary Linnaean classification:

Kingdom: Animalia
- Phylum: Arthropoda
  - Class: Insecta
    - Order: Hemiptera
      - Suborder: Auchenorrhyncha
        - Infraorder: Cicadomorpha
          - Superfamily: Cercopoidea
            - Family: Aphrophoridae
            - Family: Cercopidae
            - Family: Clastopteridae
            - Family: Epipygidae
            - Family: Machaerotidae

Insects of the order Hemiptera (sucking bugs) are characterized by a head with a beak specialized for piercing hosts and sucking fluid from their tissues. Hemiptera are divided into three big groups: Heteroptera ("true bugs"), Sternorrhyncha (aphids, scales, and their relatives), and Auchenorrhyncha, of which spittlebugs are a part. Hemiptera live by sucking juices from hosts. All Auchenorrhyncha and Sternorrhyncha are strict herbivores. They live solely on plants or fungi. Many Heteroptera live on plants, but some prey on animals, sucking blood or *hemolymph*, the insect equivalent of blood.

The Auchenorrhyncha have no established common name. "True hoppers" has been suggested (Bartlett et al. 2018) as a parallel to the "true bugs" (Heteroptera), but not all Auchenorrhyncha hop, and the term has not caught on widely. Bartlett et al. (2018) provide a nice overview of the group's diversity. It is divided into the Fulgoroidea (planthoppers), Membracidae (treehoppers), Cicadidae (cicadas), Cicadellidae (leafhoppers), and Cercopoidea (spittlebugs). The latter four groups together form the Cicadomorpha, or cicada-like insects. The name Auchenorrhyncha means throat-beaked. Auchenorrhyncha have a beak that appears to grow out of their throats, although in fact it grows out of the posterior ventral part of their heads (Fig. 1.5). In Heteroptera, the beak grows out of the front of the head. This will no doubt strike anthropocentric readers as normal, but it looks all wrong to those of us who devote our attentions to Auchenorrhyncha. Sternorrhyncha beaks are completely preposterous, appearing, as the name implies, to grow out of their chests.

The Cicadomorpha form a compact evolutionary group, a *clade*, clearly descended from a common ancestor (Chapter 2). Cicadas are distinguished by their large size and loud vocalizations, both of which contribute to their status as the closest thing we have to a charismatic megafauna among Auchenorrhyncha. They are also distinguished by their transparent forewings and by their universally subterranean nymphs. Some spittlebugs also have subterranean nymphs or transparent forewings.

Leafhoppers are the largest subgroup of Auchenorrhyncha, the one most likely to be confused, as adults, with spittlebugs. Many are similar in size and shape to spittlebugs. They can be distinguished from spittlebugs by the different arrangement of spines on their hind legs (Fig. 1.7), easily visible with a hand lens. Treehoppers are distinguished by relatively large and often extravagantly sculptured outgrowths of their *pronota* (sing. pronotum), the section of the dorsal thorax immediately behind the head (Fig. 1.4). Many have ant-tended nymphs, rewarding their tenders with drops of excreted sugar-rich phloem sap. Some species of one family of spittlebugs, the Machaerotidae, have dorsal processes that give them a superficial resemblance to treehoppers (Figs. 1.9h, 4.9e,

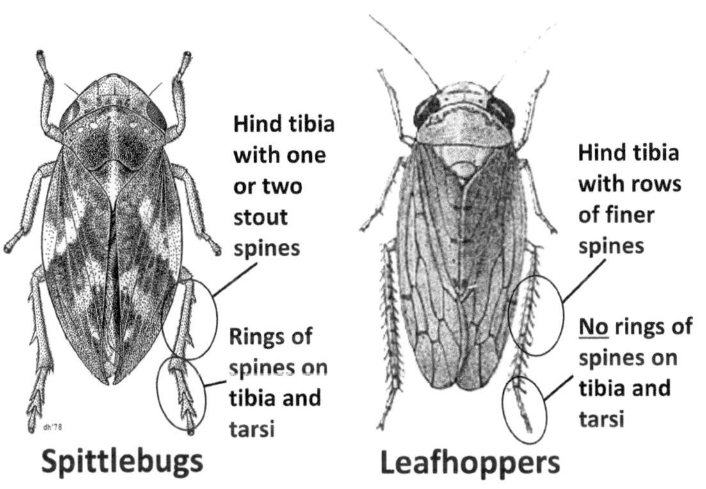

FIGURE 1.7. Key characters for distinguishing spittlebugs from leafhoppers, their closest lookalikes. Left: *Philaenus spumarius* (color form TYP, Fig. 5.1), image by Desmond W. Helmore via Wikimedia Commons reproduced under Creative Commons Attribution 4.0 International license https://creativecommons.org/licenses/by/4.0/deed.en. Right: *Bhatia olivacea*, image from Distant (1908). Note that while most spittlebugs have one or two lateral tibial spines on each hind leg, species of the machaerotid genera *Dianmachaerota*, *Tapinacaena*, and *Machaerota* (Figs. 1.9h, 4.9e, Plate 6a, 6d) have none and the genus *Sinophora* has four. (Maa 1963; Nie & Liang 2009).

8.3d, Plate 6a, 6d). Dietrich (2002) gives a good overview of the evolution of Cicadomorpha.

The taxonomy of the Superfamily Cercopoidea dates back to Linnaeus in the eighteenth century, who not only observed spittlebugs in nature, but also, in a later phase of his career, named several spittlebug species, as did his student, Fabricius. Major nineteenth and early- to mid-twentieth century contributors to spittlebug taxonomy include Amyot and Serville, Germar, Stål, Walker, Fowler, Distant, Berg, Jacobi, Schmit, Melichar, Lallemand, Synave, Doering, Nast, Maa, and Fennah (many refs. in Carvalho and Webb 2005). Late twentieth and early twenty-first century contributors include K. G. Andrew Hamilton, Ai-Ping Liang, Gervasio Carvalho, Mick Webb, Adeline Soulier, Andressa Paladini, Ulises Castro-Valderrama, and Elorde Crispolon. Together, with a host of biologists making smaller contributions (myself included), taxonomic specialists work to describe new species and hone the classification framework that guides spittlebug studies. The websites Cercopoidea Organized Online (COOL), created by Adeline Soulier of the Paris Natural History Museum (Soulier 2024), and the World Auchenorrhyncha Database, spearheaded by Dmitri Dmitriev of the University of Illinois (Dmitriev et al. 2024), have been particularly important tools for this effort, putting a wealth of spittlebug taxonomic information at people's fingertips.

Within Cercopoidea, five families are presently recognized. Two, Aphrophoridae and Cercopidae, account for a large majority of species. Aphrophoridae are found on all continents except Antarctica but are concentrated in north temperate zones, including Europe and the cooler portions of Asia, where they are dominant. Usually relatively slender, with elongated eyes (Figs. 1.6, 1.7, Plate 13), their adults resemble the adults of many leafhoppers (Fig. 1.7). Most are camouflaged or inconspicuous (for clear instances see Plates 11a, 11d, 11f, 13a, 13d, 13f, 13h) but some are flamboyant (Plates 8a, 11b, 13c). The name Aphrophoridae derives from the genus *Aphrophora*, from the Greek *áfros* and *fero*, which means froth-bearing (Buckton 1890). This is a reference to Aphrodite, the Greek goddess of love and desire, born from white foam produced by the genitals of Uranus after his son Cronus lopped them off and cast them into the ocean.

Cercopidae, the other large family, are also found on all continents except Antarctica but are concentrated in the tropics and subtropics. They tend to have relatively wider bodies and their eyes are rounded, resembling those of cicadas (Fig. 1.4). Many adult Cercopidae have striking coloration (Plates 3c, 3d, 4c, 4d, 7c, 7d, 10a, 10d, 10e, 11e, 11g, 12, 14c–h, 15a–d, 15f, 15h), thought to be associated with chemical defenses (Chapter 5). Buckton (1890) suggests that the name Cercopidae derives from the Cercopes, Mediterranean islanders who, according to the Roman poet Ovid, were turned into monkeys as punishment for dishonesty and subterfuge. By Buckton's account, most Cercopidae have "a grotesque monkey-like face, with swollen monkey-like cheeks" that probably suggested the name.

Clastopteridae (Figs. 1.9o, 8.3i, Plate 5a, 5d, 5e), named for their "broken" apical forewing membranes (as in "iconoclast"), are confined to the New World, the Philippines, and Borneo. They are small, all under five and a half millimeters in length. Most are globose, proportionately wide compared to long. The Epipygidae (Fig. 1.9k, Plate 5f) are confined to the New World tropics and are aphrophorid-like, with the exception that their wings appear to insert directly behind their eyes (Hamilton 2001). DNA evidence suggests that they are a subset of the Aphrophoridae (Chapter 2). Machaerotidae (Figs. 1.9h, 4.9e, 4.9f, 8.3d, Plate 6a, 6d) are an oddball group distributed in Sub-Saharan Africa, South and Southeast Asia, and Australia (Maa 1963). Their nymphs live in calcareous tubes (Fig. 4.9a–d, Plate 6b, 6c) and make spittle only when first hatched and in association with molting (Chapter 4). Some Machaerotidae have spine-like dorsal growths of the scutellum, the part of the thorax immediately posterior to the pronotum, giving them, as noted above, a superficial resemblance to treehoppers. A few spittlebug groups do not neatly fit these divisions (Hamilton 2015), and Cercopoidea as a whole is ripe for reexamination of family groupings (discussed in Chapter 2).

Eat, Prey, Love: Finding Food, Evading Predators, and Reproducing

Like all animals, spittlebugs have to find sustenance and reproduce, avoiding, in the meantime, falling prey to other animals. These imperatives can be summarized in a slight play of words on the title of the best-selling 2006 memoir *Eat, Pray, Love* by Elizabeth Gilbert.

Eat

Spittlebugs spend most of their time feeding, to the point that some even feed while mating (illustrated in Liang 2020). Both nymphs and adults probe plant surfaces with their beaks, looking for a suitable place to insert their stylets to pierce xylem vessels. For newly hatched first instar nymphs, finding a feeding place is a life-or-death quest. Host suitability depends in part on finding the right part of a plant. Parts that are too woody or defended by hairs are effectively inaccessible for feeding (Hoffman and McEvoy 1985a, 1985b; McEvoy 1986). Once nymphs find a good source of xylem sap, they tend to stay in place for long periods of time and often feed on the same plant for the whole nymphal cycle. They do move about, sometimes from one host to another (Whittaker 1965; Cid-Muñoz et al. 2020), but quantitative data on movement are rare. Brooks and Whittaker (1999) report that between 25% and 45% of *Neophilaenus lineatus* nymphs changed position during five- to ten-day intervals. Host choice is a matter of finding good xylem sap sources. The tips of the beak's piercing stylets have elaborate sensory receptors that provide information on both the mechanical structures being probed and the chemical nature of the sap encountered (Wang et al. 2015; Ranieri et al. 2020).

The nature and range of spittlebug hosts are extraordinarily wide. They have been recorded on more than 2,500 host plants, and a single species, the meadow spittlebug *Philaenus spumarius*, has been recorded on 1,311 hosts, a world record for herbivorous insects (Thompson et al. 2023). Patterns of feeding and host plant use are discussed in detail in Chapter 3. Here, note that spittlebugs are drawn disproportionately to nitrogen-fixing and ectomycorrhizal plants, which apparently offer especially rewarding xylem sap based on root associations with microorganisms (Thompson 1994, 2022).

Prey

While feeding and seeking mates, spittlebugs must avoid becoming prey. For nymphs, spittle itself is the primary defense against predators. There is ample evidence, detailed in Chapter 4, that spittle protects nymphs, either by concealing their presence and exact location within spittle masses or by directly deterring would-be invertebrate predators that do not want to get caught in the goo. Adults, in part, ward off predation with protective coloration. Some species are camouflaged to blend into the background. Others have warning coloration advertising effective defenses, and some have individuals of multiple color forms that form predator-confusing arrays. Defenses based on coloration are discussed in detail in Chapter 5. They are all backstopped by the ability of spittlebugs to jump, suddenly catapulting away when molested.

When it comes to jumping, spittlebugs hold another world championship. For its size, the meadow spittlebug, *Philaenus spumarius*, jumps higher than any other animal (Burrows 2003). Its hind legs are connected to two huge muscles that occupy a large portion of the thorax and account for a full eleven percent of total body weight. Preceding a jump, these muscles contract, store large amounts of energy by flexing the hardened, reinforced body wall of the thorax and then, in less than a millisecond, release the energy to snap back the folded, spine-ringed rear legs, throwing the spittlebug into the air. The effect is like the release of a crossbow, hurdling the bug with startling speed out of reach of a predator homing in for the kill (Chapter 5).

This world-class jumping mechanism is a major evolutionary innovation central to spittlebug behavior and ecology. It is so important that spittlebugs make little use of their hind legs for ordinary walking. Instead, they mostly drag them behind, locked and ready to make themselves scarce in a dicey situation with predators. The conical rings of tibial and tarsal spines that distinguish spittlebugs from leafhoppers (Fig. 1.7) represent a special adaptation to jumping. They pierce plant surfaces at the moment of takeoff, providing the frictional contact necessary for successful launches (Goetzke et al. 2019). Combined, the lavish investment in jumping muscles, the peculiar leg dragging, and the group-defining circular spinal arrays on the hind legs are a testament to the evolutionary importance of jumping in spittlebugs. Nevertheless, despite their jumping prowess and other defenses, some adult spittlebugs do fall prey to a wide variety of predators (see Chapter 5).

Love

To reproduce, spittlebugs must meet and mate. They find one another through at least two routes. The first way is vibratory signaling. Unlike their noisy cicada cousins, spittlebugs do not communicate by sound vibrations in air, but they do call to one another by vibrating the plants on which they are perched (Avosani et al. 2023). These vibrations can be recorded like sound signals. In the meadow spittlebug, both males and females make calling signals that say, in essence, "here I am." Males can respond with a courtship signal saying, "let's mate," to which females in turn can reply with a response signal ("I'm ready") or a rejection signal ("get lost"). This permits potential partners to find one another on the plant and engage in courtship that may or may not result in mating (Avosani et al. 2020). Vibrational signals are found across all spittlebug families (Moore 1961; López et al. 2001; Tishechkin 2003), created by second abdominal segment tymbals that are analogous to the cicada sound organ (Davranoglou et al. 2020). Large abdominal muscles contract, causing distinct areas of cuticle to buckle inward, producing the vibrational equivalent of sound. Aphrophoridae and Cercopidae have ribbed tymbals. The tymbals of Clastopteridae and Machaerotidae have not been well characterized (Davranoglou et al. 2020).

In addition to calling and courtship signals, many species make vibrational copulatory, territorial, and distress signals (Tishechkin 2003, 2011). Made audible to humans through electronic translation to sound, these signals can be spectacular. I recall an evening at a campground in Little Orleans, Maryland, when a colleague, Rex Cocroft, played a recording of a spittlebug of the genus *Clastoptera*. It sounded like a roaring lion and set off the dogs in the neighboring campsite. Not surprisingly, given the importance of perceiving vibrations in everyday interactions, meadow spittlebugs have a large and elaborate array of vibration-sensing mechanoreceptors built into the first abdominal segment (Ehlers et al. 2022), evidently serving as their ears to a vibratory world.

While vibratory signaling is relatively well-studied, there is evidence that spittlebugs also communicate through pheromones, or chemical signals. Meadow spittlebug males are attracted to extracts from female heads but not male heads. Females are not attracted to extracts of either sex (Sevarika et al. 2022a). This suggests that females may have pheromones that attract males. Additional evidence comes in the form of exocrine glands in meadow spittlebug (Sevarika et al. 2022b) and the males of the rice spittlebug (Fig. 8.3b, Plate 12a) *Callitettix versicolor* (Liang 2020). *Callitettix versicolor* males rub females with a gland-filled area of the flattened male sucking pump preceding copulation. Likewise, males of the New Guinian spittlebug *Augustohahnia dilatitibia* have flattened forelegs, replete with what are presumed to be male pheromone-producing glands (Liang 2003), and females of the genus *Locris* (Plate 7d) have an enigmatic abdominal gland hypothesized to produce sex pheromones (Boulard and Boulard, 1979). Females of two tube spittlebugs, *Machaerota spangbergii* (Fig. 1.9h, Plate 6a) and *Makiptyelus dimorphus*, emit fluids or odors that attract males for mating (Xie 1994; Pramanik and Liang 2019). It also appears that pheromones mediate the aggregation of *C. versicolor* nymphs in spittles (Chen and Liang 2015). If spittlebugs do communicate through pheromones, they have no shortage of potential pheromone receptors. The bases of adult spittlebug antennae (Fig. 1.8) are filled with sensory pits and organs so diverse in structure that they are a useful aid in deciphering evolutionary relationships (Liang and Fletcher 2002; Hamilton 2016). It appears, therefore, that spittlebugs communicate both by feeling the vibes and smelling the perfume of conspecifics, although the relative paucity of antennal olfactory sensilla in Auchenorrhyncha compared to some other groups suggests that vibratory signals are the key auchenorrhynchan mode of communication (Ranieri et al. 2016).

When spittlebugs do mate, they often remain in copula for hours (Xie 1994)—male and female genitalia united with the couple in a characteristic V-configuration (Fig. 1.2), the male genitalia coming from underneath. During this time, the genitalia are tightly intertwined, bringing into full function the often

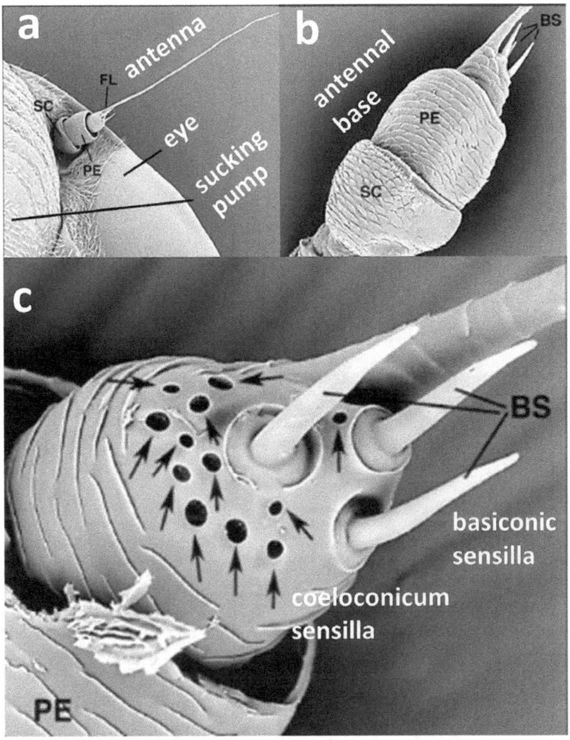

FIGURE 1.8. Scanning electron micrographs of the antenna of *Philaenus spumarius.* a) Antenna showing position on face between eye and cibarial pump. It consists of three parts: a long, mostly threadlike flagellum (FL) emerging from the pedicel (PE), which in turn is set at the base in the scape (SC). b) A closer view of the scape, pedicel, and the enlarged basal region of the flagellum bearing three elongated, conic basiconic sensilla (BS). c) Enlargement of the basal region of the flagellum showing the basiconic sensilla (BS) and pit-like coeloconicum sensilla (arrows). The sensilla are believed to be olfactory receptors, sensitive to volatile organic compounds like pheromones. From Germinara et al. (2017), reproduced, with some added annotations, under Creative Commons Attribution 4.0 International license https://creativecommons.org/licenses/by/4.0/deed.en.

elaborately sculpted *aedeagi* (penises), *parameres* (claspers), and subgenital Plates of the males. One of these matings, trapped in stone, has, in a sense, lasted for 156 million years, yet another world record (Chapter 2). Following copulation, females may deposit eggs shortly thereafter or following a delay of months, depending on the species.

How Many Kinds of Spittlebugs Are There and Where Are They Found?

There are about 2,500 described spittlebug species, encompassing a wide diversity of morphological types spread over the five families (Fig. 1.9). Described species are those that have been formally named and characterized in scientific literature. Table 1.1 shows the distribution of described species across families and the major biogeographic regions of the world. It is notably imbalanced. Over 90% fall in just two of the families, the Cercopidae and Aphrophoridae. About 75% occur in the tropics and subtropics, which is probably an underestimate given the disproportionate historical residence of taxonomists in temperate climates. This table has been assembled from multiple sources, including, among others, Metcalf (1962), Carvalho and Webb (2005), and Bartlett et al.

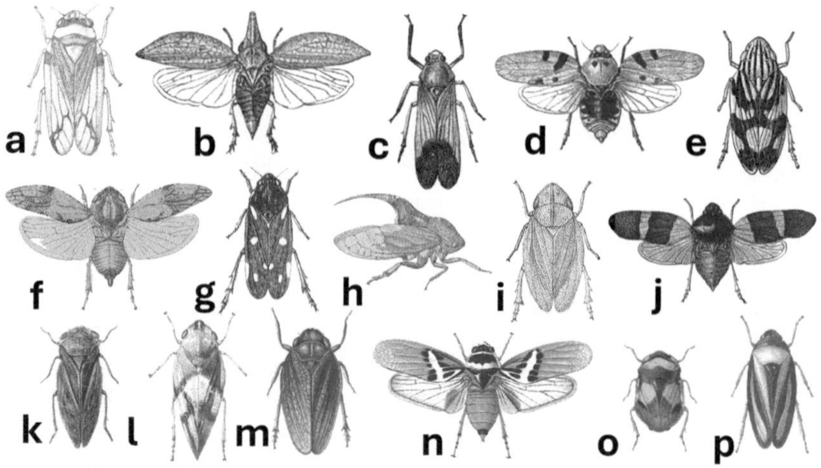

FIGURE 1.9. A diverse selection of spittlebugs illustrating variation in body form, wing venation, and markings. a) *Kanozata shillonganus* (Cercopidae), b) *Philagra fusiformis* (Aphrophoridae), c) *Callitettix braconoides* (Cercopidae), d) *Phymatostetha pudens* (Cercopidae), e) *Ptyelus sexvittatus* (Aphrophoridae), f) *Peuceptyelus trifasciatus* (Aphrophoridae), g) *Considia nitidula* (Cercopidae), h) *Machaerota spangbergii* (Machaerotidae, for color see Plate 6a), i) *Abdas nuncupatus* (Aphrophoridae), j) *Cosmoscarta heroina* (Cercopidae), k) *Eicissus decipiens* (Epipygidae), l) *Aphrophora permutata* (Aphrophoridae), m) *Poophilus adustus* (Aphrophoridae), n) *Locris areata* (Cercopidae), o) *Clastoptera compta* (Clastopteridae), and p) *Hemitriecphora haglundi* (Cercopidae). Images a, f, g, h, and i from Distant (1916). Images b, c, d, e, and j from Distant (1908). Images k and o from Fowler (1897). Images l, m, n, and p from Lallemand (1912). Note: not scaled to relative size.

TABLE 1.1. Approximate number of described spittlebug species, organized by family and geographic area

FAMILY	NEARCTIC	NEOTROPICS	PALEARCTIC	E, SE, AND S ASIA	AFRO-TROPICS	AUSTRALIA	OCEANIA	NOT FIXED	WORLD
Aphrophoridae	36	76	297	257	103	16	21	26	**832**
Cercopidae	9	473	93	694	181	13	4	35	**1,502**
Clastopteridae	37	53	0	3	0	0	0	0	**93**
Epipygidae	0	3	0	0	0	0	0	0	**3**
Machaerotidae	0	0	14	79	10	10	0	0	**113**
Total	**82**	**605**	**404**	**1,033**	**294**	**39**	**25**	**61**	**2,543**

(2018). The data have many imperfections, including some double counting of species that occur in multiple regions and the omission of some recently described species, but they accurately reflect the major taxonomic and geographical patterns of described species.

Described species are only a subset of all species. Hamilton (2001), for example includes only three described species in the newly created family Epipygidae, but he notes the existence of thirty-one total, meaning that twenty-eight of the species known to him were still undescribed. Likewise, there is ample evidence that the genus *Clastoptera* is represented by innumerable undescribed species in museum collections. Costa Rica, a small country but one of the best surveyed for insects, has about a dozen described *Clastoptera* species but at least three dozen known undescribed species in museum collections. Scaling up to the Neotropics as a whole, this suggests that several hundred *Clastoptera* species may await collection or description (Paladini et al. 2020). In general, spittlebug faunas are best known where taxonomists have been concentrated historically—areas like North America, Europe, and Japan—not in hotspots of spittlebug diversity such as Indonesia, the Philippines, and Brazil. As a result of these systemic biases and hit or miss collecting, we simply do not know how many spittlebug species exist.

There are, however, two studies that give us a hint of how many species exist. In one study, investigators set out to sample all the forest canopy Auchenorrhyncha at sites in Sulawesi, an Indonesian island in the hyperdiverse Malay Archipelago, and then determine how many were described (known and formally characterized) and how many were not. Half the spittlebugs were new to science (Hodkinson and Casson 1991). In a larger endeavor, Terry Erwin and his colleagues systematically fogged the tree canopy in hyperdiverse Amazonian rainforests with insecticides that caused insects to fall onto sheets below (Erwin 1983). I have a museum drawer-full of borrowed spittlebugs recovered from a series of these foggings in one locality in Ecuador in the 1990s. About three dozen spittlebug species were recovered. They are approximately evenly distributed among the families Aphrophoridae, Cercopidae, and Clastopteridae. About 60% are undescribed. Taken together, these results roughly suggest that only half the diversity of tropical spittlebugs has been described. Extrapolated to the world fauna and considering both the better-known state of temperate faunas and the severe undercounts of groups such as Epipygidae and Clastoptera, it appears that the total number of spittlebug species may be in the vicinity of 6,000, give or take a few hundred. This estimate tacitly assumes that morphological differentiation is more or less congruent with species differentiation. If cryptic species, morphologically indistinguishable but genetically and biologically differentiated, are common among spittlebugs, the estimate could be low.

The Meadow Spittlebug: A Recurring Species of Interest

It may have caught your attention that the meadow spittlebug, *Philaenus spumarius* (Figs. 1.2, 1.7, 8.1, Plate 11a–c), has popped up recurrently in this chapter. It is the most widespread spittlebug in the world and probably the one most often encountered, certainly in north temperate habitats. It is also the most widespread spittlebug pest in the United States and Canada, although no longer a particularly serious one (Chapter 8). It is, however, the most notorious spittlebug in Europe, where it is the only important vector of the plant pathogenic bacterium *Xylella fastidiosa*, which has killed swathes of olive trees across hundreds of thousands of hectares in the Salento (boot heel) area of Italy. This still-unfolding crisis, covered in detail in Chapter 9, has led to an unprecedented proliferation of work on *P. spumarius* biology, ecology, and behavior. As a result, *P. spumarius* has become the *Drosophila melanogaster* of the spittlebug world, the best studied exemplar and the primary organism of choice for spittlebug studies.

Looking Forward

To sum up this chapter, spittlebugs are a diverse insect group, perhaps six thousand strong, characterized by two key evolutionary innovations: the production of spittle by nymphs and the unique jumping mechanism of adults. Other overviews of spittlebug biology can be found in Peck and Thompson

2008, Bartlett et al. 2018, and the introductory sections of Hamilton 1982 and Carvalho and Webb 2005. The next chapters outline how spittlebugs evolved over a period of 200 million years and how xylem feeding has profoundly affected their ecology. Later chapters probe the nature and function of spittle and how interactions with predators have determined spittlebug coloration, both flamboyant and reserved. This sets the stage for a close examination of their importance as pests of sugarcane and tropical pasture grasses, as well as carriers of the deadly plant pathogen, *Xylella fastidiosa*. Along the way, you will meet Dr. Livingstone (of "I presume" fame) and make the acquaintance of world-class pee-ers and the world's earliest known insects in copula. Spoiler alert: They are spittlebugs.

2

TWO HUNDRED MILLION YEARS OF SPITTLEBUG EVOLUTION

This chapter presents the evolutionary history of spittlebugs based on fossils and DNA-based family trees. It gives an overview of spittlebug ancestry, their origin and diversification during the age of dinosaurs, subsequent extinction of their early forms, and the patterns of relationship among living forms.

Fossil Bugs in Fond Embrace

In November 2013, two Jurassic spittlebugs achieved improbable media immortality. A Chinese team, including one Chinese-American scientist, announced the discovery of a 165-million-year-old spittlebug mating pair, trapped in rock *in flagrante* (Li et al. 2013). Heralded as the oldest ever recorded insect copulation, the report caused a minor media sensation, complete not only with images of the couple joined face-to-face in the fossil impression (Fig. 2.1) but a full color reveal-all artist's 3-D ecological reconstruction of the living pair *in copula*. Demonstrating, as though it were necessary, that sex sells, this generated at least forty-one news articles around the world, not including China. It was a headline writer's dream. Here are a few of my favorites: "Insects Caught in the Act–in 165M-Year-Old Fossil," "Oldest Sex Fossil Shows Bugs Did it Missionary Style," and "Fossil Captures Ancient Insects Having Sex." Even the staid *New York Times* weighed in with "Case of Insect Interruptus Yields a Rare Fossil Find." In a tight contest, the prize goes to *The Register*, an online enterprise technology news publication that

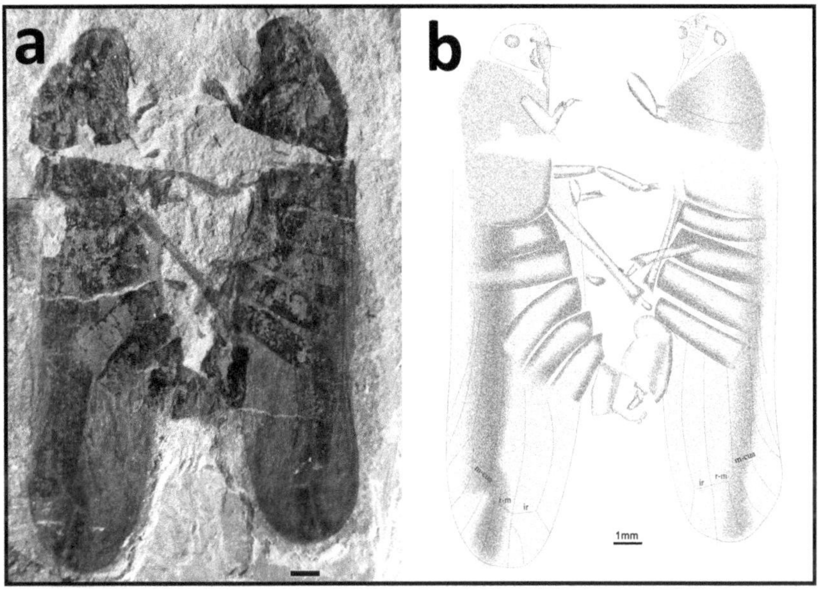

FIGURE 2.1. The world's oldest mating insects. a) Fossil impression of mating pair of the Jurassic spittlebug *Anthoscytina perpetua*, with the male on the right, (b) Interpretive line drawing of same. Scale bars = 1 mm. Reproduced from Li et al. (2013) under Creative Commons Attribution 4.0 International license https://creativecommons.org/licenses/by/4.0/deed.en, with thanks to Chungkun Shih for high-resolution images.

trumpeted "Horny Lovers FOSSILISED in Steamy RUMPY-PUMPY Session 156m [sic] yrs Ago."

Behind the hoopla was a very interesting fossil and a good backstory. The copulating spittlebugs, collected in the Daohugou locality in Inner Mongolia, China, were described as a new species, *Anthoscytina perpetua*, of the spittlebug family Procercopidae, a group known only from fossils. Fossils of *A. perpetua* are numerous, including about two hundred specimens with well-preserved genitalia (Li et al. 2013). The mating pair, apparently killed in a volcanic eruption, is the oldest example of only thirty-six known cases of copulating insects in the entire fossil record, most of the others preserved in amber.

The copulating spittlebugs were first discovered by Dr. Chungkun Shih, a Chinese-American scientist who, with local graduate students, was sorting through a couple of thousand fossil insects stored in the basement of the insect paleontology lab at Capital Normal University in Beijing. Shih, a retired chemical engineer now based in New Jersey, first encountered fossil insects in open-air Sunday fossil markets while working for international petroleum companies in Northeast China. Collecting started as a hobby, which led to a long-term relationship as a visiting professor with Capital Normal University, where, in the fine tradition of collections-associated institutions, he has been welcomed into the fold as an active collaborator with the team of Professor Dong Ren. He has visited two to four times annually and coauthored about 290 papers on fossil and amber insects since 2006. Such is retirement for those of us with an all-consuming interest in bugs.

The Nature of the Spittlebug Fossil Record

The mating spittlebugs are just one of the more spectacular examples of a long fossil record documenting the history of spittlebugs over evolutionary time. Fossils are the mineralized or otherwise preserved remains or imprints of long-dead organisms, and, over time, the only direct record of their existence. We cannot observe primordial spittlebugs, but we can learn a lot about their physical forms, their changes over geological time, and their ecological associations from the fossil record.

Insect fossils take two major forms: compression fossils and amber. Compression fossils form when the original organisms are buried in lake or ocean sediments and preserved in sedimentary rock. They are generally squished flat or nearly flat and revealed when layered sedimentary rocks are split. Because insects are small and relatively delicate, most of their fossil record comes from about forty Lagerstätte deposits, rock formations that were laid down in fine sediments that preserve exceptional detail, including, in some cases, impressions of soft parts (Grimaldi and Engel 2005). Amber inclusions preserve even more detail. They form when organisms are trapped in plant resin, which subsequently hardens and is buried, preserving insects and other small organisms in three dimensions in a clear plastic-like material. Amber deposits are rare but offer spectacularly well-preserved specimens that, in the best instances, can be studied morphologically like dead specimens of recently collected insects (Grimaldi and Engel 2005).

The fossil record is famously imperfect and incomplete. Rock formations with fossils are only formed in certain circumstances, which have occurred sporadically over Earth's history, and exposed formations are subject to destruction by erosion over time. As a result, we get only glimpses of the past, analogous to getting a page here, a paragraph there, often just a word or two or nothing, from a long series of mostly lost geological volumes of the history of life. We see glimpses from different parts of the world at different times, further complicating interpretation. For many parts of the world, for long stretches of time, we have no fossil record at all. For insects, these difficulties are exacerbated by their small size and relatively delicate constitution. It is one thing to fossilize a *Tyrannosaurus rex* femur or the hard shells of marine mollusks. It is quite another to fossilize small, soft-bodied insects. Despite these obstacles, insect fossils are abundant and reasonably well-dispersed over geological time (Labandeira 1997).

Most insect fossils are wing impressions. Wings are comparatively tough, flat structures. They stand up well to compression fossilization and, when conditions are right, the preservation of anatomical detail can be excellent. This is particularly true of spittlebug forewings, or *tegmina*, which have evolved as relatively hardened wing covers. Detailed analysis of wing vein patterns is mind-numbingly dull, and there is no consensus on nomenclature for wing vein details, even among scientists who study spittlebugs. But fossil wings can permit the assignment of specimens to family, genus, and even species. This lets paleontologists trace spittlebug taxa through evolutionary time. Wing impressions reveal nothing about nymphs, which are wingless, or about the evolution of other anatomical features.

Most of the history described here is based on compression fossils (Figs. 2.1, 2.2), largely wings, supplemented from the mid-Cretaceous on by spectacular amber inclusions that preserve spittlebugs as old as 100 million years as though they had been embedded in the last few months. The bulk of recent work on

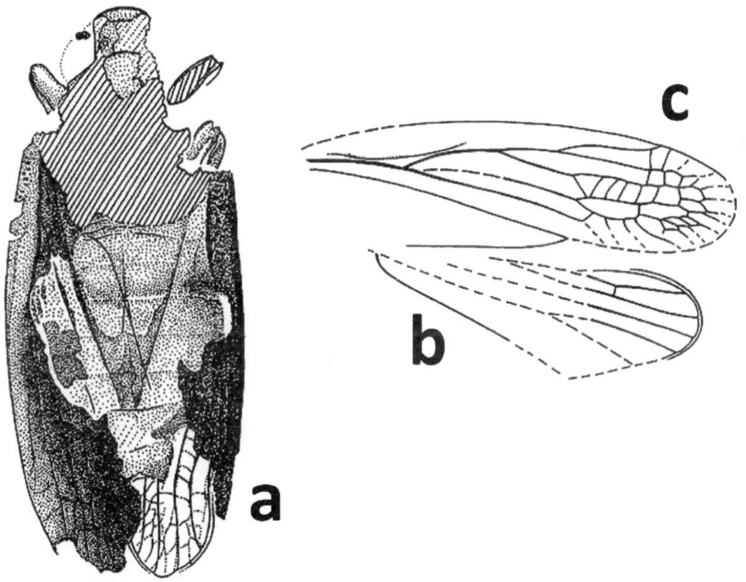

FIGURE 2.2. Interpretive drawings of *Cercopion reticulata*, the sole fossil specimen of the Cretaceous family Cercopionidae. a) Whole body, b) Reconstructed hindwing, c) Reconstructed forewing. Reproduced from Hamilton (1990) with permission of the American Museum of Natural History.

fossil spittlebugs has come from laboratories in China, building on substantial earlier work in Russia. For general background, Grimaldi and Engel 2005 provide an excellent and readable overview of insect evolution and the insect fossil record.

When and How Did Spittlebugs First Evolve?

Spittlebugs debut in the fossil record at the very beginning of the Jurassic Period in rocks from Kazakhstan, which are dated to about 200 million years ago (mya) (Chen et al. 2020). Like the famous fossil mating pair, these earliest fossils belong to the extinct family Procercopidae, which persisted for at least 99 million years, into the mid-Cretaceous Period. Their appearance in the early Jurassic suggests that spittlebugs evolved in the late Triassic, the period immediately preceding the Jurassic. Together, the Triassic, Jurassic, and Cretaceous periods make up the Mesozoic Era, or, more popularly, the age of dinosaurs.

The fossils trace a line descending from early Hemiptera that lived in the Permian Period, the last geologic period before the Mesozoic. The first Auchenorrhyncha date to the early Permian, and one Permian group, the Prosbolidae, is thought to be ancestral to the line leading to the immediate ancestors of spittlebugs (Shcherbakov 1996, 2000; Szwedo 2016). The Prosbolidae flourished during the Permian but did not survive the Permian extinction, the most intense extinction event in the history of life on Earth, which is estimated to have wiped out 90% of species then living. Before they went extinct, they gave rise to the Hylicellidae, which survived the Permian and apparently gave rise in the Triassic to cicadas, leafhoppers, and the earliest spittlebugs. Among the living Cicadomorpha, spittlebugs are morphologically closest to the Hylicellidae and, in that sense, are the most primitive living Cicadomorpha (Chen, Wang, Zheng et al. 2019).

Based on their modestly inflated sucking pumps and relatively long beaks, most of the Prosbolidae appear to have been phloem sap feeders on woody plants (Shcherbakov 1996), but some exhibit the robust sucking pumps characteristic of xylem feeders (Labandeira 2019). Shcherbakov (1996) illustrates inferred evolutionary trends in sucking pump size—smaller pumps in the phloem feeding Fulgoroidea and larger pumps in the xylem feeding ancestors of the Cicadomorpha (Fig. 2.3). The Hylicellidae had huge, swollen sucking pumps, indicating that they were xylem sap feeders (Shcherbakov 2002). This is strong evidence that the evolution of xylem feeding, a major feature of spittlebug biology (Chapter 3), preceded the evolution of spittlebugs themselves. Evidently, cicadas, spittlebugs, and leafhoppers had a xylem-feeding common ancestor among the Hylicellidae. Figure 2.4 presents a schematic

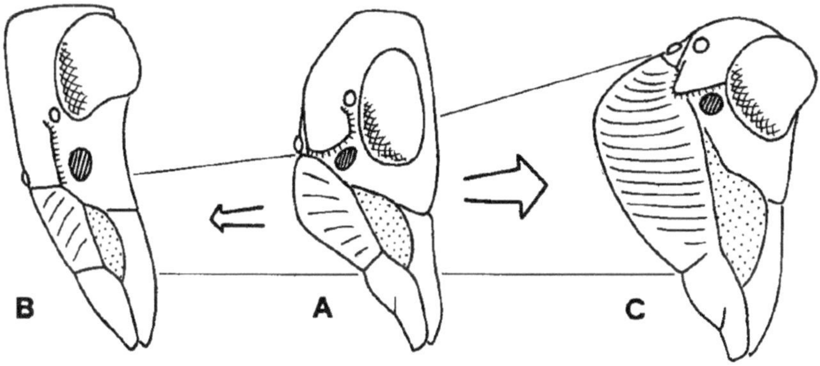

FIGURE 2.3. Hypothesized transformations of the ancestral auchenorrhynchan head (A) toward reduction of the sucking pump in phloem feeders (B) and inflation of the sucking pump in xylem feeders (C). The sucking pump (postclypeus) is striated, marking the attachment of sucking pump muscles. Reproduced from Shcherbakov (1996) with permission of the Entomological Society of America.

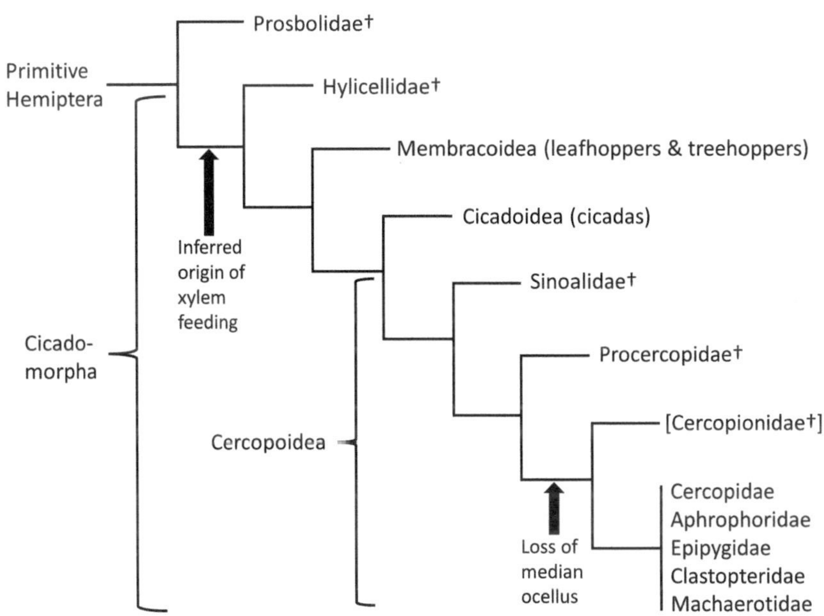

FIGURE 2.4. The ancestral line leading to present day spittlebugs as inferred from the fossil record. Prosbolidae gave rise in the late Permian to xylem-feeding Hylicellidae, which gave rise during the Triassic to the ancestral Membracoidea and the common predecessor of Cicadoidea and Cercopoidea. It appears that by the late Triassic the ancestral Cercopoidea had split into two lines, the Sinoalidae, which lasted to the mid-Cretaceous, and the Procercopidae, which are believed to have given rise to modern forms. Taxa marked with a † are extinct. Drawn after Shcherbakov (1996, fig. 2) and Wang et al. (2012, fig. 1).

illustration of the hypothesized course of spittlebug evolution from the ancestral Prosbolidae and Hylicellidae to the modern groups through Mesozoic intermediates.

The Mesozoic: Spittlebugs First Proliferated in the Age of Dinosaurs

The evolution of spittlebug predecessors sets the stage for Jurassic Bug Park, land of the copulating spittlebugs. Jurassic spittlebugs fall into two families, the Procercopidae and the Sinoalidae, which likely split in the late Triassic (Chen, Wang, Zheng et al. 2019), before the earliest spittlebug fossils. The Procercopidae (Fig. 2.5) first occur in the earliest Jurassic and appear to have evolved from a

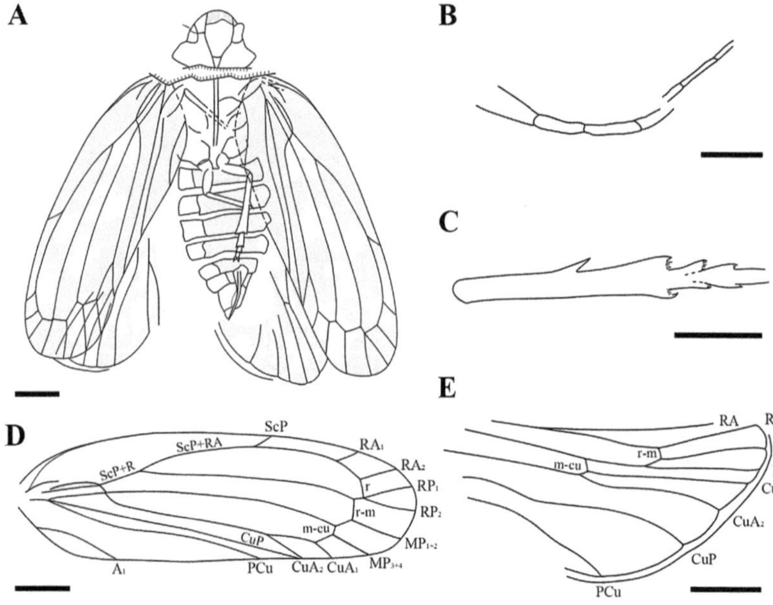

FIGURE 2.5. Drawings of compression fossils of *Anthoscytina daidaleos*, a species of Procercopidae from the Jurassic. A) Ventral view of whole body with notable mottled patterning on wings; B) Antenna, which, unlike in modern spittlebugs, is segmented; C) Hind tibia with single, sharp lateral spine, like some modern spittlebugs; D) Forewing showing detailed pattern of veins; E) Hind wing showing detailed pattern of veins. Scale bars = 2 mm in A, D, E; 1 mm in C; 0.2 mm in B. Reproduced from Fu et al. (2018) under Creative Commons Attribution 3.0 Unported License https://creativecommons.org/licenses/by/3.0/deed.en.

Hylicellidae line that gave rise to ancestral spittlebugs, cicadas, and leafhoppers. They spanned all of the Jurassic and a good part of the Cretaceous, diversifying into eight described genera and thirty-seven species (Chen et al. 2020). They are the most abundant Auchenorrhyncha fossils in both the Jurassic and Cretaceous of China, comprising about 40% of all specimens (Li et al. 2010), including the famous mating pair (Fig. 2.1). They were apparently the dominant Auchenorrhyncha in several Jurassic and early Cretaceous faunas and spanned the continental land mass known as Laurasia (Shcherbakov and Popov 2002). Two Procercopidae genera are represented by beautifully preserved specimens (Plate 3e, 3f) in mid-Cretaceous Burmese amber dating back about 99 million years (Fu et al. 2020).

The second major Mesozoic spittlebug group is the Sinoalidae (Fig. 2.6), which apparently diverged from the Procercopidae preceding the appearance of the first spittlebug fossils in the early Jurassic (Chen, Wang, Zheng et al. 2019; Chen, Wang, Zhang et al. 2019). The sinoalid fossil record extends for over 60 million years, from the mid-Jurassic to the mid-Cretaceous. They have a much narrower geographic distribution than the Procercopidae, having been recovered only from China and Myanmar, localities then located at the eastern end of Laurasia. What they lack in geographical span they make up in diversity, with twenty-nine species in nineteen genera (Chen et al. 2023), including fifteen well-preserved species in Burmese Amber (Plate 3g) (Chen, Wang, Zheng et al. 2019). Although abundant in both the Jurassic and Cretaceous, they did not survive the Cretaceous and left no descendants in the modern fauna. The Burmese amber insect fauna exhibits a peculiar combination of holdover primitive forms from the earlier, gymnosperm-dominated Mesozoic Era and more modern forms associated with a rising angiosperm flora (Labandeira 2019). Perhaps by luck, this relic fauna preserves the two oldest known spittlebug groups in exquisite morphological detail.

A third extinct Mesozoic group, the Cercopionidae (Hamilton 1990) (Fig. 2.2), likely evolved from the Procercopidae (Chen, Wang, Zheng et al. 2019). This family is known from a single compression fossil from the Brazilian Crato fossil beds, a formation from the middle of the Cretaceous, 113 mya. It appears to be closer morphologically to living Cercopoidea than any other Mesozoic spittlebugs, apparently having, for example, two ocelli rather than the cicada-like three ocelli of the Procercopidae and the Sinoalidae, and it is the only Mesozoic spittlebug specimen found outside what is now Eurasia. Its differentiation from modern forms at the family level depends on just two characters, the apparent location of the ocelli far back on the head and a single detail of hindwing venation. The fact that one of three Mesozoic spittlebug families and a whole branch

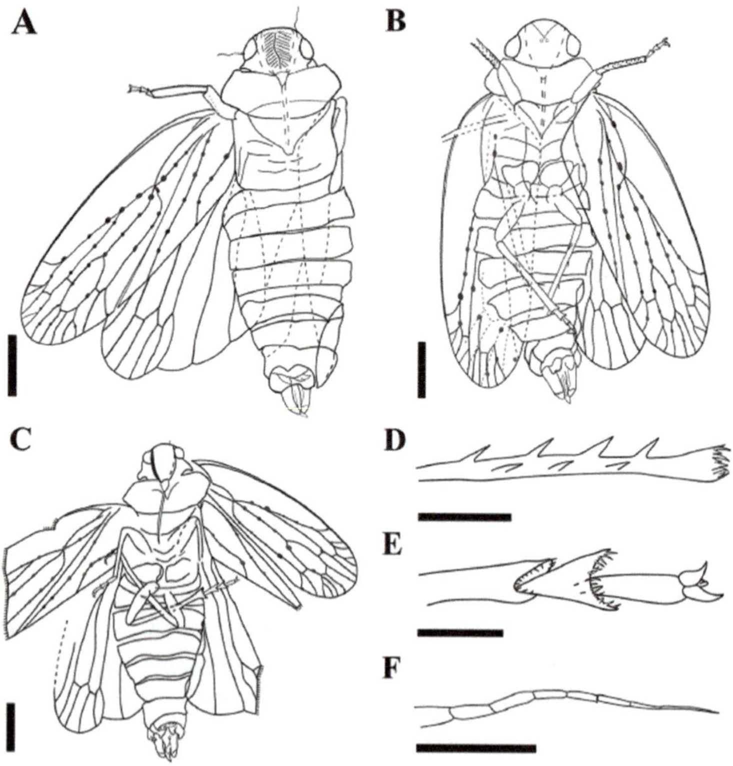

FIGURE 2.6. Drawings of compression fossils of *Stictocercopis wuhuaensis*, a species of Sinoalidae from the Jurassic. A, B, C. Ventral view of whole body of three different specimens with notable dotted patterning on wing veins; D. Hind tibia showing double row of spines, a feature that distinguishes Sinoalidae from all other spittlebugs; E. Hind tarsus. F. Antenna, which, unlike modern spittlebugs, is segmented. Scale bars: 2 mm in A, B, C; 1 mm in D; 0.5 mm in E, F. Reproduced from Fu and Huang (2018) under Creative Commons Attribution 4.0 International License https://creativecommons.org/licenses/by/4.0/deed.en.

of the Cercopoidea family tree hang on two details of a single highly imperfect specimen illustrates the often fragile nature of our understanding of the past, which is based on the exceedingly fragmentary fossil record. (As an aside, I have had the opportunity to examine this unique fossil, which presently resides in an office down the hall, awaiting repatriation to Brazil.)

It appears that the line leading to the living families of Cercopoidea evolved from the Procercopidae in the early to mid-Cretaceous (Shcherbakov 1996; Szwedo 2016). Suggested transitional forms include the early Cretaceous genera

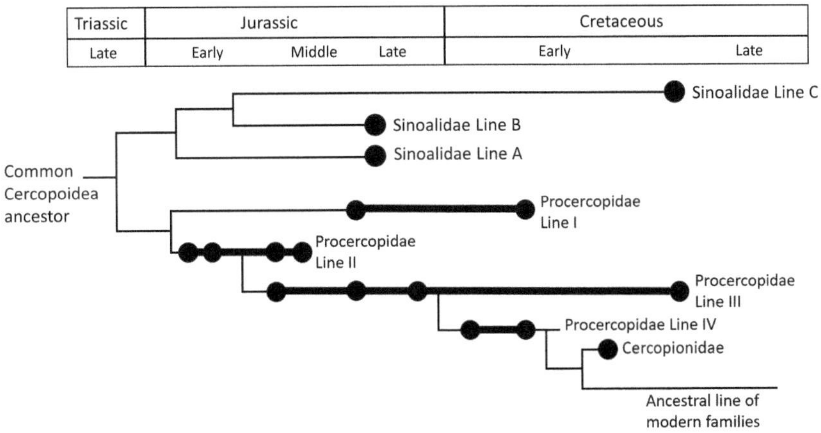

FIGURE 2.7. Schematic family tree of Mesozoic Cercopoidea as interpreted from the fossil record. Circles represent fossil occurrences. Thick lines represent inferred presence of a lineage over time between fossil occurrences. Thin lines represent speculative connections. This figure is redrawn from Chen et al. (2020, fig. 8) with the addition of more detailed information on the Sinoalidae from Chen, Wang, Zheng et al. (2019, figs. 11,12) and the incorporation of records of Procercopidae in Burmese amber (Fu et al. 2020). With the exception of Cercopionidae, each lineage encompasses multiple genera. Line designations I, II, III, and IV for the Procercopidae are taken from Chen et al. (2020, fig. 8), with Line III extended based on addition of one genus from Fu et al. (2020). Line designations A, B, C for the Sinoalidae are arbitrary.

Stellaris and *Sinocercopis* (Chen et al. 2020) that date back to about 125 mya. The Jurassic genus *Longucercopis*, dating from 175 to 190 mya, has also been suggested to be transitional (Zhang et al. 2022), but it seems to occur too early for such a role. The last recorded Procercopidae date to the late mid-Cretaceous, about 99 mya. This leaves a gap of 45 to 70 million years between the putative Procercopidae ancestors and unambiguous records of modern spittlebugs. Consequently, the fossil record is unclear as to how and when early Cercopoidea evolved into modern forms (Wang et al. 2012; Chen et al. 2020). Figure 2.7 illustrates the ranges of geological occurrence and hypothesized phylogenetic relationships among the three Mesozoic Cercopoidea families and the modern families. Both the Sinoalidae and Procercopidae appear to have spawned multiple phyletic lines, with one Procercopidae line apparently giving rise to the Cercopionidae and the ancestral line of modern spittlebugs.

The Fossil Record of Modern Spittlebugs

Modern spittlebugs are known only from the Cenozoic, the major geologic period following the mass extinction event that ended the Cretaceous. Confidently identified fossils of Cercopidae, Aphrophoridae, and Epipygidae date back to the Eocene, about 45 to 55 mya. The oldest definite Clastopteridae date to the Miocene, about 18 mya. The Machaerotidae have no fossil record (Szwedo 2016). The oldest Cercopidae species, *Eocercopidium maculata*, dates from about 55 mya in Scotland. The oldest Epipygidae come from Eocene Baltic amber, about 45 mya (Wang et al. 2012). The oldest Aphrophoridae date from early Eocene Fushun amber, about 50 mya (Szwedo et al. 2013). Notable Cercopidae fossils include vividly patterned wings from the Miocene period in Iceland (7.2–11.8 mya) (Wappler et al. 2014) and three well-preserved whole-body compression fossils of *Nangamosthethos tibetense* (Fig. 2.8) from the Tibetan Plateau (Eocene, about 39 mya) (Xu et al. 2022), also with patterned wings. Eighteen-million-year-old Dominican amber has yielded an extinct Clastopteridae genus with two species, each based on a single specimen (Poinar et al. 2014). This is notable because the present New World clastopterid fauna consists of just two genera, one represented by a single recently discovered species (Paladini et al. 2020). Bridging the ancient with the modern, four fossil specimens of the living Sub-Saharan African aphrophorid spittlebug *Ptyelus grossus* (Plate 8a) have been recovered from three-million-

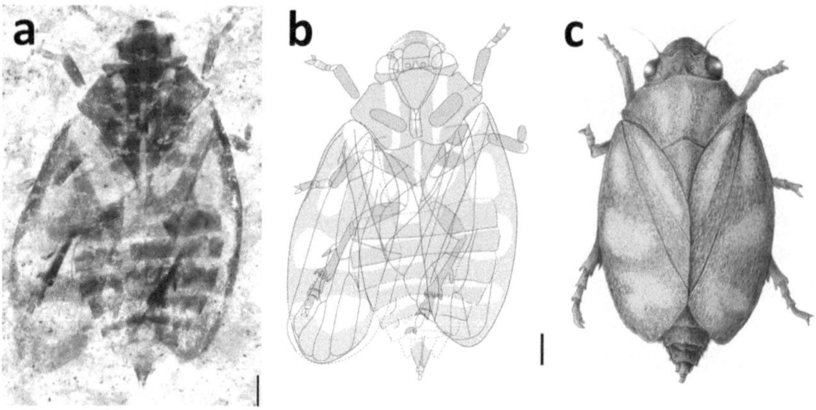

FIG. 2.8. The Eocene cercopid spittlebug *Nangamostethos tibetense*. a) Fossil impression, b) Interpretive drawing of same, c) Artist's reconstruction of living insect. Scale bars = 1 mm. Reproduced from Xu et al. (2022), with rearrangement of images, under Creative Commons Attribution 4.0 International license https://creativecommons.org/licenses/by/4.0/deed.en.

year-old Pliocene strata in Germany, establishing a preglacial distribution in a warmer Europe not yet isolated from lower parts of Africa by the Sahara (Wagner 1968).

The big pattern of spittlebug evolution is apparently unusual, with a proliferation of Mesozoic forms ending in extinction and all modern forms descending from a single surviving lineage that is, at the moment of writing, absent in the Mesozoic fossil record. It is parallel to an analogous pattern in ants (Sosiak et al. 2024), a much larger and ecologically very different group.

What Do the Fossils Tell Us About Early Spittlebug Ecology?

What were early spittlebugs eating? Based on the possession of swollen sucking pumps, the ancestral Hylicellidae, Procercopidae, and Sinoalidae are all inferred to be xylem sap feeders. Modern spittlebugs are all xylem feeders that feed on a broad range of hosts, running the gamut from ferns, herbs, and grasses to vines, shrubs, and trees—this breadth of hosts apparently permitted by xylem feeding itself (Chapter 3). Some of the presently favored host categories, like high photosynthesis C4 grasses (Thompson 2004) and nitrogen-fixing plants (Thompson 1994), did not exist until the end of the Cretaceous or later, ruling them out as Mesozoic hosts for known fossils.

Based on inferred ectomycorrhizal fungus associations, the most likely late Triassic, Jurassic, and early Cretaceous hosts are the Pinaceae (present-day pines, spruces, firs, and their relations) and early relatives of *Gnetum* (broad-leafed tropical gymnosperms) or lines ancestral to those taxa (Thompson 2022). The mid- to late-Jurassic Daohugou site, source of thousands of spittlebug fossils, including the copulating pair, has remains of at least four conifers, one of them, *Yanliaoa sinensis*, apparently quite abundant (Pott and Jian 2017), though all more closely related to Cupressaceae than Pinaceae, and therefore not clearly ectomycorrhizal. Fossil evidence of a *Gnetum* relative has been recovered from the early Cretaceous Yixian formation in Northeast China (Rydin and Friis 2010), a formation in which fossil spittlebugs are also diverse (Chen et al. 2020) and abundant (Li et al. 2010). There is inconclusive evidence that a Burmese amber sinoalid spittlebug may have fed on ferns (Poinar and Brown 2023), which are among the many hosts of the highly polyphagous modern spittlebug, *Philaenus spumarius* (Thompson et al. 2023). More information on the relationship between spittlebugs, C4 grasses, nitrogen-fixing plants, and ectomycorrhizal plants can be found in Chapter 3.

It has been suggested that the decline of ancient gymnosperms and explosive diversification of angiosperms (flowering plants) in the Cretaceous may have contributed to the extinction of the Procercopidae and Sinoalidae (Chen, Szwedo, Wang et al. 2018; Chen, Wang, Zheng et al. 2020; Chen, Wang, Zheng et al. 2019). This is unlikely. The Pinaceae expanded their footprint from the Late Cretaceous into the Cenozoic and are presently primary hosts for about 10% of all spittlebug species with known hosts (Thompson 2022). Pinaceae date back about as far as spittlebugs, at least to the beginning of the Jurassic, and were available hosts throughout the Cretaceous (Delcos et al. 2023; Thompson 2022), where they account for about 7.7% of all fossil wood samples (Peralta-Medina and Falcon-Lang 2012). Extant spittlebugs live on a diverse array of angiosperms and gymnosperms. In many species, angiosperms host nymphs and Pinaceae host adults (cf. Hamilton 1982; Cid-Muñoz et al. 2020; Thompson 2022). This makes it highly improbable that the extinct Mesozoic spittlebugs perished for lack of available gymnosperm hosts.

Although there are no published records of fossil spittlebug nymphs, several have been observed in Miocene amber from the Zhangpu formation in Southeastern China (Bo Wang, personal comm. 2023), dating from about 15 mya, when modern forms prevailed. Intriguingly, the Zhangpu amber originated from dipterocarp trees (Wang, Shi et al. 2021), a group suspected but not demonstrated to host spittlebugs, based on their ectomycorrhizal status (Thompson 2022). Since all known present day spittlebug nymphs make spittle at some point in their existence (Chapter 4), we can strongly infer that their common ancestors were also spittle producers, but we cannot determine where in deep evolutionary history the ability to make and live in spittle evolved. Well-preserved Jurassic or Cretaceous nymphs with invaginated ventral abdominal breathing tubes would provide strong evidence of spittle formation. A single nymph preserved in Late Cretaceous Canadian amber was at first interpreted to represent a new family, Jascopidae, intermediate between leafhoppers and spittlebugs, but the ventral abdomen was not included in the description (Hamilton 1971). Evans (1972) subsequently assigned Jascopidae to the leafhoppers. Shcherbakov and Popov (2002) suggest that the line leading to cicadas and spittlebugs had soil-dwelling nymphs like modern cicadas and some Cercopidae. However, there is no evidence for soil-dwelling nymphs in the more basal (early evolving) existing lines of spittlebugs, strongly suggesting that soil-dwelling in Cercopidae, where it occurs, evolved secondarily, independently of the cicadas (more on this in Chapter 3).

While it is impossible to know from fossil adults whether Mesozoic Cercopoidea made spittle, it is possible to make direct inferences about jumping. Cercopoidea evolved enlarged, reinforced, lateral forewing flanges to accommodate the force of the femur gliding over it during jumps (Wang et al. 2012). By this criterion, both the Procercopidae and Sinoalidae appear to have been jumpers like modern spittlebugs. Evidently, jumping is an ancient attribute of the Cercopoidea (more on jumping in Chapter 5).

DNA Phylogenies: An Alternative Route to Understanding Spittlebug Evolutionary History

Fossils are not the only way we can make inferences about the past course of evolution. Organisms leave traces not only in sedimentary rocks and amber but in the DNA they bequeath to their descendants. A tremendous amount of information about evolutionary history is encoded in the DNA sequences of present day species. Like the fossil record, this information is imperfect and subject to conflicting interpretations, but, by permitting detailed reconstruction of the evolutionary family trees of living groups, it provides a second, powerful window on the past. These phylogenies have been made possible by the advent of accurate, inexpensive DNA sequencing in the late twentieth century and the development of sophisticated computer programs that analyze the enormous amounts of data generated by sequencing. Together, these developments have revolutionized evolutionary biology, most especially the reconstruction of the history of life.

In spittlebugs, this permits us to pick up where the fossils leave off and make detailed inferences about the evolutionary relationships among living groups. DNA sequences consist of a four-letter code: A, T, G, and C. They determine the amino acid sequences of the proteins that underpin the structure and function of living organisms and also control the timing and cellular location of the expression of these proteins. Other portions of the DNA of most organisms, often the largest part, have no direct role in the biology of the organism. Nonetheless, all these sequences, which in spittlebugs typically include one to a few billion As, Ts, Gs, and Cs for a given species (Chen et al. 2022; Griffiths et al. 2024), are subject to mutation and change, substituting one letter for another at given points in the sequence. Over evolutionary time, independently in each lineage, these changes lead to slow and steady changes in DNA sequences. The longer species are separated in time, the greater the differences between their sequences. Making simple assumptions and applying a lot of computing power, biologists convert collections of DNA sequences into family trees and make educated guesses about how long ago the various branches on the tree separated.

We have already seen big picture family trees of spittlebugs and their close relatives inferred from the fossil record (Figs 2.4, 2.7). Those trees have a major shortcoming. They tell us nothing about how the modern families of spittlebugs diversified because the fossil record is simply too incomplete to answer that question. DNA phylogenies chart the evolutionary relationships among living lines. They also illuminate larger scale ancient patterns that, so far, cannot be determined from the fossil record, such as whether spittlebugs are most closely related to cicadas, leafhoppers, or neither. DNA approaches have a major limitation. They only permit comparison of groups with living representatives or extinct groups for which DNA is attainable from dead specimens. It is impossible, for example, to get DNA sequences for extinct lineages, such as the Sinoalidae, and impossible to know whether we have DNA sequences descended from the Hylicellidae or Procercopidae, putative ancestral groups without known living representatives.

What Do DNA Phylogenies Tell Us About the Family Tree of Living Spittlebugs?

A bit disconcertingly, three decades of sequencing and analysis have not led to consensus on major elements of the family trees of the Cercopoidea in particular or the Auchenorrhyncha as a whole. While many of the intermediate level taxonomic groupings, such as families, are clearly good *clades*, meaning evolutionary lines descended from a common ancestral species, this is not true of all named groups, and branching patterns among some groups have proven difficult to determine with certainty (Cao and Dietrich 2021).

Figure 2.9 illustrates a simple but highly contentious example. Based on morphology and physiology, spittlebugs (Cercopoidea), cicadas (Cicadoidea), and leafhoppers plus treehoppers (Membracoidea) are clearly closely related. The fossils suggest that all three groups arose from ancestral Hylicellidae, but what was the order in which they split? There are three possible branching patterns (Cryan 2005): option A, with Membracoidea branching first and the Cercopoidea and Cicadoidea separating later, making spittlebugs and cicadas more closely related to each other than to leafhoppers and treehoppers; option B, with cicadas branching first, which would make leafhoppers the closest relatives of spittlebugs; and option C, with spittlebugs branching first, making leafhoppers and cicadas equidistant relatives of spittlebugs. Morphological arguments have been advanced for each of these hypotheses (refs. in Cao and Dietrich 2021). The most recent and thorough DNA studies offer internally conflicting evidence favoring either option A or B, depending on the analytical

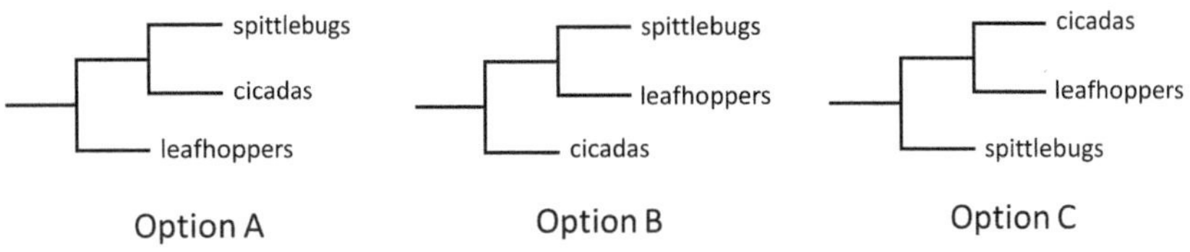

FIGURE 2.9. Three alternative trees for branching in the living Cicadomorpha. In Option A, the leafhoppers and treehoppers (Membracoidea) branch before the separation of spittlebugs (Cercopoidea) and cicadas (Cicadoidea), making cicadas the closest cousins of spittlebugs. In Option B the cicadas branch first, making leafhoppers the closest cousins of spittlebugs. In option C, the least likely, the spittlebugs branch off first, making them equally closely related to, but more distant cousins of, both cicadas and leafhoppers. Note that the relative placement of taxa at the prongs of the forks is arbitrary. In option A, for example, you could reverse the positions of the cicadas and spittlebugs and the meaning would be the same. There is no implication of closer similarity in having the cicada prong closer to the leafhopper prong. Only the branching order matters. Drawn after Cryan (2005), with the A and B options interchanged.

approach to the data. The evidence, on balance, seems to weigh in for option A, with spittlebugs and cicadas as closest cousins (Johnson et al. 2018; Skinner et al. 2020; Song et al. 2024). However, DNA evidence from *Sulcia*, a coevolving and codiversifying ancient endosymbiotic bacterium of Auchenorrhyncha (Chapter 3) is consistent with option B (Cao and Dietrich 2021). As a result, the issue remains unresolved. In Figure 2.4 I have opted, in the interim, for option A, which I believe is most likely considering the combined morphological, physiological, and DNA evidence. Li et al. (2015) make a good case for this option based on a number of morphological features.

The uncertainty over the phylogenetic relationships of spittlebugs to their nearest relatives is mirrored in efforts to discern branching patterns within the Cercopoidea. Figure 2.10 presents a simplified picture of what I believe to be the most likely large-scale family tree of living spittlebugs, based on two decades of often conflicting studies (Cryan 2005; Cryan and Swenson 2010; Bell 2016; Paladini, Takiya et al. 2018; Cao and Dietrich 2021; Crispolon et al. 2023). Family tree details depend on the particular array of species analyzed, the fine points of choosing genes for analysis and aligning DNA sequences, with oft conflicting results from equally valid, alternative analytic algorithms. The resulting trees are not written in stone. They are data-based approximations subject to change with more data, more taxa, and evolving analytical techniques.

The big picture appears to be as follows: The Machaerotidae, Clastopteridae, and a few allied groups appear to occupy the base of the phylogenetic tree.

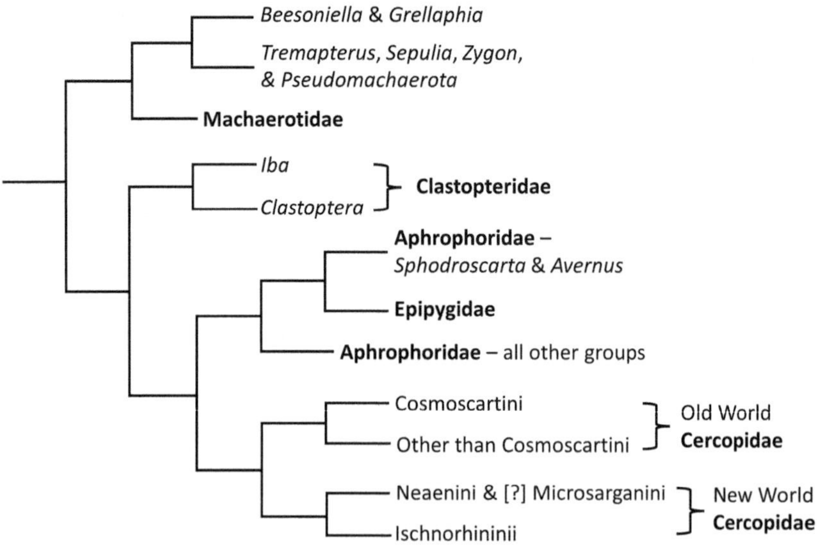

FIGURE 2.10. Family tree of living spittlebug groups as interpreted from DNA sequencing studies informed by auxiliary morphological, biogeographic, and ecological information. The five presently recognized families are in bold. It appears that the Machaerotidae and allied forms originated first among extant groups, followed by the Clastopteridae, Aphrophoridae, and Cercopidae, in that order. Epipygidae falls within the currently defined Aphrophoridae, calling into question its family rank. The placement of the Microsarganini is uncertain. For more discussion, see text.

This suggests that the other modern forms (Aphrophoridae, Epipygidae, and Cercopidae) branched from ancestors in or close to one or more of these groups. The Epipygidae appear to be a clade within the Aphrophoridae and most closely related to the aphrophorid genera *Sphodroscarta* and *Avernus*, both restricted, like the modern Epipygidae, to the Neotropics. Within the machaerotid/clastopterid group, Machaerotidae and a clutch of odd genera from Asia (*Beesoniella* and *Grellaphia*), Africa (*Tremapterus* (Plate 6e), *Sepulia*, *Zygon*), and Madagascar (*Pseudomachaerota*) seem to be most basal (primitive), but the relative position of Clastoptera and Machaerotidae are one of the major points of contention among competing molecular analyses. Depending on the study or alternative analytic methods used within the study, either Clastopteridae or Machaerotidae come out as basal. I refer the reader to several papers that grapple with these issues—Cryan and Svenson 2010, Johnson et al. 2018, Skinner et al. 2020, Cao and Dietrich 2021, Crispolon et al. 2023—with the caution that this is an area of investigation itself subject to rapid evolution.

There are important points upon which all, or almost all, DNA phylogenies agree. The Machaerotidae (Bell et al. 2014), the Clastopteridae, and the Epipygidae are all good clades, monophyletic groups descended from common ancestors. The Machaerotidae, the Clastopteridae, and their disparate odd relatives are also clearly related, a grouping that is reinforced by morphological characters (Hamilton 2014, 2015).

The Cercopidae also appear to be a monophyletic group, divided in turn into a handful of well-defined clades, or monophyletic subgroups. Old World Cercopidae fall into two big subclades. One subclade encompasses the South/Southeast Asian tribes Cosmoscartini and Phymatostethini and includes many big, colorful bugs in several genera, including *Cosmoscarta*, *Leptataspis* (Plate 12d), *Gynopygoplax*, *Oxymegaspis* (Plate 3b–d), and *Phymatostetha*. The other subclade hosts a disparate collection of Palearctic and Paleotropical genera, many of them also brightly colored, including *Cercopis* (Plate 11g), *Locris* (Plate 7d), *Callitettix* (Plate 12a), and *Eoscarta* (Plate 12c) (Crispolon et al. 2023).

With one possible exception, the New World Cercopidae are also a monophyletic group, likely descended from a single introduction to the New World after the separation of the Americas and the Old World through continental drift. They fall into two well-defined subclades, the Neaenini and the Ischnorhininii. Like the Old World Cercopidae, many are brightly colored (Paladini, Takiya et al. 2018), probably representing warning coloration that advertises a combination of reflex bleeding and quick jumping. *Reflex bleeding*, the emission of smelly haemolymph, seems to be universal among New World Cercopidae and present in at least some Old World Cercopidae, including the basal group *Cercopis* (Peck 2000a). It may be a defining feature of the family Cercopidae, a shared, evolutionarily inherited characteristic called a *synapomorphy* (see Chapter 5 for a detailed discussion of reflex bleeding and warning coloration).

Figure 2.11 presents a family tree of the major genera of the Ischnorhininii, a large and diverse group of New World Cercopidae that includes many important pests of pasture grasses and sugarcane that are discussed in Chapters 6 and 7 (Thompson 2004). There is good agreement on the general pattern of taxonomic relationships in this group (Bell 2016; Paladini, Takiya et al. 2018; Crispolon et al. 2023). Two branches, corresponding to the genus *Prosapia* (Plates 10c–e, 14c, 14h) and a clade encompassing the genera *Aeneolamia*, *Mahanarva*, *Deois*, *Zulia*, and *Notozulia* (Plates 10a, 10b, 10f, 14a, 14b, 14d–g, 15a–c, 15e–h), are composed primarily of grass feeders. It appears that these groups radiated in conjunction with the spread of C4 grasses in the mid to late Cenozoic (Thompson 2004; Paladini, Takiya et al. 2018), a relatively recent development in the history of life on Earth. Chapters 6 and 7 cover the agricultural ecology of these important Neotropical pests.

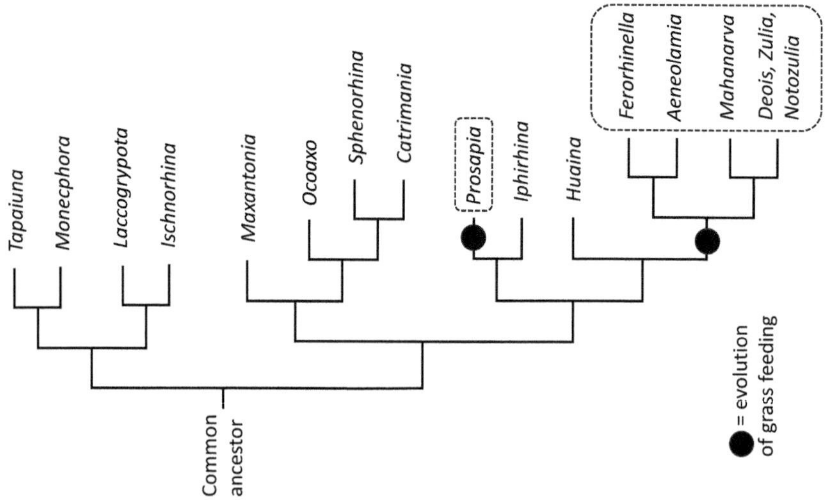

FIGURE 2.11. Simplified family tree of the Ischnorhininii, the largest branch of New World Cercopidae. With the exception of *Ferorhinella*, the grass feeding genera (enclosed in broken lines) include an important array of Neotropical sugarcane and pasture pests (Chapters 6 and 7). The tree implies that grass feeding evolved independently (dark circles) in at least two lineages and relatively recently. This is a redrawn, simplified version of a more detailed tree in Paladini et al. (2018, fig. 2) and closely approximates an equivalent tree in Crispolon et al. (2023, fig. S7)

One line of New World spittlebugs has been the subject of conflicting phylogenetic interpretation. The Neotropical genus *Microsargane* was originally assigned to Aphrophoridae based on an aphrophorid-like body and eye shape. Surprisingly, the first broadscale molecular phylogeny of Cercopoidea placed it in Cercopidae but among the Old World lines (Cryan and Svenson 2010). Subsequently, based on morphology, Hamilton (2016) placed *Microsargane* and a handful of related genera into a tribe of their own, the Microsarganini, which he judged to be allied to the tribe Neaenini. Shortly thereafter, an updated DNA phylogeny continued to place Microsargane among the Old World Cercopidae, widely separated from the Neaenini (Paladini, Takiya et al. 2018). More recently, a new molecular phylogeny, with much wider representation of Old World Cercopidae taxa, assigned *Microsargane* to the Aphrophoridae (Crispolon et al. 2023). Consequently, the issue of where *Microsargane* belongs is unsettled, indicated by the "[?]" in Figure 2.10. Interestingly, the single member of the Microsarganini tested, *Liparonotum praenitida*, is positive for reflex bleeding (Hamilton and Thompson 2007). This supports linkage to the Cercopidae, consistent with Hamilton's (2016) placement of the Microsarganini in the Cercopidae (his Cercopinae). The molecular studies place Neaenini at the base of the family tree of New World Cercopidae. None seems to have been tested for reflex bleeding. This leaves the placement and relationship of the tribes Microsarganini and Neaenini unsettled.

The taxon Aphrophoridae presents its own set of ambiguities. As noted previously, molecular studies concur that the family Epipygidae is nested within it, relatively closely related to the Neotropical aphrophorid genera *Sphodroscarta* and *Avernus*, but they differ as to whether it is in a single clade with these genera (as in Fig. 2.10) or in a more complicated nesting separating other clusters of aphrophorid lines. Either way, the validity of Epipygidae as a family-level taxonomic unit is questionable. Also, unlike the clean Old World/New World split in Cercopidae lineages (aside from, possibly, *Microsargane*), several important aphrophorid lineages overlap the Old World and North America, including the Philaenini and the genera *Aphrophora* (Plate 11f, 13a) and *Lepyronia* (Plate 13h). Consequently, the taxonomic status of Aphrophoridae is also subject to ongoing discussion (Skinner et al. 2020; Cao and Dietrich 2021; Crispolon et al. 2023).

Dated DNA Phylogenies

When did the phylogenies branch to produce the major inferred patterns in evolutionary trees? We date the ages of fossils from the ages of the rocks in which they are preserved. Alternatively, DNA sequence analysis permits us to make indirect inferences about the dates of splits in family trees. Changes in DNA sequences over evolutionary time often occur at relatively steady, predictable rates, permitting us to use those changes as a molecular clock to estimate approximately when evolutionary lines divided. Established geological dates for fossils help calibrate the molecular clock. Using these tools, it appears that the line leading to cicadas and spittlebugs diverged from the line leading to leafhoppers about 265 mya, while the cicadas and spittlebugs diverged about 225 mya, give or take 30 to 40 million years in each case (Johnson et al. 2018). In turn, modern forms started their radiation about 90 million years ago. These dates suggest an early mid-Triassic origin for the line leading to Jurassic spittlebugs and a late mid-Cretaceous origin for the line leading to modern spittlebugs, both roughly consistent with what we know from the fossil record, though still susceptible to considerable uncertainty. The same data push the split with leafhoppers back to the late Permian, which, from the fossil record, seems too early. Alternatively, Dietrich et al. (2017) estimate the divergence time with leafhoppers at 227 mya, a more plausible date, but push the cicada–spittlebug split up to the Jurassic (186 mya), which is clearly inconsistent with the fossil record. In their analysis, the separation of the lineages leading to modern spittlebugs dates back to 110 mya, a plausible estimate but with no Mesozoic fossils to test its validity. All of these estimates are subject to large margins of error.

Combing the results from fossils, DNA analysis, and morphological studies, we achieve a plausible overview of spittlebug evolution. They originated in the Triassic from the now extinct Hylicellidae, which also gave rise to the other modern groups with xylem-feeding: cicadas and leafhoppers (Fig. 2.4). Two lines, the Sinoalidae and the Procercopidae, radiated into diverse genera and species in the Jurassic and Cretaceous (Fig. 2.7). The Procercopidae ultimately gave rise to modern lines in the mid to late-mid Cretaceous, possibly through the Cercopionidae, and modern forms were established by the early Cenozoic. Later in the Cenozoic, a major Neotropical cercopid line radiated in conjunction with the development of C4 grasslands. The Machaerotidae and Clastopteridae seem to have given rise to the Aphrophoridae and Cercopidae, in that order, although many details of the phylogeny of modern groups are still unresolved. This sets the stage for the panorama of present-day spittlebug life outlined in coming chapters.

Coda: Could a Spittlebug Living Fossil Be Lurking Undiscovered?

I daydream that one day, someone, most likely somewhere in Southeast Asia, will pick a small bug with three ocelli out of a net or off a collecting sheet, marking the discovery of a surviving procercopid or sinoalid spittlebug. Although unlikely, there is precedent. Among almost 2900 known cicada species, two, in the genus *Tettigarcta*, are the sole remnants of the family Tettigarctidae, widespread and abundant in the Triassic, Jurassic, and Early Cretaceous, but now extinct everywhere except Australia (Bartlett et al. 2018). They were first described in 1845, but their fully divergent character was not documented until well into the twentieth century (Evans 1940). Cicadas are big, showy insects, making the discovery of a relict cicada more likely, but nothing rules out the survival of a lingering Mesozoic spittlebug line, one more argument for preserving large portions of the Asian tropical forests that might still sustain them.

3

XYLEM FEEDING

A Challenging Way to Make a Living

Xylem feeding is a distinctive and core feature of spittlebug biology, linked to many other spittlebug attributes, such as spittle production and broad host range. It inevitably intersects with the technicalities of xylem sap transport in plants, a daunting subject in itself. This chapter presents an overview of xylem feeding and its profound consequences for spittlebug biology.

"Dr. Livingstone, I Presume."

In December 1854, not for the first time, David Livingstone was held up by severe attacks of fever, affecting both himself and the carriers he employed for his first trans-African expedition. Sixteen years later, in 1871, he was to gain international fame as the "Dr. Livingstone, I presume" of his encounter with *New York Herald* reporter Henry Stanley, who had been sent to track him down on a later expedition to Central Africa seeking the origin of the Nile. In 1854 Livingstone was in the midst of an earlier epic journey that took him from the British Cape Colony in present-day South Africa north through the interior to Portuguese-controlled Luanda and then east via the Zambesi River to the Indian Ocean in Mozambique. His caravan, including twenty-seven warriors of the Kololo people assigned by their leader, Sekeletu, as interpreters and guides, was camped for rest and recovery at Golungo Alto, about 160 km east of Luanda in present day Angola.

Livingstone was a Scot from the Glasgow region. Born to an impoverished, working-class family, he worked at the same mill from age ten to twenty-six, putting himself, in the meantime, through medical school. He was a committed Congregationalist Christian, with a strong interest in science and natural history, and he was deeply influenced by the philosophy of Thomas Dick (1774–1857), a minister and astronomer who reconciled Christian theology with the practice of empirical science for human betterment. Livingstone pursued medicine to prepare himself for missionary work abroad (Jeal 2001).

At Golungo Alto, finally having recovered from fever and shortly before he left for places further east, Livingstone made a series of observations on "a curious insect, which inhabits trees of the fig family." He recorded that "Seven or eight of them cluster round . . . the smaller branches, and there keep up a constant distillation of clear fluid, which, dropping to the ground, forms a little puddle below. If a vessel is placed under them in the evening, it contains three or four pints of fluid in the morning" (Livingstone 1857, 415–416). He recognized the insect as a relative of the common British cuckoo-spit bug (*Philaenus spumarius*). Based on Hinde (1906), it was almost certainly *Ptyelus flavescens* (Plate 1c, 1d), a large aphrophorid spittlebug distributed widely in sub-Saharan Africa.

Livingstone knew that European scientists believed spittlebugs sucked the excreted liquid out of their host plants, and his African associates told him the same. But he was skeptical that plants circulated so much water and conjectured, instead, that the bugs were condensing moisture out of the air. Remarkably, in the midst of shepherding a large caravan across a continent still largely unexplored by Europeans, he spent three days trying to demonstrate this hypothesis experimentally.

> Finding a colony of these insects busily distilling on a branch of the Ricinus communis, or castor-oil plant, I denuded about 20 inches of the bark on the tree side of the insects, and scraped away the inner bark, so as to destroy all the ascending vessels. I also cut a hole in the side of the branch, reaching to the middle, and then cut out the pith and internal vessels. The distillation was then going on at the rate of one drop each 67seconds, or about 2 ounces 5½ drams in 24 hours. Next morning the distillation, so far from being affected by the attempt to stop the supplies, supposing they had come up through the branch from the tree, was increased to a drop every 5 seconds, or 12 drops per minute, making 1 pint (16 ounces) in every 24 hours. I then cut the branch so much, that during the day it broke; but they still went on at the rate of a drop every 5 seconds, while another colony

> on a branch of the same tree gave a drop every 17 seconds only, or at the rate of about 10 ounces 4 and 4/5 drams in 24 hours. I finally cut off the branch; but this was too much for their patience, for they immediately decamped . . .
>
> Having but one day left for experiment, I found again that another colony on a branch, denuded in the same way, yielded a drop every 2 seconds, or 4 pints 10 ounces in 24 hours, while a colony on a branch untouched, yielded a drop every 11 seconds, or 16 ounces 20 drams in 24 hours. I regretted somewhat the want of time to institute another experiment, namely, to cut a branch and place it in water, so as to keep it in life, and then observe if there was any diminution of the quantity of water in the vessel. This alone was wanting to make it certain that they draw water from the atmosphere. (Livingstone 1857, 416–417)

As it happened, Livingstone got the interpretation wrong. The spittlebugs were, in fact, sucking plant sap in enormous quantities, not condensing it from the atmosphere. The idea of cutting off sap circulation from the plant by girdling bark was clever, but, despite his evident diligence, Livingstone clearly did not manage to sever all the xylem vessels, which in woody plants are buried in the interior. E. B. Poulton (in notes accompanying Hinde 1906) called him out on the question of whether he had demonstrated condensation versus sucking. Poulton was one of the most famous evolutionists of the late nineteenth/early twentieth centuries. He pioneered the study of animal coloration and coined the concepts of aposematic selection (selection for warming coloration) and frequency-dependent selection (selection favoring rare or common forms), both of which will feature prominently in Chapter 5. His critique of Livingstone included the following prescient observation:

> The interpretation of the copious exudation is almost certainly to be found in the relatively small amount of nutriment contained in the sap, so that a great quantity must pass through the body of the insect in order to yield a sufficient supply of food. Analysis of sap drawn direct from the tree as compared with that of the fluid which has passed through the body of the insect might well yield interesting results bearing upon the physiology of insect nutrition. (Poulton in notes accompanying Hinde 1906)

Despite getting the physiological explanation wrong, Livingstone was the first person to report and scientifically investigate the prodigious fluid production of tropical spittlebugs. He was also one of the first to report a phenomenon that came to be known as "rain trees," trees that weep copious droplets of liquid due to excretion by large concentrations of spittlebugs (Decary 1928). These trees, which have been observed widely in Africa and in the Neotropics (Box 1929; Palmer and Pitman 1961; Kareem et al. 1971), appear to weep large amounts of water, sustaining a green understory and even puddles in otherwise parched dry season environments. Owen and Wiegert (1976) have suggested that the rain tree hosts may benefit from spittlebug infestation by getting access to near-surface soil nutrients that otherwise could not be accessed by roots in the dry season. Perhaps not coincidentally, many rain tree species are nitrogen-fixing legumes, which, as we will see later in this chapter, are an important category of spittlebug hosts.

Livingstone was a keen observer of nature, a supporter of economic development for Africans, and a vocal opponent of the continuing East African slave trade. He respected Africans and treated his native crews, though not his European coworkers, humanely. They reciprocated. When he died in present-day Zambia in 1873, still looking for the Nile, his African crew, led by James Chuma and Abdullah Susi, carried his preserved body in a bark cylinder 1600 km to the Indian Ocean for return to Britain. This pioneering student of spittlebugs was interred in Westminster Abbey in 1874 (Jeal 2001).

Xylem sap, as Poulton inferred, is indeed low in nutrients—uniquely low in many respects. This chapter revolves around the profound consequences of that nutritional poverty for spittlebug biology. I submit that xylem feeding determines key spittlebug attributes, from head and gut anatomy to host plant repertoire, and influences the role of spittlebugs as important agricultural pests. Excellent overviews of xylem-feeding and its consequences can be found in Raven 1983, Press and Whitaker 1993, and Redak et al. 2004. This chapter summarizes its consequences for spittlebug biology.

Evolutionary Origins and Phylogenetic Distribution of Xylem Feeding

As we learned in Chapter 2, xylem feeding is believed to have first evolved in the Hylicellidae—the line that gave rise to spittlebugs, cicadas, and leafhoppers. It appears that all living spittlebugs and cicadas are xylem feeders. This is consistent with a strong consensus that xylem feeding is ancestral in spittlebugs and cicadas (Johnson et al. 2018). Evidently, when the early Hemiptera were sorting out feeding modes in the Permian (Chapter 2), the line leading to Cicadomorpha opted for the challenges associated with xylem feeding, while the Sternorrhyncha (aphids and their kin) and the Fulgoromorpha (planthoppers) opted for the challenges associated with phloem feeding.

The situation in leafhoppers is more complicated. Only one subfamily, the Cicadellinae, or sharpshooters, are xylem feeders, with the likelihood, based on their inflated sucking pumps, that xylem feeding occurs in some related, smaller groups (Dietrich et al. 2017). The other large leafhopper subfamilies are phloem feeders or, in one case, cell content feeders (the tiny Typhlocybinae). On the other hand, fossil evidence suggests that the earliest leafhoppers were xylem feeders (Dietrich et al. 2017). Thus, it appears that at some point in their early evolution leafhoppers evolved phloem feeding, but that the line leading to the Cicadellinae reverted to xylem feeding. This is a complex scenario, but one that best fits the present combined fossil, morphological, and molecular data (Johnson et al. 2018). However, the issue is not settled. One DNA study suggests a single origin for xylem feeding, followed by subsequent loss in some lines (Bell-Roberts et al. 2019). The strongest evidence against this hypothesis is the phylogenetic position of the xylem-feeding sharpshooters. They are embedded in an otherwise phloem-feeding clade (Skinner et al. 2019), strongly implying that they arose from phloem feeding ancestors.

The Evidence for Xylem Feeding

There is overwhelming evidence, direct and indirect, for xylem feeding in spittlebugs, a feeding mode apparently first suggested by Withycombe (1926). The direct evidence comes from tracing the tracks left by spittlebug stylets when they pierce plant tissue. These tracks are marked by stylet sheathes formed by saliva egested as the stylets pierce plant tissue. Wiegert (1964b) observed that eighteen of twenty *P. spumarius* nymph stylet tracks ended in xylem vessels. The other two missed the xylem vessel and ended in "pith." Working independently in the same period, Carle and Moutous (1965) presented tracking evidence that *P. spumarius* feeds exclusively from xylem. Horsfield (1978) demonstrated that a large majority of several hundred nymphal and adult *P. spumarius* stylet probes, many branched, ended in xylem vessels, though a small number ended in phloem tissue and a larger number in *parenchyma tissue*, the living cells surrounding both xylem and phloem vessels. Byers and Wells (1966) report that *Prosapia bicincta* "feeding tunnels all end in xylem vessels." In her 1979 PhD thesis, Nickel (cited in Crews et al. 1998) traced 28 spittlebug stylet tracks to xylem conduits and two to parenchyma. Crews et al. (1998) demonstrate the direct puncture of xylem vessels in the context of adjacent probing through phloem tissue.

Working with *Clastoptera arborina*, Kuenzi and Coppel (1985a) found that most tracks, including all feeding tracks, ended in xylem tissue. Marshall and

Cheung (1975) traced stylet tracks in two Machaerotidae species (*Machaerota coronata* and *Chaetophyes compacta*) and one Cercopidae species (*Cosmoscarta abdominalis*). With a single exception leading to parenchyma cortex, every track led to xylem vessels. William and Ananthasubramanian (1991) traced stylet tracks of *Ptyelus nebulus* to xylem vessels. These studies provide strong evidence for consistent xylem feeding, with some adjacent tracks leaving a record of probes searching for xylem vessels.

In a recent comprehensive study, Cornara, Garzo et al. (2018) tied electrical penetration graph (EPG) evidence of xylem feeding directly to sectioned olive stems, showing that the stylets of feeding *P. spumarius* individuals penetrated xylem vessels (Fig. 3.1). They also coordinated xylem feeding with excretion. Phloem feeding waveforms, which are distinctive (Chuche et al. 2017), were absent, which is consistent with chemical tests of the excreta. Other workers provide strong, indirect evidence for xylem feeding based on analysis of the

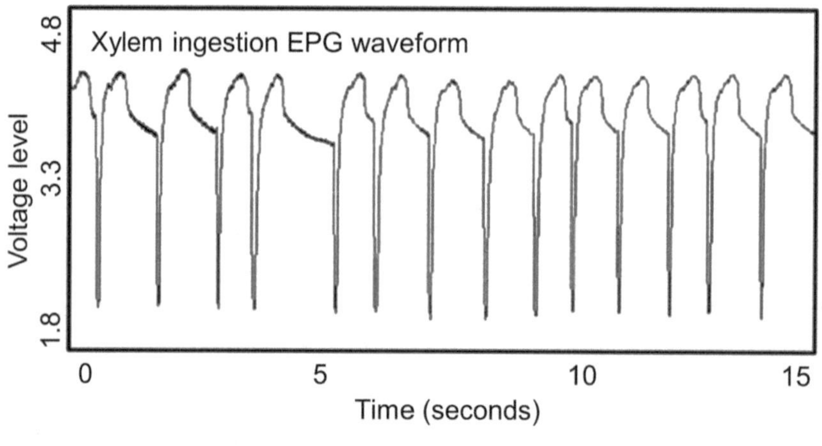

FIGURE 3.1. Electrical Penetration Graph (EPG) waveform pattern generated by a *Philaenus spumarius* adult during xylem sap ingestion. The peaks correspond to retraction of the cibarial muscles, filling the cibarium with sap. The plateaus correspond to relaxation of the cibarial muscles, with collapse of the cibarium and emptying of sap into the foregut. In this example, there are 15 peaks in 15 seconds, corresponding to a pumping frequency of 1 Hz (one per second). Other EPG waveform patterns correspond to stylet insertion and probing, to xylem vessel contact, and to resting or feeding interruption. Such waveforms, combined with direct observation of the stylet paths of the individuals monitored, are direct evidence for feeding from xylem vessels. From Fig. 3c of Cornara, Garzo et al. (2018), with added annotations, under Creative Commons Attribution 4.0 International license https://creativecommons.org/licenses/by/4.0/deed.en.

excreted fluid (Cordoba et al. 1990; Crews et al. 1998; Malone et al. 2002; Ponder et al. 2002; Buxton et al. 2007).

Two discordant studies merit comment. Hagley and Blackman (1966) examined feeding tracks produced by the spittlebug *Aeneolamia varia* (Plate 15b) in sugarcane (see Chapter 6). For nymphs feeding on roots, 260 of 322 reached xylem vessels. Fourteen ended in parenchyma and 35 in phloem. Taking probing tracks into account, this is consistent with xylem feeding as the primary mode. In contrast, for adults feeding on leaves, 230 of 307 tracks ended in parenchyma and only 14 in xylem. From this they concluded that adults, in contrast to nymphs, feed on parenchyma cells. In a follow-up experiment, Hagley (1966) fed fifth instar nymphs on leaves and found that 108 tracks of 217 ended in parenchyma and only 63 in xylem, implying that, like adults, nymphs feeding on leaves prefer parenchyma. There are two serious problems with these experiments. First, *A. varia* nymphs generally live on roots. Leaves are not a natural nymphal food source. Second, both experiments with leaves used cut four-inch leaf blades set in water at one end as a food source, instead of intact plants. In all likelihood, this disrupted the xylem vessels, compromising the validity of the outcome, as did the use of cut stems in another spittlebug/host system in Costa Rica (Thompson 1997). Consequently, and counter to Fewkes (1969a), these experiments do not provide convincing evidence of parenchyma feeding. Malone et al. (1999) argue convincingly that feeding on parenchyma cells could not account for the immense volume of fluid passing through spittlebugs.

Working with another spittlebug sugarcane pest, *Mahanarva fimbriolata* (Plate 14g), on intact plants, Garcia et al. (2007) report that while the leaf-feeding adults and root-feeding nymphs feed on xylem vessels, 35% of third instar nymphs also "reached the root phloem" and, at least in one case, penetrated phloem vessels. However, while Garcia et al. tested spittle for pH, they did not test for the tell-tale presence of sugars, which could have confirmed phloem feeding. Occasional mix-and-match xylem feeding is known to occur in some phloem feeders and might occur in xylem feeders (see this chapter's section "Why Feed on Xylem at All?"). The extravagant volume of spittlebug excreta, accompanied by the absence of any significant sugar content, suggests that xylem feeding is the rule across all life stages among all spittlebugs studied.

Xylem Sap Is a Very Challenging Source of Nutrition

By any normal standard, xylem sap is a terrible nutrient source. First, it is extremely dilute. Insects have to imbibe an enormous amount to get a little bit of food. Second, the nutrients it does contain are highly imbalanced. Not only is there a paucity of food content, but the nutrients present are not the right mixture to sustain protein synthesis. Third, to add insult to injury, the sap itself is hard to extract because negative pressure inside the xylem elements pulls against the insects as they feed. Consequently, xylem feeders have to expend a lot of energy for a paltry return. It is a small miracle that any organism can live on xylem sap at all. Nevertheless, at least three groups, including spittlebugs, have evolved to do so. Here is how they do it.

The First Challenge: A Very Dilute Food Source

To begin, here is a little elementary plant plumbing. Vascular plants, which include all terrestrial plants more complicated than mosses, have two basic sorts of vessels to circulate water and nutrients. In the simplest telling, phloem vessels bring the products of photosynthesis from the leaves to the roots. As a result, phloem sap is nutritionally rich, full of sugars and amino acids. Xylem elements bring water and minerals from the roots to the leaves, supporting photosynthesis. The most important minerals drawn from the soil and transported via xylem sap are nitrogen, phosphorus, and potassium. This is the paradigm I learned in high school. Were this the end of the story, all xylem feeders would starve for want of any organic material in the sap. However, xylem sap does have organic content. Much or most of the nitrogen in xylem sap is transported as amino acids. This is because, depending on species, a lot of nitrogen from the soil enters plants as toxic ammonium ions, which are immediately converted to amino acids for xylem transport from the roots. The result is a dilute but reliable stream of organic compounds in the xylem sap, laying the basis for the existence of xylem feeders. The carbon in the xylem-transported amino acids comes from photosynthates carried to the roots by the phloem. At least 13% of carbon assimilated by plants is recycled up the xylem in this fashion (Raven 1983), so ultimately, xylem-feeding insects are sustained by the photosynthesis of their host plants. In some plants, the primary source of soil nitrogen is nitrate, sometimes assimilated into amino acids in the roots but often transported to the leaves in mineral form, in which case it is not a source of organic carbon for xylem feeders (Thompson 2004; Andrews and Raven 2022).

Quantitatively, the organic carbon level of xylem sap is two to three orders of magnitude lower than that of phloem sap (Raven 1983). Table 3.1 presents comparative levels of xylem and phloem content for red oak (*Quercus rubra*), a nymphal host of the spittlebugs *Lepyronia gibbosa* in Michigan (Dunn et al. 2002), and *Lepyronia coleoptrata* (Plate 13h) in New Hampshire. Red oak xylem sap carries about three hundred times less organic carbon than its phloem sap,

TABLE 3.1. Comparative levels of xylem and phloem sap constituents in *Quercus rubra* (red oak)

SAP COMPONENT	XYLEM SAP	PHLOEM SAP	PHLOEM TO XYLEM RATIO
Dry matter g/l	1.4	220	157
Sucrose mM	0.4	409	1,105
Reducing sugars mM	1.7	211	127
Organic carbon mM	26.0	7,580	292
Nitrogen mM	3.9	470	121
Energy for feeding organism kJ/l	7.8	2,280	292

Source: Data from Raven 1983.
Note: g = gram, l = liter, mM = millimole, kJ = kilojoule

resulting in a 300-fold disadvantage in available food energy per unit of sap for xylem feeders compared to phloem feeders, such as aphids. In the castor oil plant, on which Livingstone conducted his experiments, amino acids occur at a concentration of 4.9 millimolar (mM) (Peuke 2010), approximately the same as the very dilute 3.9 mM concentration in red oak (Table 3.1). Despite the dilute food source, the castor oil plant sustains spittlebugs at pest levels (Chapter 8).

Big Sucking Pumps and Tremendous Throughput: Spittlebugs Are World Class Imbibers and Pee-ers

Extracting xylem fluid requires a lot of pumping, which spittlebugs, both nymphs and adults, carry out with an organ called the *cibarium*, which, along with the muscles operating it, takes up a large portion of the head. It is basically a piston. With the pharynx (throat) closed, the cibarium muscles contract to expand the cibarium, creating a low-pressure void into which xylem fluid is drawn through the precibarial valve from the stylets inserted in xylem vessels. When the precibarial valve, first described by Rodman (1921), then shuts, the cibarial muscles release, the pharynx opens, and xylem fluid enters the foregut leading to the digestive tract. These movements are synchronized and the cibarium pumps at a rate of 0.7 to 1.7 Hz, or about 1 cycle per second (Malone et al. 1999; Cornara, Garzo et al. 2018; Beckett et al. 2019; Bergman et al. 2021). Cibarial muscle contraction and filling of the cibarial chamber corresponds to peaks in the characteristic EPG xylem ingestion waveform (Cornara, Garzo et al. 2018; Fig. 3.1). Muscle relaxation and cibarium collapse that pushes xylem fluid toward the foregut corresponds to plateaus in the EPG waveform (Fig. 3.1).

The cibarial muscles are anchored to the exoskeleton on the inside of the *postclypeus*, or frons, as it is sometimes called in spittlebugs (Figs. 1.5, 3.2). This

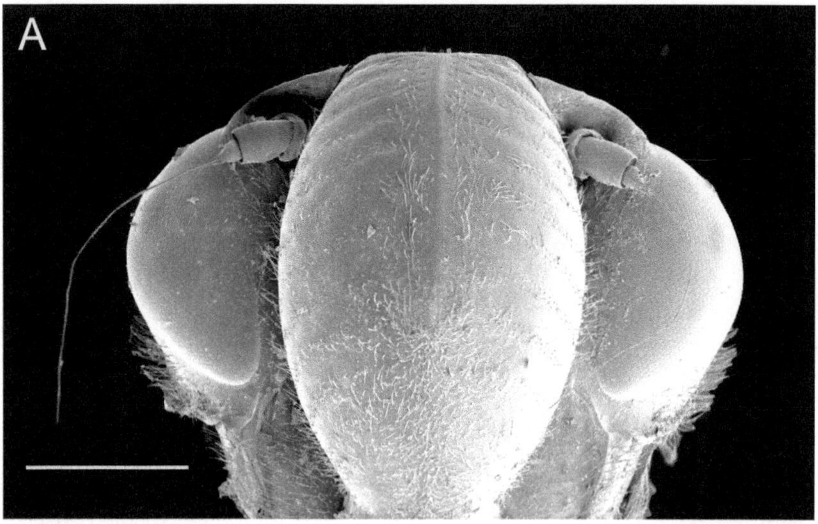

FIGURE 3.2. Scanning electron micrograph of face of the Chinese rice pest spittlebug, *Callitettix versicolor* (Fig. 8.3b, Plate 12a) showing the large inflated postclypeus, which holds a powerful sucking pump. The transverse striations on either side of the median ridge (carina) mark the attachment points of the cibarial muscles that power the pump. Scale bar = 0.5 mm. From Zhu et al. (2019) under Creative Commons Attribution 4.0 International license https://creativecommons.org/licenses/by/4.0/deed.en.

is an expanded area of the face between the eyes and antennae and generally reaches to and a bit over the dorsal side of the head, where, in adults, it is set off by cuticle sutures and called the *tylus*. Expanded postclypei are a hallmark of spittlebug external morphology. They are typically crossed by traverse striations, often more darkly pigmented than the ground color. These mark the points of cibarial muscle attachments, thirty total in *P. spumarius*, lined up in fifteen pairs (Clarke et al. 2023).

Spittlebugs seem to feed round the clock (Withycombe 1926; Fewkes 1969a; Newby 1980; Malone et al. 2002). In proportion to body weight, they imbibe and excrete enormous amounts of liquid, just as Livingstone observed in his African explorations. Throughput is measured in the form of excretion rates, since ingestion rates are difficult or impossible to measure, and it is reasonable to assume that very little volume is lost in transit through the digestive tract. Some measured spittlebug excretion rates appear in Table 3.2, with data for a few sharpshooters for comparison. Rates lie between 0.27 and 1.80 microliters/minute (μl/min). The unit term microliters suggests that throughput is

TABLE 3.2. Excretion rates for spittlebugs and selected sharpshooter leafhoppers

SPECIES	STAGE	EXCRETION RATE (μL/MIN)	SOURCE
Spittlebugs			
Philaenus spumarius	adult	1.20 (> 2.4 max)	Malone et al. 1999
Philaenus spumarius	adult	0.57 (max 1.0)	Bergman et al. 2021
Philaenus spumarius	adult	0.43	Wiegert 1964
Philaenus spumarius	adult	1.2-1.5	Horsfield 1978[a]
Philaenus spumarius	4th instar	0.52	Wiegert 1964
Philaenus spumarius	5th instar	1.8	Horsfield 1978[b]
Aeneolamia varia	adult	1.4	Withycombe 1926
Aeneolamia reducta	adult	0.14-1.11	Cordoba et al 1990
Notozulia entreriana	adult male	0.07	Valério & Nakano 1987
Notozulia entreriana	adult female	0.14	Valério & Nakano 1987
Mahanarva spectabilis	adult	0.007–0.50[c]	Silva et al. 2017
Ptyelus flavesens	adult	0.34	Livingstone 1857[d]
Ptyelus grossus	5th instar	12.0	Boulard 1969
Clovia bigoti	5th instar	4.8	Boulard 1969
Clastoptera arborina	5th instar	0.27	Kuenzi & Coppel 1985b
Sharpshooter leafhoppers			
Homalodisca vitripennis	adult	2.39–3.95	Brodbeck et al. 1993
Homalodisca vitripennis	adult	1.43–2.18	Brodbeck et al. 1995
Homalodisca insolita	adult	1.13–4.71	Brodbeck et al. 1993
Cuerna costalis	adult	1.90–4.38	Brodbeck et al. 1993
Graphocephala atropunctata	adult	approach 3.33	Mittler 1967

Note: All rates have been translated into μl/min from disparate units of measurement in the original works.
[a] As interpreted by Bergman et al. 2021.
[b] As interpreted by Kuenzi & Coppel 1985b
[c] Range over nine different pasture grass hosts, one of which was extremely unpalatable
[d] Based on an estimated 8 individuals producing 3.5 British pints (1.99 liters) over an estimated 12 hours. Species identity follows Hindc 1906.

low, but there are a lot of minutes in the day, and the ultimate quantities excreted are large proportional to the body weight. Feeding *P. spumarius* (adult weight 10 mg) excrete 150 to 284 times body weight per day depending on life stage and sex (Horsfield 1978). This is equivalent to a 68 kilogram (150 pound) human peeing 10,000 to 19,000 liters a day, an astonishing rate of throughput. In general, adult *P. spumarius* females excrete more than males, which is partly a function of greater body weight, and there is evidence in at least one species that feeding rates are greater on pasture grass species less resistant to spittlebug attack (Cordoba et al. 1990). Livingstone's observations on liquid output, which require a few assumptions to cast in modern comparative form, fall right in the range of modern measurements (Table 3.2), a remarkable accomplishment for an explorer in the African hinterland more than a century and a half ago.

In nymphs, much of the excreted liquid is incorporated into spittle masses (the focus of Chapter 4). In adults the liquid is usually flicked away in tiny droplets, one at a time, at a rate of about seven per minute in *P. spumarius* females and a bit lower in males (Malone et al. 1999; Cornara, Garzo et al. 2018). It is not clear why spittlebugs (and sharpshooter leafhoppers, which are named for the phenomenon) have evolved this behavior. It may be advantageous to avoid making obvious puddles of excreta that could alert predators to their presence. Whatever the selective advantage, there is fascinating (and counterintuitive) evidence that production of flicked drops is energetically more favorable than the production of simple jets of liquid (Challita et al. 2023). It is also true that even in adults the excreted liquid sometimes puddles and runs on the adjacent surface of plants, as evident in a photograph of *Ectemnonotum fruhstorferi* taken in Vietnam (Plate 12f)

Filter Chambers and Extraordinarily Efficient Gut Absorption of Nutrients

Hemiptera are the only insects able to live entirely on plant sap, xylem or phloem, which is likely a result of the evolution of specialized midguts that enable them to deal with large amounts of fluid food (Terra et al. 2023). Xylem feeding insects feature a well-developed, water-shunting morphological contrivance called a *filter chamber* that diverts about 90% of the water entering the foregut directly to the hindgut, bypassing the midgut where nutrients are absorbed. As a result, the midgut receives and processes nutrients concentrated tenfold. In three xylem feeding leafhopper species, this results in absorption of a phenomenal 99% of the total ingested carbon and 74–93% of all the incoming organic nitrogen (Brodbeck et al. 1993). Since *P. spumarius* and other spittlebugs excrete little or no organic waste (Licent 1912), their efficiency must be about the same. By comparison, one measured aphid species extracts only 55% of incoming amino acids and a mere 5% of the sugars (Goodchild 1966). The filter chambers of phloem feeders are distinctly less complex than those of xylem feeders (Goodchild 1966; Lindsay and Marshall 1980).

Xylem feeder filter chambers and the accompanying alimentary tract plumbing form elaborate structures (Figs. 3.3, 3.4), first described in detail by Licent (1912) and Kershaw (1914). The posterior midgut loops back to and zig zags around the surface of a diverticulum of the beginning of the anterior midgut.

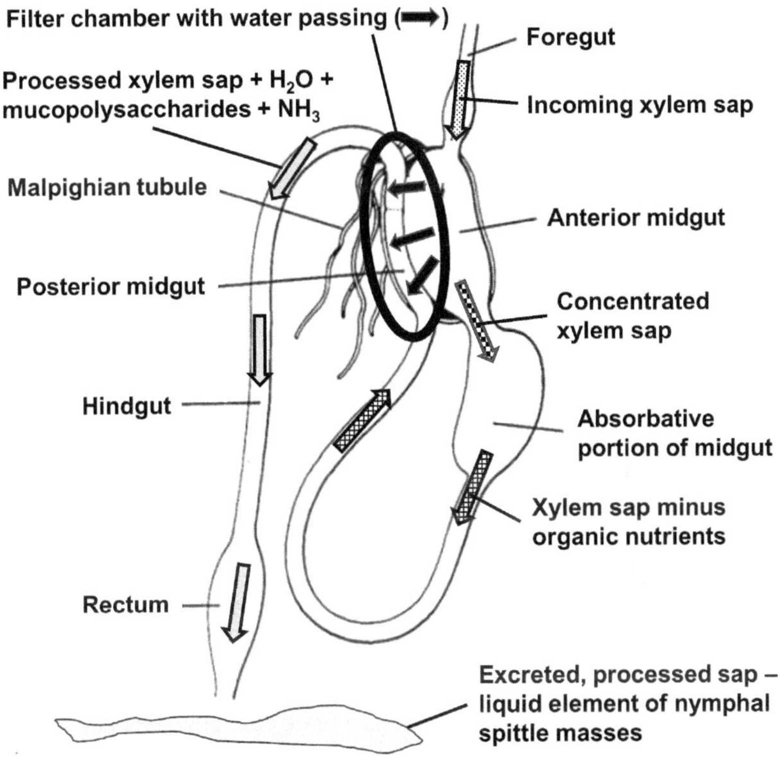

FIGURE 3.3. Schematic diagram of a spittlebug nymph alimentary tract. Sap enters the foregut from the cibarial chamber through the esophagus. Ninety percent of the water passes directly to the posterior midgut (black arrows) through the filter chamber. The remaining, concentrated sap is processed in the midgut to remove organic nutrients. The depleted liquid is shuttled back to the filter chamber, where the diverted water is added back to the mix. Coming out of the filter chamber, mucopolysaccharides and ammonia are received from the Malpighian tubules. The resulting processed fluid then passes through the hindgut and is excreted. For simplicity this diagram omits some important details. The posterior midgut zig zags and winds around the filter chamber to maximize the contact area that permits direct passage of water. Also, the Malpighian tubules (four) are proportionately larger than shown and their distal ends reach and attach to the exterior of the rectum (see Marshall 1964a and Boulard 1991 for alternative schematic representations). The same illustration also applies to adults, absent the addition of the mucopolysaccharides, and with the excreted liquid flicked away in droplets rather than added to spittle masses.

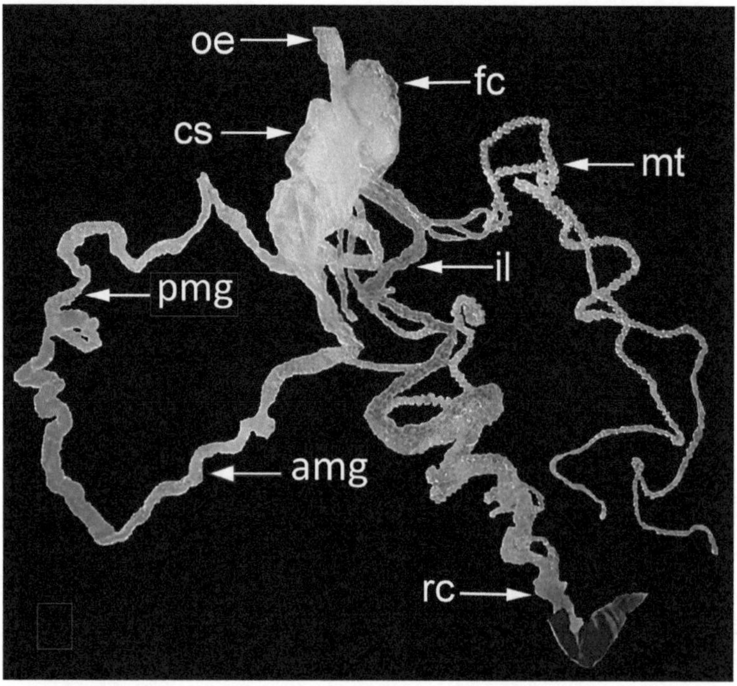

FIGURE 3.4. Photomicrograph of the alimentary tract of the spittlebug *Lepyronia coleoptrata* (Plate 13h). After initially exiting the filter chamber through the conical section, the gut eventually loops back around the exterior of the filter chamber and, re-exiting, intersects with the Malpighian tubules before ending at the rectum. Only two of the four Malpighian tubules are visible. oe = esophagus, fc = filter chamber, cs = conical section, amg = anterior midgut, pmg = posterior midgut, il = ilium, mt = Malpighian tubule, rc = rectum. A similar photo appears in Zhong et al. (2013), who discuss the anatomy in detail. Original photograph by H. Zhong, courtesy of Yalin Zhang, annotations added.

This creates an interface where water passes directly from the front of the digestive tract to near the end, apparently through the mediation of a class of water channel proteins (*aquaporins*) common to all xylem feeders (Le Caherec et al. 1997). Water transfer appears to be driven by passive osmosis, with the fluid in the receiving posterior midgut hyperosmotic (more concentrated) in relation to the very dilute incoming xylem sap (Cheung and Marshall 1973; Hubert et al. 1989). Peristaltic contractions of the well-developed muscles around the filter chamber apparently also play a role in pushing water through the interface (Marshall and Cheung 1974). The physical complexity of filter chambers and variation in structure among taxa (Goodchild 1966) suggests that their comparative morphology could shed light on the unsettled question of whether xylem feeding evolved only once, in the ancestral Cicadomorpha line, or re-evolved independently in the sharpshooter leafhoppers (Zhong et al. 2015).

Living Like a Spittlebug on a Diet of Caffeine

To put the nutrient challenges of xylem feeding in everyday perspective, consider the case of an ordinary human, the author for example, trying to get sustenance from a solo diet of the caffeine in black coffee. Humans require, on average, about 2,250 kcal per day in food intake. One 8-oz (237-ml) cup of coffee has about 95 mg of caffeine (at a concentration of about 2.0 mM, reminiscent of the molarity of organic compounds in xylem sap). Caffeine has about 90 kcal/mol (Huq 2006), which, at a molecular weight of 194 g/mol translates into 0.44 kcal per cup. However, 82% of the caffeine is metabolized to paraxanthine, which is excreted in urine (Institute of Medicine 2001). Subtracting the paraxanthine, only 18% of caffeine calories, or 0.08 kcal per cup, support metabolism. This would require human consumption of 1,750 gallons (6,650 liters) of black coffee per day to sustain life on the proceeds of caffeine alone, in the range of the human equivalent of 10,000 plus liters per day of xylem sap processed by *Philaenus spumarius*, taking body weight into account.

The Second Challenge: An Imbalanced Diet and an Excess of Nitrogen

The sheer poverty of available food energy for xylem feeders is compounded by the tremendous imbalance in organic compounds transported by xylem sap. Xylem sap organic solutes have very little sugar or other carbohydrates and are usually dominated by just one or a few amino acids of the twenty necessary for protein synthesis. The details differ by plant species and vary over the course of a day and seasonally, but in a wide variety of plants the dominant amino acids are asparagine/aspartic acid, glutamine/glutamic acid, and arginine (Anderson et al. 1989, 1992, 2003; Brodbeck et al. 1993; Pfautsch et al. 2015). Both xylem and phloem sap also have proteins but at low levels (Rodríguez-Celma et al. 2016), and a study of proteases in the digestive tract of the Brazilian sugarcane pest *Mahanarva fimbriolata* (Plate 14h) suggests that xylem sap proteins do not play a significant role in its nutrition (Pimentel et al. 2020). One recent study found that a substantial portion of the dilute organic xylem content in olive, citrus, and grape consists of sugars and alcohols (Surano et al. 2024). These observations are different enough from earlier work with other plant species to merit further examination,

Like all animals, insects lack the biosynthetic pathways for ten amino acids necessary for protein synthesis. Xylem sap feeders are chronically short on these ten essential amino acids (Koga and Moran, 2014), a problem common to all sap feeders, xylem or phloem (Douglas 2006). This exacerbates the poverty of xylem sap as a source of nutrients. It is also possible that the very large intake of xylem fluid leads to an overdose of some minerals. Nichol and Locke (1990) observe that *Philaenus spumarius* sequesters unusually large amounts of iron in the form of intracellular ferratin, an iron storage protein, and suggest that this is an adaptation against the potential for iron (and possibly copper and sulfur) overload due to large throughput xylem feeding.

Symbiotic Partners Provide Essential Amino Acids

Sap-sucking insects make up for the dietary deficiency of essential amino acids by hosting symbiotic bacteria (or in a few cases fungi) that can synthesize them from the nonessential amino acids present in sap. These symbioses are ancient and probably date back to the Permian origin of the Auchenorrhyncha (Koga et al. 2013; Bell-Roberts et al. 2019). Most spittlebugs host two intracellular bacterial symbionts with extremely reduced genomes: *Sulcia muelleri* and *Zinderia insecticola*. Together, they synthesize complementary sets of the ten essential amino acids, seven from *S. muelleri* and three from *Z. insecticola*. In the tribe Philaenini, including *Philaenus spumarius*, *S. muelleri* is replaced by a bacterium closely related to *Sodalis glossinidius*. This *Sodalis*-like symbiont has a substantially less-reduced genome and carries genes that may permit exploitation of the abundant nonessential amino acid glutamine for energy production, possibly contributing to the enormous host range of *P. spumarius* and to polyphagy in other Philaenini (Koga and Moran 2014; Thompson et al. 2023).

The spittlebug symbiotic partners are housed in specialized, paired abdominal structures called *bacteriomes*, generally with the complementary symbionts physically separated. The bacteriomes are often brightly colored, easily visible in nymphs through the translucent abdominal wall (Plate 2a, 2b, 2c, 2f, 2g) and in dissected adults. In *Clastoptera proteus*, the *Sulcia*-containing parts of the bacteriomes are bright red, and the *Zinderia*-containing portions are yellow (Ankrah et al. 2020). Red and yellow abdominal bacteriomes are also conspicuous in the first four instars of *Aphrophora* species nymphs (cf. Thompson 2021). On the other hand, bacteriomes are not always prominently visible, and this can be used to differentiate nymphs of similar species, such as *Clastoptera testacea*, in which they are prominent, from *Clastoptera querci*, in which, past the first instar, they are not (Thompson et al. 2020).

Xylem Sap Feeders Excrete Nitrogen as Diluted Ammonia, Conserving Carbon

Most herbivores suffer a deficiency of nitrogen (Mattson 1980). Despite the very low absolute concentration of nitrogen in their diet, xylem feeders have the opposite problem. When they metabolize amino acids for their carbon skeletons, they produce excess nitrogen, which they excrete as toxic ammonium ions. In contrast, most terrestrial organisms convert nitrogen to nontoxic urea or uric acid for excretion. Xylem feeders dilute toxic ammonia with large amounts of liquid and excrete it directly, a phenomenon called *ammonotelism*, otherwise confined to aquatic insects (Anderson et al. 1989). This permits relatively carbon-starved xylem feeders to avert the loss of precious carbon. Unlike urea and uric acid, ammonia has no carbon and no calories (Brodbeck et al. 1993). Paradoxically, despite living on an ultralow nitrogen diet, xylem feeders ingest excess nitrogen in relation to carbon, and as a result they are measurably "nitrogen-replete" (Ankrah et al. 2019); this is not surprising for creatures that subsist almost entirely on amino acids. Ammonotelism is an adaptation to both the abundant liquid throughput and the relative carbon starvation inherent in xylem feeding. Consistent with ammonotelism, Mello et al. (1987) found no uric acid in *Deois* spittlebug foam. Turner (2000) suggested that the formation of spittle foam may itself speed the diffusion of ammonia by increasing surface area and allowing it to volatilize more quickly.

The Third Challenge: A Food Source Under Negative Pressure and Energetically Costly to Extract

In addition to being dilute and nutritionally unbalanced, xylem sap is characteristically under negative pressure. Insects have to suck hard to get it out of the plant, an energetically costly exercise (Raven 1983). This is because in the ordinary course of events xylem sap is pulled up through the xylem tissue by transpiration in leaves, creating negative pressure. In contrast to phloem feeders, which encounter positive pressure, xylem feeders suck against a pressure equivalent to several times the pressure in an automobile tire, but negative. Xylem elements have thick walls comprised primarily of *lignin*, a tough, complex polysaccharide that helps them resist collapse under negative pressure but makes them all the harder for xylem feeders to penetrate.

Background: The Cohesion-Tension Theory of Xylem Sap Ascent

Discussion of spittlebug xylem feeding leads inevitably to one of the most important and most contentious theories in modern science, the cohesion-tension theory of xylem sap ascent (Brown 2013). Trees move xylem sap from roots to leaves with no expenditure of internal energy and no organs equivalent to the various sorts of pumps that circulate fluids in animals (Tyree and Zimmerman 2002; Brown 2013). They do so via a mechanism that has no equivalent in human technology. Thin, unbroken threads of water are pulled up through the xylem vessels by transpiration in the leaves, powered ultimately by the sun and the difference in water potential between the atmosphere, the interior of the leaves, and, ultimately, the soil around the roots (Tyree and Zimmerman 2002). This is possible because water, counterintuitively, has very high tensile strength, about one-tenth the tensile strength of steel, and because the structure of xylem elements permits these tiny water columns to reach unbroken from the leaves to the roots. The negative pressure against which xylem feeders pump is actually the internal tension on water in the xylem elements (see sidebar on Tensions and Pressures). Measured daytime tensions run from about –0.2 to –9.0 MPa (Stroock et al. 2014). MPa stands for *megaPascals*, a unit of pressure equal to about 145 pounds per square inch, 10 kilograms per square centimeter, or 10 atmospheres.

Tensions and Pressures: What's the Difference?

Technically, zero pressure corresponds to a void, and it is impossible to have pressures less than zero. As a result, the negative pressure-like pull associated with liquids is described as *tension*, analogous to the force pulling on a rope and employing the same units of measurement used to describe pressures, historically measured in atmospheres, with one atmosphere equal to 14.7 pounds/inch^2 (psi) or 10.2 kg/cm^2. Presently, the standard international pressure measurement is the Pascal, with one megapascal (MPa) equal to 145 psi, or approximately 10 atmospheres. Thus, when spittlebugs feed against negative tensions up to –1.5 MPa, they are feeding against pressure equivalents up to 215 psi, an extraordinary feat for small insects. By comparison, mosquitoes suck blood at 0.01 MPa, elephants suck water at 0.02 MPa, and bedbugs suck blood at 0.08 MPa (Kim 2013; Lambert 2021).

The cohesion-tension theory was first advanced in a remarkably lucid and succinct paper by Dixon and Joly (1894) in the Proceedings of the Royal Society of London. John Joly was a versatile physicist who later contributed to the development of radiotherapy for cancer treatment and the estimation of the age of geological periods based on radioactive decay. Henry Dixon was a twenty-five-year-old botanist. Both were teaching at Trinity College Dublin (Brown 2013). They were immediately attacked by those who thought it impossible that tiny threads of liquid water could exist under such high implied tension, let alone be pulled intact to such heights. Botanist Francis Darwin, son of Charles Darwin, called it a "violent disturbance of our current views" and complained that

> to believe that columns of water should hang in the tracheals [xylem elements] like solid bodies, and should, like them, transmit downwards the pull exerted on them at their upper ends by the transpiring leaves, is to some of us equivalent to believing in ropes of sand. (Darwin 1896)

Francis Darwin appears to have lifted the sand metaphor, perhaps unconsciously, from the Victorian novelist and commentator Samuel Butler. In a letter to Francis two decades earlier, Butler (1877) announced that his soon to be published book, *Life and Habit* (Butler 1878), would side with the evolutionary ideas of French zoologist Jean-Baptiste Lamarck's ideas over those of Charles Darwin and that "as for 'natural selection' frankly to me it now seems a rope of sand as in any way accounting for the 'origin of species.'" Evidently, Butler's metaphor was memorable enough to be resurrected by Francis Darwin in an entirely different controversy years later.

Although the cohesion-tension theory is widely accepted as the only compelling explanation of xylem sap transport (Brown 2013; Strook et al. 2014; Jensen et al. 2016), skepticism about some of its tenets persists. As recently as 2004, 110 years after Dixon and Joly's revelatory paper, forty-five scientists from seven countries signed a letter to the journal *New Phytologist* (Angeles et al. 2004) repudiating a published review that cast doubt on the theory (Zimmerman et al. 2004). Researchers rarely attack scientific adversaries in such strong, public terms, a measure of the vehemence of ongoing contention.

Powerful Sucking Ability Overcomes High Negative Xylem Tension

Spittlebugs managed to stick their beaks into the cohesion-tension controversy in the most literal possible way. In 1998, as part of an attack on the theory, Martin Canny (1998) proposed that the very fact that spittlebugs are able to feed from xylem elements is evidence of low xylem tension. Based on calculations in

Raven 1983, which he neglected to credit, Canny asserted that, at most, spittlebugs could draw against a maximum tension of –3 atmospheres (–0.3 MPa), much lower than tensions spittlebugs are predicted to encounter by the theory. A year later, a British group (Malone et al. 1999) returned fire with a paper mustering an array of direct and indirect evidence that *P. spumarius* feeds directly from xylem elements under high, cohesion-tension theory-compatible negative pressures. Kim (2013), making explicit reference to Raven's calculations, echoed this debate without taking a stand. John Raven himself told me in 2023 that the first version of his 1983 paper was rejected, largely because the reviewers felt he overestimated the strength of negative xylem tension. A later dinner party encounter with an Academic Press editor eventually provided an avenue to invited publication.

Using microCT scans of the piston-like cibarial pump and the muscles driving it, Bergman et al. (2021) revisited the ability of *P. spumarius* to suck against negative tension. They calculate that *P. spumarius* can feed at pressures up to about –1.5 mPa, easily encompassing the range of xylem tensions it would encounter in its loçal hosts and many other plants (Stroock et al. 2014) and consistent with earlier work by Andersen et al. (1992) showing that the xylem-feeding leafhopper *H. vitripennis* can feed at tensions up to about –2.0 MPa. For perspective, ordinary positive pressure in an automobile tire is about 0.25 MPa.

Based on measurements of metabolic rates in feeding and resting insects, Bergman et al. (2021) calculate that *P. spumarius* can extract net positive energy from the dilute xylem sap, even at high negative tensions. This work demonstrated that spittlebugs solve the challenge of negative xylem tension with powerful sucking ability. It also got considerable press. Science News labeled meadow spittlebugs "the super-suckers of the animal world" (Lambert 2021). Phillip Matthews, the coauthor who supervised the work, told the *New York Times* that the large cibarial muscles were "like a huge bicep on their head," adding that their sucking powers were such that sitting on the top of the torch of the Stature of Liberty a spittlebug could drink through a straw from a glass of water on the ground (with the caveat that it be an unbroken column of water) (Imbler 2021). The lead author, Elizabeth Bergman, who performed the work as a master's thesis, described the bugs as "awesome sucking machines" (Lambert 2021).

Why Don't Xylem Feeders Induce Catastrophic Cavitation? A Remaining Mystery

Though widely accepted, the cohesion-tension theory has an Achilles heel that bears directly on xylem feeding insects. At 20°C, water in a vacuum has the potential to boil at or below positive 0.0023 MPa (Tyree and Zimmerman 2002; Kim 2013). As a result, any negative tension renders xylem sap fundamentally unstable. At a given moment, the water in the xylem elements of actively transpiring plants (as well as nontranspiring plants over 10 m in height) is primed to vaporize, or cavitate, generating bubbles that can expand to form an embolism that interrupts the water column, destroying the continuity of the ropes that elevate the water. If rampant and unchecked, cavitation and embolism would be a disaster for plants, and the natural world as we know it could not exist.

Supporters of the cohesion-tension theory have recognized this issue since its inception (Dixon and Joly 1894). They point out that while water under xylem-level tension is thermodynamically unstable, it is, in theory, kinetically stable up to tensions of more than –100 MPa (Strook et al. 2014; Jensen et al. 2016), tensions far beyond any that would occur in a plant in nature. In the absence of preexisting small bubbles or external damage that leads to infiltration of bubbles, xylem water will continue indefinitely in metastable condition, held together by the internal cohesion of hydrogen bonds and by adhesion of water to the walls of the xylem conduits. The relationship of tiny bubbles to cavitation is analogous to ordinary boiling, in which steady trains of bubbles emanate from small gas nuclei trapped on the wall of a pan (Tyree and Zimmerman 2002; Kim 2013; Brown 2014).

Problems arise when the water column is disrupted by outside forces. Freezing and thawing, for example, can precipitate bubbles that embolize and break the water columns, and mechanical damage to xylem tissue can let in air. Drought can raise negative tensions so high that air leaks in through portions of the xylem column walls otherwise impervious to such breaches. A lot of the current literature on cohesion-tension theory centers on whether there are mechanisms that prevent the growth of tiny cavitation bubbles (Yang et al. 2020) and whether and how embolisms caused by cavitation might be reversed (Strook et al. 2014; Jensen et al. 2016).

For spittlebugs, the vulnerability of xylem vessels to cavitation raises a fundamental question (Raven 1983). Xylem-feeding insects must poke holes in xylem conduits to feed. Why doesn't this lead to embolisms that block feeding altogether? The answer apparently lies in the ability of spittlebugs to form strong, thick salivary sheaths as their stylets advance through plant tissue. Crews et al. (1998) demonstrate that unlike aphid sheaths, which end at the outer surface of phloem vessels, the *P. spumarius* sheath penetrates and spreads out inside xylem vessels, suggesting that it forms a seal impenetrable to gases. They found no evidence of embolisms. This is all the more impressive because *P. spumarius* punctures are "quite brutal" (Carle and Moutous 1965), causing a lot of damage to living cells and piercing xylem vessel walls with "particularly spectacular" openings, "occupying almost one-quarter of the circumference of the vessel" (Crews et al. 1998; and see Withycombe 1926). Ironically, one of the Crews study coauthors is the same Martin Canny who, as noted previously, used the fact of spittlebug feeding to question the very existence of strong negative xylem tensions (Canny 1998). A later experiment using X-ray videography of stylet penetration of grapes by glassy-winged sharpshooters provides corroboration for the *P. spumarius* results, showing no evidence for air in the stylet tubes or in punctured xylem vessels and no feeding-induced embolism (Backus and Lee 2011). This work, never formally published, needs confirmation.

Mittler (1967) observed that xylem-feeding insects must establish an unbroken water column under full xylem tension in their food intake canal and wondered how they do this without precipitating cavitation. Barbara Nickel (cited in Crews et al. 1998) suggests that egestion of salivary fluid down the stylets at the moment of penetration serves the additional purpose of maintaining a fluid continuum from the unbroken water column to the insect head. If so, the transmission of the bacterial pathogen *Xylella fastidiosa* with egested fluid (Chapter 9) may be, in part, a side effect of a mechanism that preserves a column of uninterrupted xylem sap at full negative tension right into the head. This issue merits further investigation.

It is possible that cavitation resistance has direct consequences for economic damage by spittlebugs. In a study of sugarcane resistance to damage by spittlebug nymphs feeding on roots, Melo et al. (2018) found that a susceptible commercial strain had roots with large internal air chambers (aerenchyma) surrounding the xylem and phloem vessels, while a resistant wild cane species did not. It is plausible that the presence of air conduits so close to the xylem vessels raises the probability of embolisms caused by feeding punctures. Perhaps more directly pertinent, there is evidence that differential resistance to cavitation and the ability to repair cavitation-caused embolisms may influence the susceptibility of olive cultivars to *P. spumarius* vectored *Xylella fastidiosa* (Sabella et al. 2019; Walker et al. 2023; and see Chapter 9).

Why Feed on Xylem at All?

Given the litany of negative factors associated with xylem feeding, why feed on xylem sap at all? The same host plants offer an alternative, much richer food source in the form of phloem sap under positive pressure. The likely answer is that phloem sap offers its own steep set of challenges (Douglas 2006). Like xylem sap, phloem sap is nutritionally unbalanced and short on essential amino acids. Unlike xylem sap, it is also relatively short on nitrogen compared to carbon. As a consequence, phloem feeders have to process excess carbohydrates to get enough amino acids and nitrogen, resulting in sugar rich excreta. Worse, the very high concentration of organic solutes in phloem sap creates a tremendous osmotic challenge. High molarity phloem sap passing through the digestive tract will tend to pull water out of body tissues, desiccating an insect as it feeds. In effect, phloem feeders are at risk of dying of dehydration while feeding on a liquid diet, whereas xylem feeders are at risk of starving to death for lack of sufficient dietary carbon while processing enormous amounts of sap. One commentator called spittlebugs "stupid xylem parasites" in contrast to "smart phloem feeders" (Turner 2000); given the evident evolutionary tradeoffs, this is a one-sided declaration. Xylem feeders have thrived for over 200 million years. They must be doing something right.

Xylem versus phloem feeding is not an absolute dichotomy. Quite a few phloem feeders, including leafhoppers (Chuche et al. 2017), aphids (Pompon et al. 2011), and psyllids (Ebert et al. 2018), tap into xylem sap now and then, likely for purposes of maintaining water balance. There seem to be no direct reports of xylem feeders tapping into phloem sap. This is a bit of a mystery. One might suppose that tremendous selective advantage would accrue to xylem feeders that could supplement their meager xylem carbon resources with a bit of carbohydrate rich phloem sap now and then. The absence of evidence for such behavior (Cornara, Garzo et al. 2018; Markheiser et al. 2023) invites evolutionary explanation. Perhaps evolution from xylem to phloem feeding is a one-way street (Bell-Roberts et al. 2019,) and phloem feeding ability, once evolved, is so dietarily lucrative that the balance shifts decisively in its favor, with the relict ability to tap into xylem a vestige of evolutionary history.

Fluctuations in Xylem Tension: Consequences for Spittlebug Feeding

Negative Xylem Tension Diminishes at Night

Xylem sap is not always under negative tension. At night, in the absence of transpiration, xylem tension comes into equilibrium with soil water potential but does not equal that potential, which is due to gravity increasing negative tension in the water column by 0.1 MPa for every 10 meters in height. For herbs and shrubs this effect is inconsequential. For trees of 100 m, nighttime treetop negative tension would be a substantial 1.0 MPa (Jensen et al. 2016). This suggests that, outside elevated tree canopies, spittlebugs feeding at night will encounter low xylem tensions. On the other hand, they will also face a static xylem stream with flow rates close to zero (Jensen et al. 2016). How these conflicting factors play out in feeding behavior is unknown. In this context, it may be of interest that *P. spumarius*, despite its very broad host range, is underrepresented on plants with crassulacean acid metabolism (CAM) (Thompson et al. 2023). CAM species shut off daytime transpiration as an adaptation to arid environments, resulting in a static column of sap that may be an unattractive and unrewarding food source.

Root Pressure Sometimes Leads to Positive Nighttime Xylem Pressure

When water potential in the soil exceeds that in roots, root pressure can sometimes lead to positive nighttime xylem pressure (Schenk et al. 2021). Theoretically, spittlebugs feeding at night on plants with positive root pressure have the xylem sap pushed into them, analogous to aphids feeding on positive pressure phloem sap. This would provide an energetically free source of sap for a substantial portion of the diurnal cycle. As an example, positive root pressure apparently permits bamboos to refill embolized xylem vessels from the roots (Cao et al. 2012). This might account, in part, for the pest status of the spittlebug *Aphrophora notabilis* on *Phyllostachys meyeri*, a bamboo observed to exhibit positive nighttime root pressure (Chapter 8).

Positive nighttime root pressure is sometimes manifested in a phenomenon called *guttation*, wherein plants bleed drops of water from their leaf tips or edges (Singh 2016). Guttation was first observed in taro, *Colocasia esculenta* (Schenk et al. 2021). Perhaps not coincidentally, in Vietnam, I have observed the spittlebugs *Eoscarta borealis* on *Colocasia gigantea* and *Ectemnonotum fruhstorferi* on *Alocasia odora* (Plate 12f), both close taro relatives. Guttation has also been observed in many plants that host spittlebug pests, including C4 grasses (see sidebar: C4 Photosynthesis and C4 Plants), strawberries, and rice (Chapter 8) (Glen and Takeda 1989; Kerstetter et al. 1998). In Costa Rica, I have waded through pastures wet with morning guttation droplets and heavily infested by the spittlebug pest *Zulia vilior* (Plate 10f). These anecdotal observations suggest that positive xylem pressure in plants exhibiting guttation may contribute to their attractiveness as spittlebug hosts. Grass roots also exhibit nighttime guttation (Schenk et al. 2021), a particularly intriguing observation, given the importance of spittlebugs with root feeding nymphs as pests of Neotropical pasture grasses (Chapter 7).

Root Feeding is Common but Far from Ubiquitous in Spittlebugs

One of the corollaries of the cohesion-tension theory is that negative tension should be greatest at the apex of a plant and least in the roots (Kim 2013). As pointed out by Raven (1983), this should give an incentive to xylem feeders to tap into roots, which is exactly where all cicada nymphs feed. Root feeding is common but far from ubiquitous in spittlebugs. It occurs most often in Cercopidae nymphs. It is the rule, for example, in the grass feeding New World pest genera *Prosapia*, *Aeneolamia*, *Deois*, *Zulia*, and *Notozulia* (Plates 10, 14, 15). It is also common but not universal in the sugarcane/grass pest genus *Mahanarva* and appears to be a prominent mode of spittlebug feeding in the surface root mat on the Amazon forest floor (Penny and Arias, 1982).

Nymphal root feeding may have been present among the New World Cercopidae from their origin. In the earliest branching clade, the Microsarganini (Hamilton 2016; Paladini, Takiya et al. 2018), the only species with known nymphal ecology is *Liparonotum praenitida* (formerly *Zuata praenitida*). It lives on roots but not in soil, making spittles on the trunk-hugging roots of plants epiphytic on large rainforest trees (Hamilton and Thompson 2007). In *Ocoaxo assimilis*, representing a later branching clade, early nymphal instars feed on the roots of multiple hosts (Cid-Muñoz et al. 2020). The root-feeding New World grass pest Cercopidae appear to have evolved even later, around the late Miocene when C4 grasses first became common (Thompson 2004; Paladini et al. 2015; Paladini, Takiya et al. 2018).

Root-feeding nymphs have been observed in four genera of Old World Cercopidae: *Cercopis* (Plate 1b), *Haematoloma*, *Callitettix*, and *Abidama*. They all feed on grass roots: *Cercopis* spp. and *Haematoloma dorsata* on wild grasses (China 1927; Biedermann 2002), *Callitettix versicolor* and *Abidama liuensis* on the roots of rice (Chen and Liang 2012; Wang et al. 2021). *Callitettix versicolor* nymphs also feed on roots of sweet potatoes and soybeans (Lei et al. 1992). *Cercopis* lies at the base of one branch of the Old World Cercopidae (Crispolon et al. 2023). *Callittettix* and *Abidama* are very closely related genera (Liu et al. 2014) embedded in a clade further out on the same branch. The presence of nymphal root feeding in *Cercopis* suggests that it might be an ancestral character of Cercopidae, shared with the New World cercopids. Resolution of this issue will require refined phylogenies and more extensive study of the nymphal ecology of Old World Cercopidae.

A few cases of root-feeding nymphs are known in the Aphrophoridae. *Philaenarcys killa* nymphs feed on the roots of the grass little bluestem, sometimes in the company of nymphs of the cercopid *Prosapia ignipectus* (Thompson and

Carvallo 2016). In western North America, *Philaronia abjecta* and several species of the genus *Aphrophora* are reported to feed on the roots and/or root crowns of herbaceous plants, as well as higher up on some hosts (Hamilton 1982).

Tellingly, there are no reports of root feeding in the Clastopteridae or Machaerotidae, the groups that appear to have evolved earliest among living spittlebugs (Chapter 2). This argues strongly against root feeding as a primitive trait, as does the fact that root feeding may correlate with the loss among root feeders of the Batelli glands, which seem otherwise to be universal in Cercopoidea (Marshall 1965b). Additionally, Raven's hypothesis suggests that the benefits of root feeding would be most prominent among tree feeders, like cicadas, rather than among feeders on the grasses and herbs, a category including many root-feeding spittlebugs. Taken all together, it appears that root feeding evolved in spittlebugs in response to selective factors other than relatively low negative xylem tension in roots, perhaps in adaptation to positive nighttime root pressure, which, as noted previously, may make some hosts attractive to root-feeding nymphs. There appears to be only a single root-feeding leafhopper, *Xestocephalus desertorum* (Rakitov 2000), a phloem feeder.

An Unusual Instance of Positive Xylem Pressure Associated with Unique Aquatic Spittlebug Nymphs

Nymphs of two Costa Rican spittlebug species, *Mahanarva costaricensis* and *Mahanarva insignita*, can live in water-filled *Heliconia* flower bracts (Plate 4), using their ventral breathing tubes as a snorkel at the water surface (Thompson 1997). The plants extrude water into the bracts, independently of rain, strongly implying at least intermittent positive xylem pressure. Presumably, the feeding spittlebugs benefit from this positive pressure as well as making protective use of the mini-aquatic habitat provided by the flower bracts. The recently described Mexican species *Mahanarva jurael* probably shares the same niche (Castro-Valderrama et al. 2018). Along with nymphs of a few odd African cicadas (Boulard 1991), these are the only known aquatic Auchenorrhyncha. Their evolution could perhaps have been anticipated since spittlebug nymphs are preadapted to micro-aquatic habitats in the form of spittle (Chapter 4). The aquatic lifestyle is not obligatory. The nymphs can also live like ordinary spittlebugs on leaves of *Heliconia* that are not in flower.

Xylem Feeding and Body Size: Is Bigger Better?

On average, xylem feeding spittlebug, cicada, and leafhopper adults have larger body sizes than phloem and cell content-feeding leafhoppers. In two tropical/

subtropical faunas, India and Vietnam, all of the largest species and none of the smallest are xylem feeders, although the mean sizes of phloem- and xylem-feeding leafhoppers are statistically indistinguishable (Novotny and Wilson 1997). Initially, calculations of fluid flow through narrow stylet channels and the cibarial muscle mass necessary to pump such flow suggested that small xylem feeders are energetically disadvantaged (Novotny and Wilson 1997). However, recent analysis indicates that body size has no effect on the xylem-feeding capacity of third through fifth instar *Philaenus spumarius* nymphs (Bergman and Matthews 2024). Size seems not to matter, but, in retrospect, the fact that first instar nymphs, often only a millimeter or so in length (Plate 2a, 2b), feed perfectly well on xylem sap indicates that tiny size is not a serious impediment to xylem feeding.

This leaves the question of apparent minimum adult body size for xylem feeders up in the air. The smallest spittlebugs tend to be globose, wide for their length, making them relatively more robust and perhaps fortifying their sucking power relative to length. *Clastoptera* (Plate 5a, 5d, 5e), ranging from about 3.0–5.5 mm in length, fit this pattern, as do related genera like *Iba* in the Philippines and *Tremapterus* (Plate 6e) in southern Africa. However, some *Clastoptera* and *Tremapterus* species are associated with hosts likely to have high negative xylem tension (see below), calling into question the suggestion (Novotny and Wilson 1997) that diminutive size constrains small xylem feeders to feed on plants with low xylem tension. I also note that first instar nymphs have proportionately large heads (Plate 2a, 2b), consistent with the notion that relatively larger heads facilitate xylem feeding in smaller individuals.

How Xylem Feeding Influences the Spittlebug Host Plant Repertoire

Polyphagy: A World Championship for Spittlebugs

Raven (1983) seems to have been the first to observe that xylem feeders tend to feed on a multiplicity of diverse host plants, a phenomenon known as polyphagy. Spittlebugs in particular are associated with an extraordinarily wide range of hosts, thought to be a direct reflection of feeding on xylem sap, a poorly defended resource common in character across diverse plant taxa (Press and Whittaker 1993; Thompson 1994). Phloem sap has a relatively higher concentration of proteins, some of which are likely but not proven to be involved in defense against sap feeders (Kehr 2006). In contrast, xylem sap has very dilute defense proteins, which seem to be aimed at microbial pathogens (Rodríguez-Celma et al. 2016). I maintain a database of host associations. As of late 2024, it included

TABLE 3.3. Spittlebug primary hosts by nature of specialized nitrogen acquisition

PRIMARY HOST PLANT CATEGORY	NUMBER OF SPITTLEBUG SPECIES	PERCENTAGE OF SPITTLEBUG SPECIES
Rhizobial N-fixation	35	7.0
Actinorhizal N-fixation	11	2.2
Associative N-fixation	112	22.4
Ectomycorrhiza	89	17.8
Ericoid mycorrhiza	3	0.6
Xylem parasite[a]	6	1.2
No special N acquisition	262	52.5
Total	499	103.7

Source: Thompson 2022.
Note: Because a few species were counted in two categories, the percentages total is a bit over 100.
[a] Xylem parasitic plants tap into the xylem elements of host plants, diverting xylem stream nutrients.

about 3,000 spittlebug host plant species worldwide. This approaches 1% of the total number of known vascular plants, about 370,000 (Christenhusz and Byng 2016). These numbers apply only to spittlebugs with known hosts, just 500 or so of the 2,500 described species (Tables 1.1, 3.3).

The most spectacular manifestation of polyphagy is the host repertoire of *Philaenus spumarius*, the world champion herbivorous insect polyphage. It has 1,311 recorded host species in 117 plant families, about twice the host number of any other insect (Thompson et al. 2023). This is particularly impressive given that *P. spumarius* apparently originated from a monophagic lily-feeding ancestor as recently as 7.9–3.7 mya (Maryańska-Nadachowska et al. 2010).

Many other spittlebugs have broad host ranges. Their extreme host plant versatility permits them to move onto completely unfamiliar, exotic plant taxa. Two notable examples are the movement of multiple species of the New World spittlebug genus *Clastoptera* onto introduced species of *Casuarina* (refs. in Thompson et al. 2020; Thompson 2022) and the movement of native spittlebugs onto introduced *Eucalyptus* species at pest levels in Brazil (Ribeiro et al. 2005), China (Pang 2013), and Africa (Lepesme 1947). Host versatility is also reflected in the movement of dozens of native New World spittlebug species onto introduced Old World grasses, a much smaller leap taxonomically, but one with enormous economic consequences (Chapters 6 and 7).

A Predilection for Nitrogen-Fixing and Ectomycorrhizal Hosts

Within their very broad host range, spittlebugs favor two notable host categories: nitrogen-fixing plants and ectomycorrhizal plants. In nitrogen-fixing plants, symbiotic bacteria catalyze otherwise unassimilable atmospheric nitrogen and pass it to the plant as ammonium ions. Plant roots incorporate the ammonium into amino acids, which are transported from root to shoot in xylem sap available to xylem-sucking insects (Thompson 1994, 1999, 2004). In ectomycorrhizal symbioses, root-associated fungi provide access to otherwise inaccessible nitrogen sequestered in complex soil organic matter (Thompson 2022). As in nitrogen-fixing associations, this nitrogen is passed to the plant roots as ammonium ions, incorporated into amino acids, and transported as xylem-borne organic nitrogen compounds available to xylem feeders. In both cases, plant photosynthesis provides the amino acid carbon skeletons via phloem transport of organic compounds to the roots.

Judging by host associations, augmented transport of organic compounds makes the symbiotic host plants especially attractive to spittlebugs. Half of the 500 or so spittlebugs with known hosts (50.4%) feed primarily on nitrogen-fixing or ectomycorrhizal plants (Table 3.3) (Thompson 2022). Three more have ericoid mycorrhizal hosts, which are physiologically equivalent to ectomycorrhizae, and six have hosts that are *xylem parasites*, plants that tap into the xylem stream of other plants. One of these xylem parasite spittlebugs, *Clastoptera distincta*, lives as nymphs on a mistletoe parasite of pines, making it a xylem hyperparasite (Thompson 2011). The remainder are spittlebugs associated with *Santalum album*, a commercial tree that produces aromatic heartwood and parasitizes roots of other trees. Most reports of *S. album* spittlebugs come from old studies screening for potential vectors of sandalwood spike disease (Chatterjee and Bose 1933) and would benefit from confirmation. Notably, *S. album* is the only recorded host of the primitive clastopterid/machaerotid line spittlebug *Beesoniella sylvestris* (Fig. 2.10), which was first described from specimens collected from *S. album* (Lallemand 1933; Subra Rao et al. 1988). As a coda to these examples, *Neophilaenus lineatus* (Plate 11d) placed in direct competition with the plant xylem sap parasite *Rhinanthus minor* experienced elevated mortality, evidently from the plant parasite's diversion of amino acids (Bass et al. 2010).

About two-thirds of the spittlebugs linked with nitrogen-fixing plants feed on grasses that exhibit associative nitrogen fixation. They trade carbohydrates for nitrogen from free-living root zone bacteria or from symbiotic bacteria living in plant tissues other than root nodules. With the exception of rice and bamboo, these grasses have C4 photosynthesis (see sidebar C4 Photosynthesis and C4 Plants), an especially efficient type of photosynthesis that promotes high levels of carbohydrate production (Thompson 2004). Most importantly, C4 grasses host the sugarcane and pasture pests discussed in Chapters 6 and 7. While associative nitrogen fixation has been widely reported among C4 grasses, its quantitative importance to nitrogen nutrition is not as well established at that in legumes and other nodulated nitrogen-fixing plants (Thompson 2004;

Chalk 2016). In recent years, research on associative nitrogen fixation has focused on bacteria living within host plant tissues (cf. Carvalho et al. 2014; Dong et al. 2018; Baldani et al. 2024), especially in sugarcane, the spittlebug host for which the importance of associative nitrogen fixation is best established (Thompson 2004).

C4 Photosynthesis and C4 Plants

C4 plants produce unique, 4-carbon photosynthetic metabolic intermediates, compounds involved in the assimilation (fixation) of atmospheric carbon dioxide (CO_2) using energy from sunlight. Along with specialized leaf anatomy, this enables C4 plants to concentrate intracellular carbon dioxide, making the critical photosynthetic enzyme RuBisCo far more efficient than in comparable plants with ordinary C3 photosynthesis. With adequate water, C4 plants gain a competitive edge and higher productivity in warm, sunny habitats. Most C4 plants are grasses and sedges. Excepting rice and bamboo, all cultivated tropical grasses, including important hosts of spittlebug pests, have C4 photosynthesis, which, in turn, is often linked with associative nitrogen fixation, probably because plants with high photosynthetic efficiency make enough surplus carbon compounds to feed associated nitrogen-fixing bacteria (Sage 2004; Thompson 2004).

As in associative nitrogen fixation, the quantitative contribution of ectomycorrhizal fungus-derived nitrogen is hard to pin down. These fungi are taxonomically diverse and exhibit tremendous interspecific variability in their ability to scavenge organic nitrogen. Roots of individual ectomycorrhizal trees or shrubs, moreover, may host dozens of fungal species, mightily complicating analysis (Thompson 2022). In both cases, the quantitative uncertainties make it difficult to know precisely the extent to which these specialized forms of nitrogen acquisition ultimately contribute to amino acids in spittlebug-sustaining xylem sap. We do know that the spittlebugs gravitate to nitrogen-fixing and ectomycorrhizal hosts, suggesting that their contributions are important.

Many Other Factors Influence Spittlebug Host Choice

About half of all spittlebugs feed primarily on plants that do not fix nitrogen or engage in ectomycorrhizal symbioses (Table 3.3). Host accessibility is governed in part by simple mechanical considerations. Hoffman and McEvoy (1985a, 1985b) have demonstrated that hairy and hardened stems of herbs deter feeding by *P. spumarius* nymphs, particularly early instars, and Thompson et al. (2023) note that when *P. spumarius* nymphs feed on trees and woody shrubs they invariably do so at the new growing tips of branches or on adventitious shoots. Adult spittlebugs are more versatile in woody host choice, and some can even feed on tree trunks, such as *Suracarta tricolor* in Sabah (Plate 12b) and several species of cloud forest Cercopidae in Costa Rica (K. Nishida observations).

Beyond these physical considerations, spittlebugs seem to follow the trail of xylem-borne amino acid availability. In lab experiments with cocklebur, Horsfield (1977) demonstrated that *P. spumarius* nymphs selected leaves with higher amino acid concentrations and experienced higher mortality on leaves with low amino acid concentrations. He also demonstrated that both nymphs and adults processed more xylem sap when amino acid levels in leaves were higher, presumably stimulated to feed more. By far the most extensive set of observations of xylem feeding in relationship to amino acid concentrations was carried out over the course of two decades by Peter Anderson, Brent Brodbeck, and Russel Mizell at the University of Florida. Working with the xylem-feeding leafhopper pest *Homalodisca vitripennis*, they demonstrated that its feeding preferences track variations in the amino acid content of sap diurnally, seasonally, and among hosts (Brodbeck et al. 1990, 1993). In all likelihood, spittlebugs do the same but have not been studied in equal detail.

Are Any Plants Off Limits to Spittlebugs?

While the xylem sap of most plants is not chemically defended, tobacco xylem transports relatively high concentrations of the insect poison nicotine. This spurred Raven (1983) to suggest that spittlebugs do not feed on tobacco, an idea supported by the absence of tobacco from the most comprehensive host list of the extreme polyphage *P. spumarius* available at the time (Weaver and King 1954). However, there are observations of *P. spumarius* (Beirne 1972), the neotropical spittlebug *Huaina inca* (Maes and Tellez 1988), and the East Asian spittlebug *Callitettix versicolor* (Lei et al. 1992) feeding on and even damaging tobacco, indicating that spittlebugs can breach what has the appearance of a formidable defense. Spadavecchia et al. (2024) have demonstrated experimentally that spittlebug adults can feed on tobacco but experience substantially higher mortality rates (93%) than on five alternative hosts (7–39%). It has also been suggested that *P. spumarius* avoids feeding on the legume forage plant *Coronilla varia* (synonym *Securigera varia*) and species of the milkweed genus *Asclepias*, though neither case is certain, and there are recorded *Asclepias* hosts for other spittlebugs (details in Thompson et al. 2023). Spittlebugs are also pests of related giant milkweeds in South Asia (Chapter 8).

Woody mangroves do appear to be off limits to spittlebugs. The eighteen taxonomically diverse genera of tree and shrub mangroves grow with roots in salt water (Tomlinson 2016). None are recorded as spittlebug hosts. Pulling against the salt water osmotic gradient requires overcoming very high xylem tensions, measured on the order of –2 to –5 MPa (Stroock et al. 2014; Jensen et al. 2016), well beyond the apparent –1.5 MPa suction limit of *P. spumarius* (Bergman et al. 2021). However, in Puerto Rico nymphs of *Clastoptera signifera* (*C. brevis*) live on the fern *Acrostichum aureum* (Ramos 1957), a nonwoody, facultative mangrove species found in freshwater swamps as well as saltwater habitats.

Ecological Hints on Where to Look for Champion Spittlebug Sucking Power

In terms of sucking power, the champion spittlebug is probably yet to be found. *Clastoptera* (Plate 5a, 5d, 5e) are a promising group. Several species, including *C. lineatocollis*, *C. delicata*, and *C. brunnea*, are common on arid habitat shrubs in the American southwest (Doering 1928, 1942), suggesting that they feed successfully against the high negative xylem tensions characteristic of plants in dry soils. *Clastoptera distincta* nymphs feed in even more extreme conditions on a mistletoe parasite drawing xylem sap from pines during the height of the annual monsoon climate drought (Thompson 2011). In addition, several junipers host other *Clastoptera* species in the American West (Thompson 2023), juniper having recorded xylem tensions in the –4 to –5 MPa range (Stroock et al. 2014). Despite their small size, these *Clastoptera* are all good candidates for sucking power effective against tensions in excess of –2.0 MPa. Species of the genus *Tremapterus* (Plate 6e) inhabiting arid habitats in southern Africa are also good candidates.

Conclusions

Xylem feeding is an unusual nutritional mode. It poses special challenges, which have been elaborated in this chapter, and has profound consequences for spittlebug morphology, ecology, host selection, evolution, and agronomics. The agronomic consequences will be explored in Chapters 6 through 9. The next chapter focuses on the signature product of xylem feeding in spittlebugs, the spittle masses themselves.

Plate 1. Spittles. (**a**) Exposed *Philaenus spumarius* molting chamber revealing newly emerged (teneral) adult and shed fifth instar nymphal skin, Austria. © Gernot Kunz. (**b**) *Cercopis* sp. nymph in spittle mass on root, Austria. © Gernot Kunz. (**c**) *Ptyelus flavescens* adult, South Africa. Marlies Craig. (**d**) *P. flavescens* nymphs in communal spittle mass, South Africa. Marlies Craig.

Plate 2. Nymphs. (**a**) *Philaenus spumarius* first instar nymph. Picciotti et al. 2021 *https://creativecommons.org/licenses/by/4.0/*, edited. (**b**) *Neophilaenus campestris* first instar nymph. Picciotti et al. 2021, *https://creativecommons.org/licenses/by/4.0/*, edited. (**c**) *Cercopis vulnerata* early instar nymph, Austria. © Gernot Kunz. (**d**) *Clastoptera funesta* nymphs on *Ruellia simplex*, Costa Rica. © 2024-Kenji NISHDA. (**e**) Fourth instar *Aphrophora moscoviciae* nymphs on *Carpobrotus edulis*, California. V. Thompson. (**f**) *A. moscoviciae* fourth instar nymph in ventral view, from specimen in ethanol. V. Thompson. (**g**) *Aphrophora salicina* fifth instar nymph, Austria. © Gernot Kunz.

Plate 3. A striking spittlebug from the Philippines and fossils from the age of dinosaurs. (**a**) Late stage fifth instar *Oxymegaspis* sp. nymph in "nubby" spittle mass, Luzon. Adeline Soulier. (**b**, **c**, and **d**) Emerged *Oxymegaspis* sp. adult with magnificently flared pronotum: head on view, dorsal view, and lateral view. @Elizabeth Huber. (**e**) *Burmocercopis lingpogensis*, and (**f**) *Paranthoscytina xiai*, two Cretaceous Procercopidae species stunningly well preserved in amber. Courtesy of Diying Huang. (**g**) *Jiaotouia burmitica*, a Cretaceous Sinoalidae species in amber. Courtesy of Jun Chen.

Plate 4. *Mahanarva costaricensis* spittlebugs with aquatic nymphs, Costa Rica. (**a**) *Mahanarva costaricensis* spittle on *Heliconia wagneriana* flower bract, Costa Rica. Kristy Morley https://creativecommons.org/licenses/by/4.0/, edited. (**b**) Spittle inside water-filled bract, Costa Rica. V. Thompson (**c**) Aggregated *Mahanarva costaricensis* adults, Costa Rica. © 2024-Kenji NISHDA. (**d**) *Mahanarva costaricensis* adult, Costa Rica. © 2024-Kenji NISHDA. (**e**) Diving aquatic *Mahanarva costaricensis* fifth instar nymph with ventral abdominal breathing tube at water/air interface. V. Thompson.

Plate 5. Spittle capsules, *Clastoptera* and Epipygidae. (**a**) Costa Rican *Clastoptera* sp. that makes capsules rather than ordinary spittle. © 2024-Kenji NISHDA. (**b**) Costa Rican *Clastoptera* sp. capsules arrayed on a stem. © 2024-Kenji NISHDA. (**c**) Similar capsules made by nymphs of an unknown Philippine spittlebug. Elorde Crispolon. (**d**) *Clastoptera achatina*, Georgia. E. Christina Butler, CC BY 4.0, Attribution 4.0 Int. Deed, edited. (**e**) *Clastoptera cimicoides*, Brazil. Note prominent false eye spots. Thales Santos. (**f**) *Eicissus* sp. (Epipygidae), Costa Rica. © 2024-Kenji NISHDA.

Plate 6. Tube spittlebugs. (**a**) *Machaerota spangbergii*, India. © Judhajit Dasgupta. (**b**) Multiple nymphal tubes of an unidentified *Machaerota* sp. Note small early instar tubes at bases of larger later instar tubes, China. Jiang Guobin. (**c**) *Machaerota philippinensis* nymphal tube on *Premna serratifolia* (Lamiaceae), Philippines. @Elizabeth Huber. (**d**) Newly emerged *M. philippinensis* adult, nymph having made spittle at tube tip preceding molting (nymphal skin visible under bubbles). @Eliazabeth Huber. (**e**) *Tremapterus* sp., South Africa. © Cecile Roux. (**f**) Tubes built by *Tremapterus* nymphs on *Grewia* (Malvaceae), Botswana. Tony Benn © Botswanabugs.

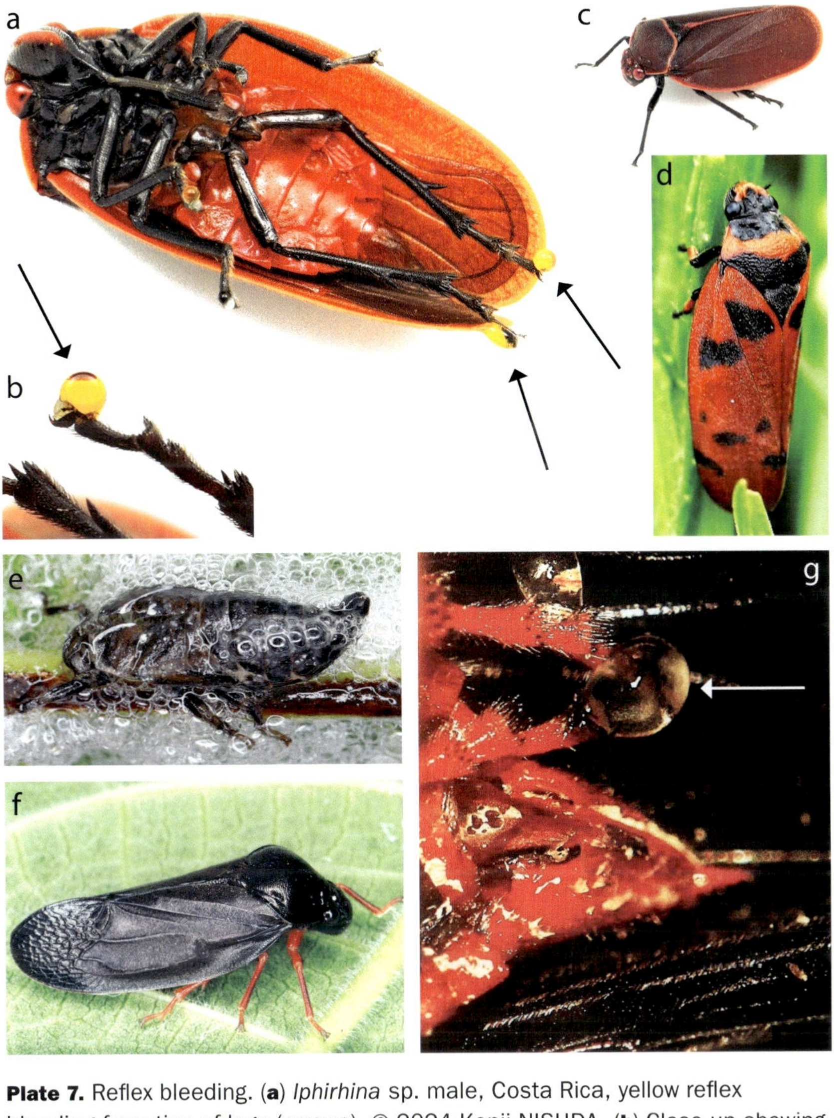

Plate 7. Reflex bleeding. (**a**) *Iphirhina* sp. male, Costa Rica, yellow reflex bleeding from tips of legs (arrows). © 2024-Kenji NISHDA. (**b**) Close-up showing bleeding from tarsal pads. © 2024-Kenji NISHDA. (**c**) Dorsolateral view, same species. © 2024-Kenji NISHDA. (**d**) *Locris arithmetica*, South Africa, a genus demonstrated to reflex bleed. Marlies Craig. (**e**) *Serapita* sp. nymph, Philippines. @Elizabeth Huber. (**f**) *Serapita* sp. adult, Philippines. @Elizabeth Huber. (**g**) *Serapita* sp. female, ventral close-up of posterior end showing reflex bleeding from tarsi (arrow, clear droplet). Maxime Le Cesne.

Plate 8. A raintree spittlebug and fly larvae that coinhabit spittle. (**a**) *Ptyelus grossus* adult, most conspicuous of several color forms. Charles J. Sharp, Wikimedia https://creativecommons.org/licenses/by-sa/3.0/deed.en, cropped. (**b**) *P. grossus* nymphs in communal spittle, Zambia. V. Thompson. (**c**) Fifth instar *P. grossus* nymphs with associated fly larvae, probably of genus *Leucophenga*, South Africa. Marlies Craig. (**d**) *Clastoptera* nymph with *Cladochaeta* fly larva, Costa Rica. © 2024-Kenji NISHDA.

Plate 9. Predation on spittlebugs. (**a**) Robber fly (Asilidae) preying on *Phymatostetha perspicillaris*, Philippines. Jerome Constant. (**b**) Jumping spider preying on male *Mioscarta obscuripennis*, Philippines. @Elizabeth Huber. (**c**) *Prosapia bicincta* male reflex bleeding in response to funnel-web spider attack, Kansas. Andy Williams (from Thompson and Carvalho 2016). (**d**) Running crab spider preying on nymph, probably *P. spumarius*, Austria. © *Gernot Kunz.* (**e**) Predaceous fly larva (probably Syrphidae) probing spittle of *Serapita* sp. for nymph, Philippines. Elorde Crispolon.

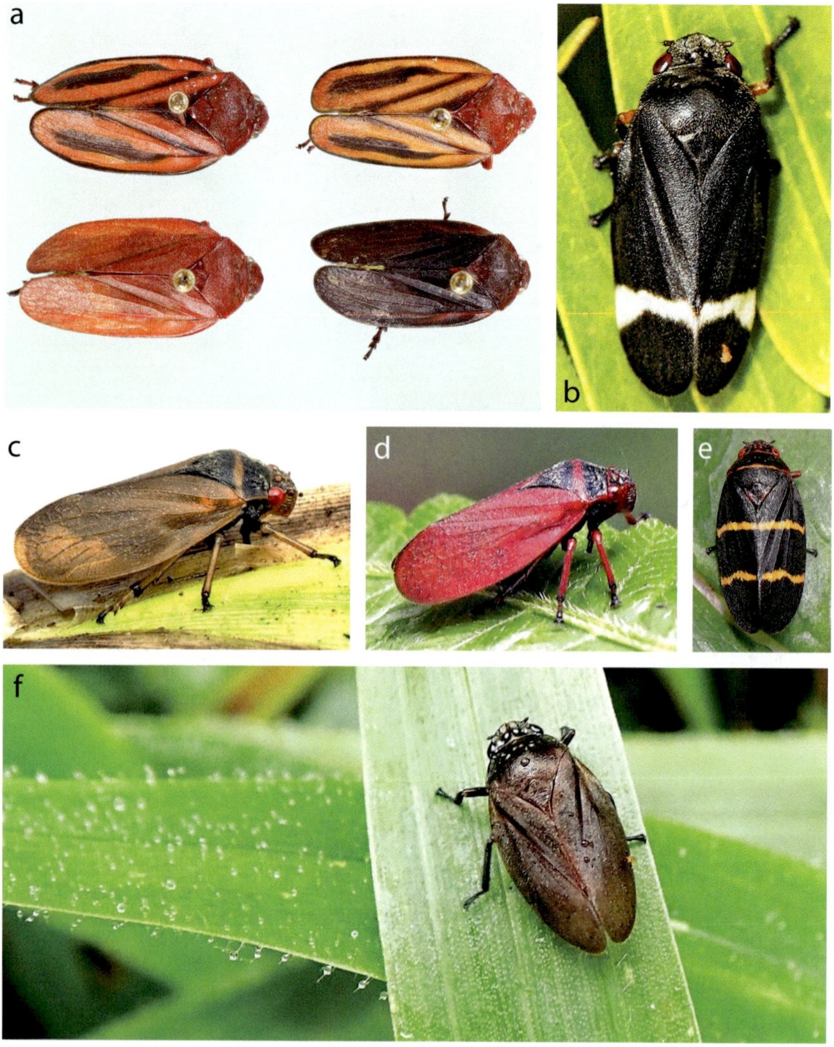

Plate 10. Sugarcane and pasture pests. (**a**) Four co-occurring *Mahanarva spectabilis* color forms, Paraguay. AMNH pinned specimens. Agnieszka Pierwola. (**b**) *Notozulia entreriana*, Brazil. © Ísis Medri. (**c**, **d**) *Prosapia plagiata*, brown and red color forms, Costa Rica. © 2024-Kenji NISHDA. (**e**) *Prosapia bicincta*, USA. © Rob Van Epps. (**f**) *Zulia vilior* on pasture grass blades exhibiting guttation droplets, Costa Rica. V. Thompson.

Plate 11. European spittlebugs. (**a**) *Philaenus spumarius* color form TYP. © Gernot Kunz. (**b**) *P. spumarius* color form MAR. © Gernot Kunz. (**c**) *P. spumarius* color form TRI. © Gernot Kunz. (**d**) *Neophilaenus lineatus*, Austria. © Gernot Kunz. (**e**) *Haematoloma dorsata*, Germany. © Gernot Kunz. (**f**) *Aphrophora alni*, Austria. Tobias Schönberg. (**g**) *Cercopis vulnerata*, Austria. © Gernot Kunz.

Plate 12. Southeast Asian Cercopidae. (**a**) *Callitettix versicolor*, Thailand. © 2024-Kenji NISHDA. (**b**) Feeding *Suracarta tricolor*, Sabah. Note the stylets inserted perpendicularly into the tree host. © Jamiun Micheal. (**c**) *Eoscarta borealis*, Vietnam. © Gernot Kunz. (**d**) *Leptataspis ophir*, Sabah. © Jamiun Micheal. (**e**) Unidentified Cercopidae sp., Sabah. © Jamiun Micheal. (**f**) *Ectemnonotum fruhstorferi* feeding on midrib of *Alocasia odora* leaf (Asian taro), Vietnam. Note the copious excreted fluid. Ruth Moscovitch.

Plate 13. Old World Aphrophoridae. (**a**) *Aphrophora corticea*, Austria. © Gernot Kunz (**b**) *Philagra quadrimaculata*, China. C.H. Dietrich. (**c**) Unidentified black and yellow Aphrophoridae species, Sabah. © Jamiun Micheal. (**d**) *Clovia conifera*, Vietnam. © Gernot Kunz. (**e**) *Lallemandana rarotongae*, Rarotonga. Gerald McCormack. (**f**) *Ariptyelus ikalihani* on *Pinus kesiya*, Philippines. Note the extremely effective camouflage. @Elizabeth Huber. (**g**) *Poophilus costalis*, Taiwan. Aerobird. © Hsiung Sheng Chih. (**h**) *Lepyronia coleoptrata*, Austria. Thomas Oswald, https://creativecommons.org/licenses/by/4.0/, edited.

Plate 14. More sugarcane and pasture pests. (**a**) *Zulia pubescens*, Suriname. David Kohl. © djhiker. (**b**) *Zulia carbonaria*, Colombia. © Juan Manuel de Roux. (**c**) *Prosapia simulans*, Costa Rica. © Gernot Kunz. (**d**) *Aeneolamia albofasciata*, Mexico. © J. Refugio Lomeli Flores. (**e**) *Aeneolamia reducta*, Panama. © Hubert Szczygieł. (**f**) *Aeneolamia contigua*, Costa Rica. © Gernot Kunz. (**g**) *Mahanarva fimbriolata*, Brazil. PUCRS pinned specimen, Gervasio Carvalho. (**h**) *Prosapia tilaranensis*, Costa Rica. © Federico Figueroa Cabezas.

Plate 15. Sugarcane, pasture, and pine pests. (**a**) *Deois flavopicta*, Brazil. Thales Santos. (**b**) *Aeneolamia varia*, Colombia. © David Villanueva. (**c**) *Mahanarva liturata*, Brazil. Thales Santos. (**d**) *Ocoaxo cardonai*, Mexico. © Héctor Naranjo. (**e**) *Mahanarva tristis*, Trinidad. Davis Gunn © davisgunn. (**f**) *Aeneolamia* colon, Brazil. Thales Santos. (**g**) *Deois incompleta*, Brazil. Jens N. Lallensack. © agujaceratops. (**h**) *Aeneolamia lepidior*, Panama. Kai Squires. https://creativecommons.org/licenses/by/4.0/, edited.

Plate 16. Some spittlebug scientists of the late twentieth and early twenty-first centuries. (**a**) Participants in session on Cercopoidea, 10th International Auchenorrhyncha Congress, Cardiff, Wales, September 1999. Left to right: John Whittaker (who missed group photo), Sakis Drosopoulos, Robert Biedermann, John Badmin, Hani Adul-Nour, Vojtech Novotny, Daniel Peck, Olli Halkka, Vinton Thompson. (**b**) An informal spittlebug conclave, Wonalancet, New Hampshire, August 2011. Left to right: Andressa Paladini, K.G. Andrew Hamilton, Daniel Peck, Vinton Thompson, Jason Cryan, Adam Bell.

4

SPITTLE

A Foamy Adaptation Unique to Spittlebugs

Spittle masses are the most conspicuous element of spittlebug biology, evoking the English common name and forming a unique setting for nymphal development. This chapter examines the formation and composition of spittle and its role in the biology of spittlebugs, particularly as a defense against predators and parasites of nymphs.

Spittle is a Biological Foam and a Biochemical Balancing Act

Spittle formation is a direct adaptation to one consequence of xylem feeding: the excretion of very large amounts of processed xylem sap (Chapter 3). In nymphs, this excreted liquid, largely water, is augmented with organic compounds that stabilize small bubbles, permitting the formation of a white foam that can range in consistency from gooey to frothy, depending on species and life stage. *Philaenus spumarius* nymphs devote about 34% of their ingested food energy to spittle production (Weigert 1964a), an enormous metabolic investment that testifies to the magnitude of spittle's evolutionary importance.

Biofoams are rare in nature, probably because they are energetically expensive to make (Cooper and Kennedy 2010). When they do occur, they are generally associated with reproduction, including, as in spittlebugs, the protection of delicate juvenile stages. The best studied examples are the foam nests of frogs constructed on water surfaces in the Neotropics or on overhanging objects in the Old-World tropics (Cooper and Kennedy 2010; Cooper et al. 2017). Biofoam production is a biochemical balancing act. It requires *surfactants* (bubble formation enablers) with contradictory properties. They must overcome bubble-bursting surface tension, but they must not disrupt animal membranes. This rules out molecules like those of the most familiar human-created surfactants, detergents, and soap. Biofoams also require stabilizers that lengthen the life of bubbles to several days and antimicrobial compounds that protect long-lived foams from degradation by microorganisms. Frog foam has a complex mixture of proteins, carbohydrates, and lipids that have evolved to support these functions, though the details are still obscure (Cooper et al. 2017). Knowledge of spittlebug foam is even less developed. This chapter presents our current understanding of the production, composition, and ecological functions of spittle, a unique biological foam.

Fooled by Frog Foam

The similarity between frog and spittlebug foam is close enough to fool a knowledgeable observer. While conducting fieldwork at the Organization for Tropical Studies La Selva Biological Station in Costa Rica in the early 1990s, I stumbled over what I thought at first might be the world's largest spittle mass, spread over a water-inundated patch of grasses. It was a false alarm. I had happened upon a large clutch of foam-covered frog eggs, likely laid by *Leptodactylus pentadactylus*, the smoky jungle frog. In its immensity, it looked for all the world like one or more giant nymphs had made an enormous spittle.

Spittle and Bubble Formation

Gruner (1901) gives an excellent account of the early history of observations of spittle and attempts to account for its origin. The first close study of spittle production seems to have been that of Charles De Geer (1773), the Swedish entomologist we met in the first chapter. His depictions of spittles on a leaf and a nymph within a spittle appear in Figure 4.1. Though later entomologists (Ball 1901) credited Morse (1876) with the earliest description of spittle formation, Morse himself cited De Geer's observations in his first substantial work on the question (Morse 1900). Morse (1876) did produce what were probably the first modern published illustrations of spittle masses with nymphs feeding and making bubbles (Fig. 4.2).

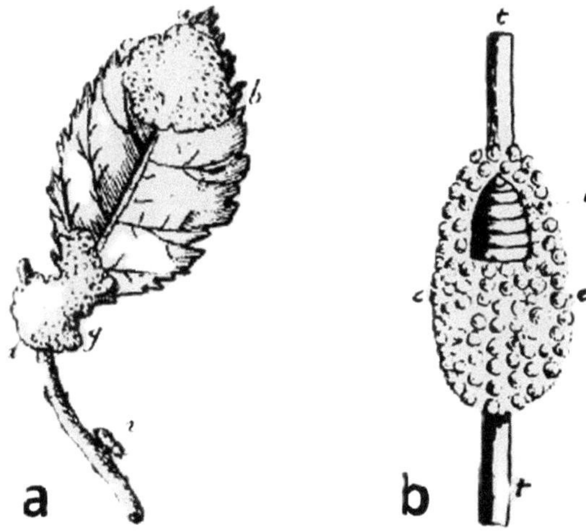

FIGURE 4.1. Early illustration of spittles by the Swedish entomologist Charles De Geer. a) Spittles on the petiole and tip of a leaf, with a tiny nymph on the stem below. b) Abdomen of a nymph exposed in a spittle. From De Geer (1773).

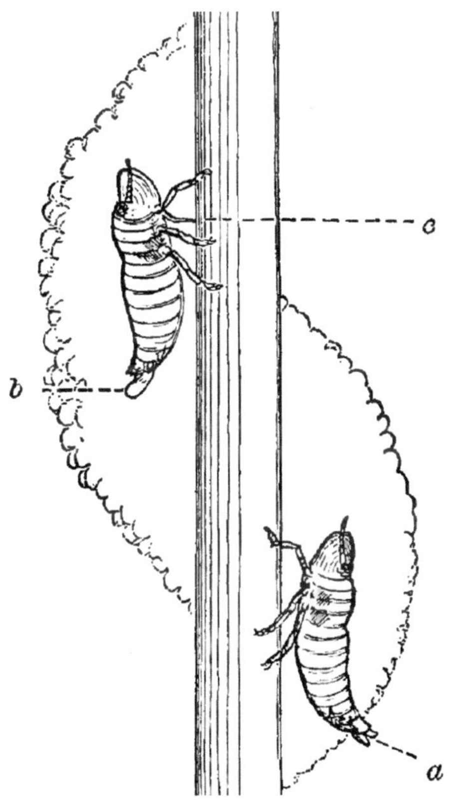

FIGURE 4.2. First known American illustration of nymphs in spittle. a) Tip of abdomen extended outside spittle to get air. b) Tip of abdomen releasing bubble inside spittle mass. c) Beak piercing grass stem. The striations on the sucking pumps are incorrectly drawn longitudinally. From Morse (1876).

Edwin Sylvester Morse was a polymath New Englander with an extraordinary range of interests (Kingsley 1926). His first short description of spittle production appeared in a zoology textbook for beginning college students (Morse 1876). He describes the production of bubbles in association with the tip of the abdomen and the abdominal breathing tube but gets one detail wrong, saying the initial fluid comes from pores. The book came between stints as a zoology professor at Bowdoin College and Tokyo Imperial University, where he amassed a world-class collection of Japanese pottery. In 1886 he served a term as president of the American Association for the Advancement of Science. Ever busy, he turned out a nonstop stream of publications, including a memorable illustrated piece titled "Latrines of the East" (Morse 1893). In 1900, he revisited spittle production in a seven-page article in *Popular Science Monthly* titled "A Bubble-blowing Insect."

The primary advance in Morse's 1900 paper is the unequivocal observation that the foam in spittle results from the repetitive grasping and releasing of air bubbles from an "air-grasping pocket" (Guilbeau 1908) at the tip of the nymphal abdomen. The bubbles do not come from within the body of the nymph. Instead, they originate at the opening of a ventral abdominal breathing tube formed by laterally extended abdominal segments. In a charming (and accurate) description of bubble formation by an early twentieth-century English commentator, the breathing tube is likened to a gentleman's attire.

> The larva [nymph] whose life is spent in a condition of fluid surroundings, is clothed in such a way that will suit its environment. Insects, as we know, breathe through a series of pores or spiracles opening on the lateral aspects of their bodies. If these spiracles were exposed to contact with fluid, they would become obstructed, and the larva would be asphyxiated. To provide against this, Nature has so clothed Philaenus' abdomen with a double integument, corresponding to the waistcoat and jacket of a man's attire. The spiracles open upon the surface of the underlayer, the waistcoat; the outer layer, corresponding to the jacket, is closed along the centre of its ventral aspect in the same way a coat is buttoned up, and protects the spiracles from occlusion by the fluid froth in which the insect lives. On the under surface of the last abdominal segment a V-shaped space is to be seen . . . [which, during froth formation] has poured upon it a watery film, closing the aperture, much as the orifice of the bowl of a clay pipe is closed by a fluid film of soapsuds when a youngster is about to blow bubbles. This film is provided by fluid poured out from the extremity of the animal's bowel, and when in position, is blown out into a bubble by a peculiar

> pumping action provided by the contraction to and fro of the outer integuments. By one set of contractions air is sucked in as so to occupy the space between the inner and outer coats, and then by a subsequent contraction–a bellows action, in fact–is forced out through the V-shaped space, distending the fluid film into bubble form. This bubble in due course frees itself and forms one of the many entering into the formation of the froth. (Rodman 1921)

The process of bubble formation begins a few minutes after a nymph starts feeding and bathes its body in excreted fluid. Bubbles accumulate from the ventral side and cover the nymphs within two to three minutes (Hoch et al. 2024), with the tip of the abdomen "moved alternately to the right and left in perfect rhythm, so that the bubbles are distributed to both sides of the body . . . till the entire fluid is filled with bubbles, and the froth thus made runs over the back and around the [plant] stem" (Morse 1900). This process continues at about eighty bubbles per minute, with the volume of air entrapped exceeding the volume of the insect itself by four to six times. Figure 4.3 shows Morse's illustration of the foam building process, which is also excellently illustrated by China (1927) (Fig. 4.4), described in detail by Weaver and King (1954), and shown in engaging video clips accompanying Hoch et al. (2024, supplementary movies 1–4).

Morse observed that nymphs make bubbles in drops of plain water, though these do not last, for want of the bubble-stabilizing ingredients of excreted nymphal fluid. However, evidently having expectorated for the occasion, he observed that nymphs make more persistent bubbles in saliva on a glass slide—spittlebugs living, at least briefly, in real spit. Working in parallel, Porta (1899) and Gruner (1901) made similar observations on the nature of spittle formation and added chemical evidence that spittle liquid lacks sugar but contains a mucilaginous material, calcium carbonate, and a salt of potassium with an organic acid. Porta studied *Philaenus spumarius*. Gruner studied *P. spumarius* and *Aphrophora salicina*. Morse refers to the species he studied simply as "*Aphrophora*," which could include a number of eastern North American aphrophorid species (his implication that it lives on grass suggests *Neophilaenus lineatus*). Together, Morse, Porta, and Gruner initiated modern studies of spittle and spittle formation.

The ventral breathing tube is lined with a waxy hydrophobic cuticle to keep it dry (Rodman 1921; Decary 1928; Messner and Adis 1992), forming a snorkel at the tip that permits nymphs to breathe while immersed in a semi-aquatic environment (Plate 4e). Spiracles opening two per abdominal segment convey air from this tube to the tracheae, which disperse it throughout the body. While the gist of Morse's work was to show that the bubbles do not emanate from inside the insect, the word "bubble-blowing" in his title might suggest that the

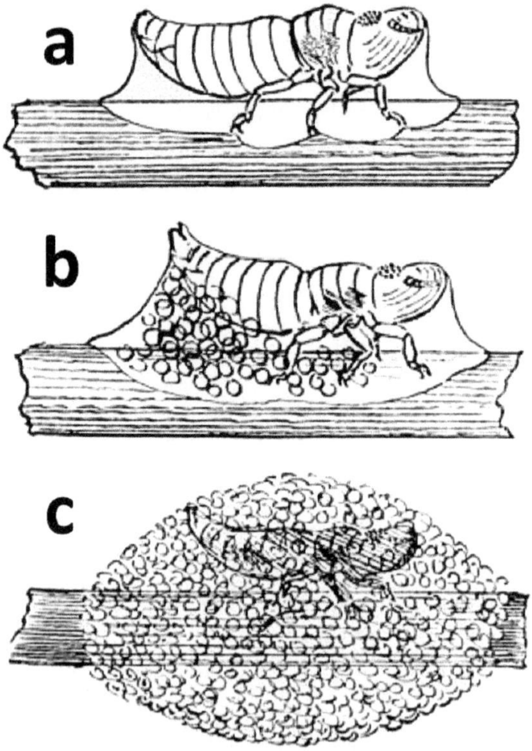

FIGURE 4.3. Progression of spittle mass formation. a) Feeding nymph covers itself with clear excreted fluid. b) Tip of nymph's abdomen introduces bubbles into excreted fluid one at a time, at rate of about 80 bubbles per minute, starting at the rear. c) Nymph covered by foam composed of innumerable small bubbles introduced one-by-one. The striations on the sucking pumps are incorrectly drawn longitudinally. From Morse (1900).

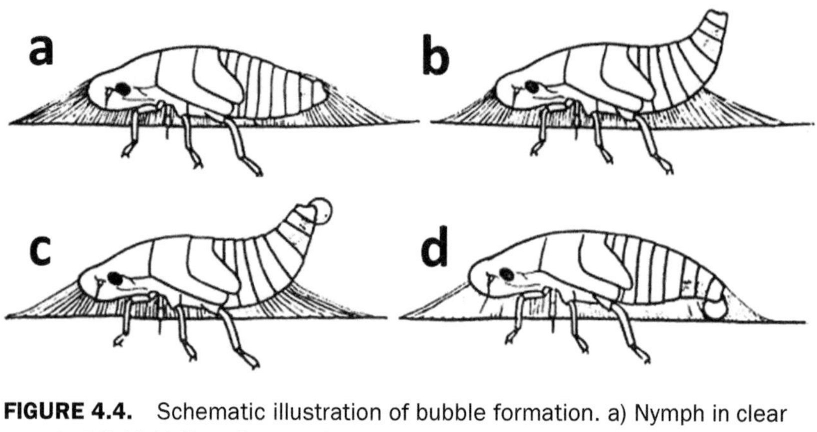

FIGURE 4.4. Schematic illustration of bubble formation. a) Nymph in clear excreted fluid. b) Nymph extends tip of abdomen out of fluid. c) Nymph forms bubble at tip of abdomen from captured pocket of air. d) Nymph brings bubble into fluid and releases it. Adapted from China (1927).

air comes from the tracheae, the internal "breathing" network, a misconception that has been perpetuated over time (cf. Crews et al. 1998). Instead, bubble formation occurs at the tip of the abdomen when it is extended out of the spittle, with a bellows-like push from the abdominal muscles, mediated by a pair of flaps (the *stigmacalypts*) on the last (ninth) abdominal segment that open and close access to the ventral air canal.

Hamilton (2001) has suggested that spittlebugs of the genus *Clastoptera* produce deficient, "primitive" spittles with small and sparse bubbles that do not protect against parasitoids and predators, condemning the *Clastoptera* to small size. In my experience, while some *Clastoptera* spittles, such as those of *C. obtusa*, are relatively viscous, with fewer bubbles, many more, such as those of *C. xanthocephala*, are as bubbly and foamy as those of the Cercopidae and Aphrophoridae that Hamilton cites in contrast. In fact, there is tremendous variation in the texture of spittles (Urich 1913), some of it no doubt related to bubble size, which can be regulated by nymphs changing the size of the air-grasping pocket at the tip of the abdomen (Guilbeau 1908). Since larger nymphs have larger air-pockets, they have the option of making larger bubbles. I have encountered spittles ranging from very viscous, made by Costa Rican rain forest and pasture Cercopidae, to very foamy, made by aggregated nymphs of *Cephisus siccifolius* in Costa Rica.

Spittle also changes in texture as nymphs develop, tending to become foamier in later instars (cf. Zhang et al. 2024). An experienced naturalist can often infer species and life stage from the appearance of spittles (Guilbeau 1908; Kuenzi and Coppel 1985a). Likewise, the foam in individual spittles also changes with age. It starts out with spherical bubbles immersed in liquid (*kugelschaum*), which tend to coalesce with time, lose liquid to gravity, and take on an irregular polyhedral structure (*polyederschaum*) (Cooper and Kennedy 2010). Traces of this process can be seen in photographs and cryo-SEM images in Hoch et al. 2024. Spittles can also take on odd and unexpected forms. In Luzon, some spittlebugs make "nubby spittles" with a pronounced bumpy surface. Plate 3a illustrates a nubby spittle holding a nymph of the genus *Oxymegaspis*, which was successfully reared to adulthood (Plate 3b–d).

The Chemical Nature of Spittle

Oddly, despite over a century of work, the issue of where the bubble-stabilizing component of spittle comes from is still unsettled (Rakitov 2002). The early workers were divided on this question, with some advocating a source in the Malpighian tubules via the gut and some advocating for the glands of Batelli, located in segments near the tip of the abdomen. Weaver and King (1954) and

Rakitov (2002) summarize the evidence in dispute. Guilbeau (1908) cauterized the glands of Batelli of multiple nymphs (probably of *Lepyronia quadrangularis*), which thereafter went through protracted abdominal bubble-making motions but failed to form bubbles. He interpreted this to demonstrate that functional Batelli glands are necessary to make spittle foam. On the other hand, Marshall (1965a) surveyed nine spittlebug species from three families and found that the glands of Batelli were absent in the two Cercopidae species examined, *Aeneolamia varia* and *Locris* sp. This indicates that the glands are unnecessary for spittle production in these species. In contrast, the glands of Batelli *are* present in the Machaerotidae species *Makiptyelus dimorphus* and *Chaetophyes compacta*, which make spittle only briefly in the course of a lifetime. Marshall's observations argue against a critical role for these glands in spittle production. It is likely that their primary role is production of waxy secretions that help create the hydrophobic surface of the ventral breathing tube (Hoch et al. 2024).

Studies of spittle chemistry date back to Porta (1899) and Gruner (1901), who both observed that spittle lacks sugars and has a mucilaginous component. Several early workers observed that the Malpighian tubules produce a mucopolysaccharide, which is excreted with the processed xylem sap and becomes a major constituent of the organic component of spittle. Mucopolysaccharides, known in biochemistry as *glycosaminoglycans* (GAGs), are long polymers of *disaccharides*, double sugars, in which one of the two component sugars has a nitrogen-bearing amino side group (DeAngelis 2002). Figure 4.5 illustrates GAG's molecular structure. Interestingly, a closely related compound, galactosaminogalactan, plays a major role in the infection of insects by the entomopathogenic fungus

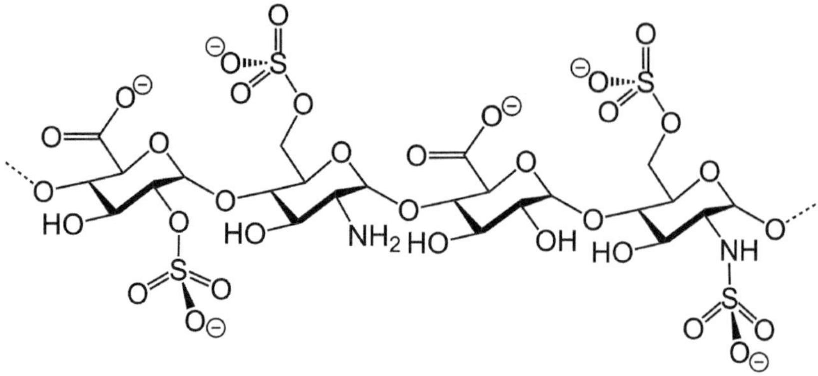

FIGURE 4.5. Repeating molecular disaccharide unit of glycosaminoglycans (GAGs or mucopolysaccharides), a major component of spittle. By Jü via Wikimedia Commons, reproduced under Creative Commons CC0 1.0 Universal Public Domain Dedication.

Metarhizium (Mei et al. 2021), which is used successfully to control spittlebug infestations in sugarcane (Chapter 6).

Spittlebug mucopolysaccharides have been best characterized in now classic cytochemical and morphological studies by Marshall (1966a, 1966b). They are secreted by a long, defined segment of the proximal part of the Malpighian tubules, shortly before the junction of the four tubules at the beginning of the hindgut coming out of the filter chamber (Figs. 3.3, 3.4). This is clearly the primary function of these segments, in which 80 to 90% of the bulk of the cells consists of mucopolysaccharide. The correlation of mucopolysaccharide production with spittle production is perfect. Production occurs only in nymphs making spittle. It disappears in adults and in the life phases of Machaerotidae nymphs when they are not making spittle (Marshall 1964b). Li et al. (2015) provide more recent evidence for the close correlation between active secretion by Malpighian tubules and spittle production.

The excreted product is a sulfated, highly charged, acid mucopolysaccharide, which is produced in abundance. It appears to be made up of four subunits: glucuronic acid, acetylglucosamine, glucose, and rhamnose. Marshall (1966a) concludes that "a highly charged water-soluble mucopolysaccharide could behave, in spittle, as a very efficient surfactant and be responsible for bubble stability. It may be such that a mucopolysaccharide could also combine with the protein evidently present in spittle, resulting in increased cohesion and surface viscosity properties of the bubble film."

Alan Marshall's Major Contributions

From 1961 to 1979, Alan Marshall, sometimes with his student Wilkin Cheung, produced a remarkable series of works on the nature of spittle, spittle production, and the tubes of Machaerotidae. He started this work in England as a graduate student when his advisor Richard Southwood suggested he investigate the production and stabilization of the nymphal foam of cercopids, which was inhibiting effective pesticide use against the Trinidad sugarcane pest *Aeneolamia varia* (see Chapter 6). He then moved to academic positions in Hong Kong and Australia, where he had the good fortune to work in areas with abundant Machaerotidae. The work petered out when he turned to other, more fundable projects, but he also later faced a prolonged drought in Australia that had profound negative effects on local insect abundance and the availability of experimental material. His pioneering work has served for decades as a bedrock of spittle and machaerotid tube studies.

Along the same lines, Mello et al. (1987) provide evidence that spittle of a species of the genus *Deois* (Cercopidae) is rich in glycoproteins, which probably combine with acid glycosaminoglycans to form acid proteoglycans. They suggest that the bubble fibrils they observed by electron microscopy are composed of acid proteoglycans. These observations, which have been confirmed in other Brazilian Cercopidae species (Mello et al. 1996), are consistent with Marshall's hypothesis that acid glycosaminoglycans (mucopolysaccharides) combine with proteins to play a major role in determining the properties of spittle. Guanidinium chloride treatment, which specifically targets glycosaminoglycans, disrupts spittle bubbles and causes loss of their characteristic optical qualities, indicating that glycosaminoglycans play a significant part in the bubbles' ordered structure (Mello 1987).

There has been surprisingly little advance in our understanding of spittle chemical composition and function since Marshall and Mello's pioneering work. Part of this may be attributable to the difficulty of analyzing GAGs due to their large size, heterogeneity, and variable molecular weight. In view of these limitations, analysis of the glycosyltransferase genes involved in their synthesis may offer an alternative route to understanding their evolutionary history (DeAngelis 2002). Zhang et al. (2024) have made the first steps in this direction. Working with the Asian rice pest cercopid *Callitettix versicolor*, they demonstrate that it exhibits a proliferation of genes associated with carbohydrate and lipid metabolism, including some specifically GAG-related functions. More directly, they compare genes expressed in nymphal Malpighian tubes with those of nonspittle making adults and find differential expression of genes governing spittle-related metabolites. This is the first direct evidence for genes related to spittle production.

Recent Studies of Spittle Chemical Composition (Still a Work in Progress)

In recent years, several groups have made stabs at characterizing the composition of spittle in more detail. Using gas chromatography-flame ionization detection (GC-FID) and gas chromatography-mass spectrometry (GC-MS) on *Mahanarva fimbriolata* spittle, Tonelli et al. (2018) measured palmitic acid and stearic acid at levels of 2.5 and 2.8 μg/ml of spittle, respectively. They also measured carbohydrates at 0.6 μg/ml of spittle and protein at 320 μg/ml. The relative quantities strongly suggest that their analysis of carbohydrates did not include polysaccharide polymers and simply measured very low levels of free sugars.

In a GC-MS study, Szterk et al. (2024) report that *Aphrophora alni* spittle dry matter content is composed primarily of carbohydrate polymers, up to 92% of dry weight. In concurrence with Tonelli et al. (2018), they report the presence of fatty acids and alcohols that might contribute to bubble stabilization. They found low levels of protein, only about 1% of dry weight. They found no free sugars and suggest, without evidence, that symbiotic microorganisms might synthesize the carbohydrate polymers (presumably equivalent to the mucopolysaccharides of earlier workers).

Campo et al. (2011) analyzed *Aphrophora cribrata* spittle for compounds that might be repellent to predators. Using dichloromethane extraction and GC-MS combined with nuclear magnetic resonance (NMR) spectroscopy, they identified five classes of suspect deterrent compounds: fatty-acid-derived alcohols, c-lactones, 1-monoacylglycerol, polyol pinitol, and poly-3-hydroxybutyrate. They found that both natural spittle and a synthetic mixture representative of the identified compounds repelled the aggressive insectivorous ant *Formica exsectoides*, apparently due to distastefulness. In a similar study of *Poophilus costalis* spittle using GC-MS and other methods, Sahayaraj et al. (2021) report the presence of several long chain fatty acids and an assortment of other organic compounds, including two minor compounds connected to defense in other organisms: benzo(h)quinoline, 2–4 dimethyle-, which is related to quinoline, a strong defensive compound released by a Peruvian stick insect (Eisner et al. 1997); and 2,4,6-cycloheptatrien-1-one,3,5-bis-trimethylsilyl, said to be microbicidal and nematocidal.

As of this writing, there is still no definitive, detailed elucidation of the chemical nature of spittle. It would not be surprising to find that composition varies among species, over the course of nymphal development, or in response to specific host plants or environmental variation. We also still do not know how spittlebugs stabilize bubbles for such an extended period. Spittles remain in a foamy state for three or four days after nymphs exit (Kershaw 1913), which is a long time for a liquid foam.

The Physical Nature of Spittle

In addition to the studies on chemical composition, there is a small body of work on the physical nature of the foam itself. Using a combination of polarization and fluorescence microscopy, Mello et al. (1987) present evidence that *Deois* bubbles incorporate a layer of thin filaments. It is tempting to speculate that these filaments might be related to the "mucofibrils" that anchor the structure of Machaerotidae dwelling tubes (discussed later in this chapter). These fibers, which have been observed only in Machaerotidae, are hypothesized to be composed of an acid mucopolysaccharide in intimate association with a protein (Marshall 1965b, 1968).

Hoch et al. (2024) have examined the physical properties of *Aphrophora alni* spittle. It is sticky, gets rapidly stickier with evaporation, and evaporates more quickly than pure water. It also has greater wetting capacity than pure water, an attribute that may help it cling to plants. When first formed, round bubbles are suspended in fluid (the kugelschaum). As the foam dries, they pack and ultimately leave pentagonal borders with converging edges in a thin filament pattern (the polyederschaum). In contrast to their finding on evaporation, Turner (1994) reported that *Aphrophora cribrata* spittle liquid evaporates more slowly than water but calculated that the increase in surface area from bubble formation should essentially nullify this effect. The discrepancy in measured evaporation rates might be due to differences in experimental conditions or to interspecific differences in spittle composition or structure (Hoch et al. 2024).

Spittle Masses Often Host Multiple Nymphs, Sometimes Multiple Spittlebug Species

Individual spittle masses often harbor two or more spittlebugs. This is most common when nymphs are abundant and/or concentrated on limited areas of a host plant and probably results from coalescence of spittle masses as nymphs develop in close proximity (cf. Yaghmaee 2008). Working in Corsica, Chartois et al. (2023, fig. S1) found an average of 1.24 *P. spumarius* nymphs per spittle, similar to the 1.7 per spittle reported by Halkka and Halkka (1990). De Geer (1773) reported the phenomenon of three, four, or five nymphs of the same species per spittle. The highest recorded number of nymphs per spittle in *P. spumarius* seems to be 70 on *Angelica archangelica* (Halkka and Halkka 1990). Biedermann (2003) presents thorough data for *Neophilaenus albipennis* in Germany. Most spittles were single occupancy, but some harbored up to four nymphs, and multiple occupancy went down from a high of 63% in the first instar to a low of about 9% in the fifth instar. In dense *P. spumarius* populations in alfalfa, Mangan and Wutz (1983) report mean densities of 1.4–3.0 nymphs per spittle. At the high end of that range, a majority of spittles had two or more nymphs. In another high-density *P. spumarius* population, McEvoy (1986) found that 63% of spittles on *Potentilla erecta* were multiply occupied, with up to twelve nymphs per spittle. There was no evidence in these works that density per spittle or per plant had any effect on development, suggesting that even at high densities there is effectively no competition for resources among nymphs sharing spittles on the same plant. This likely contributes to the ability of many spittlebug pest species to build up enormous populations. As long as individuals tap into separate xylem vessels, there is plenty of nourishment to go around. Shared spittles may give individual nymphs an advantage when they have to stop feeding to molt or restore spittles after rainstorms (Wise et al. 2006), which is consistent with measured lower mortality rates in aggregated *N. albipennis* nymphs (Biedermann 2003). Work by Matsumoto (1990) with the East Asian pine pest *Aphrophora flavipes* (Chapter 8) suggests that the benefits of aggregation diminish at the highest densities.

At least two spittlebug genera have evolved large, communal spittles. Three species of the African/Madagascan genus *Ptyelus* inhabit large, foamy aggregate spittle masses, which can harbor up to 150 individuals (Plates 1d, 8b), contributing to the phenomenon of "rain trees" (Palmer and Pitman, 1961) discussed in Chapter 3. When disturbed, *Ptyelus* nymphs desert the foam *en masse* and disperse along the surrounding branches (Decary 1928). The Neotropical species *Cephisus erythrocephala* and *Cephisus siccifolius* have similar communal masses with more than one hundred nymphs (Tsacas and Couturier 1993). Like their *Ptyelus* equivalents, *C. siccifolius* nymphs react to disturbance by running out of the foam (Peck 2000b). *Ptyelus* and *Cephisus* are both large tropical aphrophorids, suggesting that they might be related, but molecular phylogenies indicate that they evolved in separate lines (Cryan and Svenson 2010; Crispolon et al. 2023), making their congruent behavior an example of convergent evolution.

Cohabiting nymphs are not limited to the same species. Halkka et al. (1977) document the ecological co-occurrence of four spittlebug species on thirty-nine host plant species in southern Finland. They report fifty-three instances in which nymphs of different species shared the same spittle mass, fifty-one involving two species and two involving three, compared to 243 instances of nymphs of the same species sharing a spittle. Biedermann's (2003) *Neophilaenus albipennis* cohabited with *P. spumarius* in four of 368 audited spittle masses. Badmin (2007) presents multiple instances involving *P. spumarius*, *A. alni*, and *A. salicina* in Britain, and I have observed *P. spumarius* and *Aphrophora moscoviciae* nymphs sharing spittles in California. DeLong and Severin (1950) report that *P. spumarius*, *Aphrophora permutata* (Fig. 1.91), and *Aphrophora angulata* share spittles on occasion in California. Such instances are probably common where spittlebug species at high densities share hosts.

In Some Spittlebugs the Last Spittle Mass Develops a Molting Chamber

In *P. spumarius*, the spittle of very late fifth instar nymphs takes on a special, drier consistency, losing about half of its water content (Markey 1965), within which an internal air space facilitates transformation into the adult stage (Weaver and King 1954; Peck 1998a). The emerged adult remains in this characteristic capsule until its cuticle hardens and assumes adult coloration (Plate 1a). This phenomenon is widespread in Aphrophoridae and Cercopidae. The root-feeding nymphs of *Aeneolamia varia*, for example, make molting spittles at the ground surface or above (Urich 1913; Williams 1921). At least some *Aphrophora* species emerge as adults outside spittle (Doering 1931, 1941; Kelson 1964; Wilson 1987), and it is uncertain whether molting chambers are universal among the Cercopidae. They have not been observed in the Clastopteridae or Machaerotidae and thus appear to be a novel evolutionary development dating back to a common ancestor of the Aphrophoridae-Cercopidae lineage, with loss in at least one aphrophorid lineage. The mechanism of creation of the air space has not been determined (Beckett et al. 2021). Markey (1965) reports that multiple newly emerged *P. spumarius* adults are sometimes congregated in one last stage spittle mass, a unique observation that requires confirmation. In popular media, illustrations of newly emerged adults exposed in molting chambers (such as Plate 1a) risk giving the impression that adults as well as nymphs live in spittle.

Spittle is a Source of Drinking Liquid for Some Animals

A handful of insects have been reported to sip from spittle. In Virginia, the butterfly *Limenitis arthemis astyanax* has been observed to drink repeatedly from *Clastoptera xanthocephala* spittles (Wise 2010). Similarly, the western North American butterfly *Zestusa dorus* is reported to live on mud, sap, and spittlebug secretions (Scott 1986). Yellow jacket wasps (*Dolichovespula maculata* and *Vespula* spp.) have also been reported to drink spittle. *Vespula* also feed spittle to their larvae (Akre et al. 1976; Schmidt 2014). In Wisconsin, the ant *Formica montana* occasionally ingests spittle and also makes extensive use of spittle mixed with soil in the construction of protective tent-shelters for tended aphids (Henderson et al. 1990) In Italy another ant, *Crematogaster scutellaris*, has been observed to drink from spittle in the lab (Balzani et al. 2023). The prize for exotic imbibement goes to the Madagascan lemur *Phaner furcifer*, which has been observed to spend hours going from twig to twig on the shrub *Rhopalocarpus lucidus*, licking the exudation from machaerotid tubes (Petter et al. 1971).

As a matter of course, humans do not drink spittle, but Rodman (1921), who gave it a try, describes *P. spumarius* spittle as practically tasteless, an observation supported by Turner (2000) and one I can vouch for personally. On the other hand, Meena et al. (2014) report that spittle of an unidentified *Poophilus* species living on a shrub used to cultivate lac insects "has an acrid taste, deterring predators." It is possible that some hosts plants have xylem sap properties that influence the taste of spittle.

Spittle Provides Protection and a Buffer Against Ambient Environmental Conditions

Spittle protects or buffers nymphs against a number of challenges posed by the physical environment.

Desiccation

In his 1732 Lapland diary (Chapter 1), Linnaeus remarks that spittle preserves the delicate flesh of nymphs from the sun's strong heat (Linné 1913). De Geer (1773) concurred that spittle serves to counter the ardor of the sun as well as the effects of heavy rain. He also noted, as have many observers, that nymphs die quicky if driven from or deprived of spittle in dry environments (cf. Decary 1928). Markey (1965) quantified time to death experimentally, subjecting *Philaenus spumarius* nymphs to dehydration over potassium hydroxide at 18°C. For groups of twenty-five-nymphs of each instar, first through fifth, mean time to demise was, respectively, 110, 275, 210, 430, and 490 minutes. Control nymphs maintained on damp filter paper all survived to the termination of observations at 24 hours. Whittaker (1970) demonstrated that humidity gradients led to systematically smaller *Neophilaenus lineatus* spittles higher up grass stems in the lab and rush stems in nature, a function of greater distance from soil moisture and, in the field, more wind.

While spittle undoubtedly protects nymphs from desiccation, some observers (cf. Turner 1994, 2000) have questioned whether spittle is necessary for this function. They point out that xylem feeders excrete voluminous liquid, ostensibly obviating the need for an additional wet layer of foam. On the other hand, it is possible that spittle evolved in part to prevent drowning in excess fluid while preserving a moist environment (Rodman 1921), a challenge that xylem feeding leafhoppers (Challita et al. 2023) and adult spittlebugs (Fewkes 1969a) overcome by constantly flicking away emerging droplets of excreta. In two unusual instances involving *Aphilaenus nigripectus* in Japan and *Leocomia cinchonae* in Jamaica, nymphs induce an extra layer of protection by causing the formation of leaf-roll galls around their spittle masses, giving them an added defense against desiccation (Pickles 1930; Sugiura and Yamazaki 2003). In *L. cinchonae*, which deposits its eggs in spittle, this also provides effective defense against hymenopterous egg parasites (Pickles 1930), parasitoids covered in Chapter 5.

Light

Chen, Meyer-Rochow et al. (2018) demonstrate that brighter light leads to more light avoidance by nymphs, with the effects much more pronounced in the absence of spittle and stronger in early instars. They show that spittle is highly reflective at all wavelengths, bouncing back more than 85% of incident light, thus its bright white color, a challenge to spittle photography. This suggests that spittle is an effective barrier against the harmful effects of visible and UV light.

Fluctuating Temperature

Spittle also serves as an insulator against fluctuating temperatures. Tonelli et al. (2018) report that the temperature within spittles of *Mahanarva fimbriolata* nymphs feeding on surface roots was similar to that of the adjacent soil and remained relatively constant at about 25°C despite daytime temperatures, which reached 29°C. Sahayaraj et al. (2021) found that in hot ambient conditions (31–35°C), temperatures inside *Poophilus costalis* spittles were consistently cooler than outside, often by several degrees. Spittles appear to offer substantial opportunity for evaporative cooling (Whittaker 1970), but this has not been subject to investigation (Tonelli et al. 2018).

Oxygen Deprivation

Beckett et al. (2019) studied oxygen consumption and spittle oxygen levels in *P. spumarius*. Forcibly submerged nymphs breathed from consolidated air bubbles around the abdominal tip, depleting oxygen. Unconfined nymphs snorkeled, opening their breathing tubes outside the spittle mass. With the exception of transient lower oxygen levels in the molting chambers of late last instar nymphs, levels throughout spittles generally conformed to outside air levels. This disproves earlier speculation that nymphs use foam as a respiratory organ (Turner 1994). It is possible that nymphs sometimes use spittle bubbles as a reserve air supply in, for example, cases of short-term waterlogged soil (Withycombe 1926).

The Spittle Microbiota and Putative Antimicrobial Effects of Spittle

Wilson and Dorsey (1957), using dilution plating on laboratory media, measured spittle bacterial counts of 18 to 108 million/ml. They identified isolates of *Flavobacterium* and *Achromobacter* from *P. spumarius* spittle but found no inhibitory effect of fresh *P. spumarius* spittle on fifteen tested bacterial species and twelve of thirteen species of fungi. Working with *Haematoloma dorsata* nymphs, which live in soil on grass roots, Pelagatti et al. (1995) isolated sixteen species of bacteria from spittle, a few of which exhibited modest antagonism toward tested soil fungi. Using molecular techniques, Nencioni et al. (2023) found multiple α-Proteobacteria species in *P. spumarius* and *L. coleoptrata* spittle, and Tonelli et al. (2020) report a diverse bacterial flora, including 257 putative species, in *M. fimbriolata* spittle.

Chang et al. (2019) report that concentrated *Poophilus costalis* spittle represses the growth of the potentially plant pathogenic soil fungus *Fusarium oxysporum*, though not the growth of the similarly pathogenic fungus *Botrytis cinerea*. They attribute the negative effect on *F. oxysporum* to spittle chitinases of unknown origin and suggest that the observed antifungal action might be linked to the observation that spittles in nature are never moldy, despite appearing to offer a mold-friendly substrate. Chitinases might appear to be detrimental to nymphs, but the outer lipoprotein layer of the nymphal integument protects them against chitinase attack.

Sahayaraj et al. (2021) also tested *Poophilus costalis* spittle against five bacterial species for antimicrobial activity. In each case there was evidence for suppression of microbial growth correlated with the concentration of foam, but the nature of the spittle content applied, including sterilization, if any, is not specified, making the results difficult to interpret. Li et al. (2015) report that the Malpighian tubules of early instar nymphs of two Chinese spittlebug species are bacteria-free, while the tubules of their adults host numerous bacteria, suggesting that nymphal Malpighian secretions may inhibit microorganisms. However, tubules of fifth instar nymphs of two Italian species, including one of the two Chinese species studied, have been found to harbor bacteria (Nencioni et al. 2023).

The take home lesson from these studies is that spittle microbiology is still in its infancy. Spittle is rich in potentially assimilable polysaccharides. Not surprisingly, as shown above, it is full of bacteria. It also hosts protozoans (Urich 1913; Williams 1921; Hoch et al. 2024). Its fungal flora has not been explored. The studies to date suggest that spittle might have properties that suppress the growth of undesirable microbes and repel some predators, but more work is needed to substantiate these claims.

Crusts, Tubes, and Capsules: Alternatives to Ordinary Spittle Masses

Every member of the Cercopoidea with a known life history makes spittle at some time during nymphal development. The Machaerotidae construct nymphal tubes of calcium carbonate with an embedded organic matrix (Marshall 1965b).

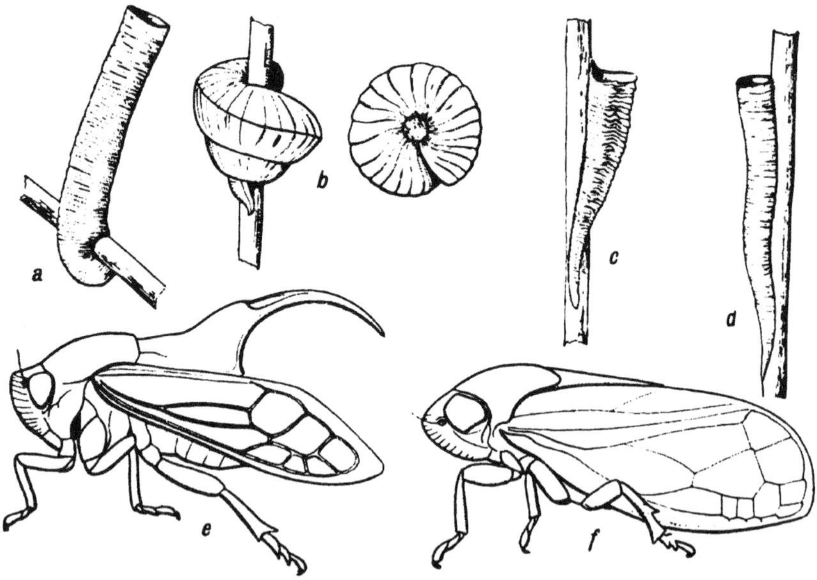

FIGURE 4.9. Nymphal tubes and adults of spittlebugs of the family Machaerotidae. a–d) Calcareous tubes of multiple species; note the variety of shapes, which are genus specific. e) Adult of Sri Lankan species, *Machaerota punctatonervosa*; note the prominent scutellar spine, which gives this group of machaerotids a superficial resemblance to treehoppers (Membracidae) and the absence of a lateral tibial spine, an unusual characteristic of the genus. f) Adult of Australian species, *Pectinariophyes stalii*. From China 1927.

These tubes vary greatly in conformation from species to species (Fig. 4.9a–d, Plate 6b–d). Nymphs live inside, head down, immersed in fluid, feeding on plant sap through the base. They make spittle only at the earliest stages of the first instar (Lomer et al. 1993), upon emerging from the tube for the final adult molt, or, sometimes, when moving from an old tube to construct a larger one during intermediate nymphal molts. They breathe and excrete excess fluid through the tip of the tube.

While tube building appears to be universal among the Machaerotidae, in many lineages the nymphal stage has not been observed. During field work in Borneo in 2009, I collected a single newly emerged adult from spittle on a rain forest understory plant. Based on its teneral coloration, I initially took it to be one of the local Cercopidae, but it matured into an adult machaerotid, tentatively an undescribed species of the genus *Hindola*. No tube was in evidence near the spittle. Whether this represents a tubeless machaerotid or a wandering nymph that had left an undiscovered tube to molt in an outside spittle awaits further observation. Known species of the genus *Hindola*, if that is the correct identity, do construct nymphal tubes and have been observed to have "emerged from nymph tubes . . ., settling on top of a mass of newly produced spittle until the cuticle hardened (Lomer et al. 1993)."

The African genera *Tremapterus* and *Sepullia* make similar tubes, with the mineral covering comprised mostly of calcium carbonate (Plate 6f). Many nymphs of the California spittlebug *Clastoptera mineralis* make mineral covered spittles, also consisting primarily of calcium carbonate, though absent the standardized structure of machaerotid and *Tremapterus* tubes and covering only part of the nymphal life cycle (Thompson 2023a). *Clastoptera arborina*, a related species living on a similar host, has also been observed, on occasion, to make partial crusts (Kuenzi and Coppel 1985b).

Recently, some spittlebug nymphs have been discovered to live in gel-like capsules. In Costa Rica, biologist and photographer Kenji Nishida discovered nymphs of an undescribed species of *Clastoptera* that live inside liquid-filled gelatinous capsules (Plate 5a, 5b), breathing through fleeting openings in the capsules using a modified abdominal tip. In Central Luzon, Elorde Crispolon has observed nymphs, possibly of the genus *Poeciloterpa*, living in similar capsules (Plate 5c), and I have observed nymphs, family unknown, living in opaque capsules in Northern Luzon.

A final enigmatic case involves the African aphrophorid genus *Cordia*. Two grass-feeding species, *C. peragrans* and *C. rotundiceps*, are reported to "surround themselves with a jelly-like substance, instead of the typical froghopper spittle mass" (Fewkes and Buxo 1964). The full life cycle has not been recorded in either of these species, leaving open the possibility of more ordinary spittle formation at points so far unobserved. Overall, the variations on nymphal shelter, including mineral crusts and tubes, gel capsules, and jelly-like material, as well as ordinary spittle, demonstrate striking evolutionary plasticity in the nymphal ecology of Cercopoidea.

Are There Any Truly Spittleless Spittlebugs?

There is one reported instance of spittleless spittlebug nymphs and one instance in which they have been inferred to exist based on circumstantial evidence. Yaghmaee (2008) describes an unusual system involving the spittlebug *Poophilus costalis* (under the synonym *Poophilus nebulosus*) and its host, the camel thorn plant (*Alhagi persarum*, Fabaceae), source of the Persian traditional medicine hedysarum (or taranjabin) manna. In the Iranian habitat studied, *P. costalis* has three generations per year that occur from April to autumn. Manna is produced by second- and third-generation nymphs. First-generation nymphs produce stem-long coalescing spittles of the conventional kind. Remarkably, it appears that the second- and third-generation nymphs do not make spittle but instead produce crystalline manna, though the mechanism for doing so is not described. In one illustration (Yaghmaee 2008, fig. 5), the manna appears to accumulate on stems in gel-like form, but the relationship between the nymphs and the gel—for example, whether they excrete it and/or live in it—is not recorded.

Poophilus costalis (Fig. 8.3f, Plate 13g) is one of the world's most widely dispersed and common spittlebugs. It occurs across Africa and Asia from Nigeria to Taiwan and has been reported as a minor pest of rice and a major pest of sorghum (Chapter 8). By all other accounts, it is a conventional spittlebug and it is difficult to believe that nymphs could survive the intense summer heat in an arid area without some form of protection. Given this uncertainty, Yaghmaee's observations require confirmation. Interestingly, the primary ingredients of hedysarum manna appear to be modified long chain polysaccharides (Hamedi et al. 2015) with some resemblance to the polysaccharide portions of the mucopolysaccharides that form the primary basis of ordinary spittle.

The second case of purported spittleless spittlebugs rests on loose circumstantial evidence. Although no one has ever reported nymphs of the family Epipygidae, Hamilton (2001), under the subheading "No spittle," says it is "doubtful" that epipygid nymphs produce spittle. He bases this in part on the assertion that spittle masses are easy to find and the fact that nymphs have been reported from the Neotropics for "all the other known major groups." This is true for families but not for every large subfamilial group. Many spittlebugs, for example, have root-feeding nymphs (Chapter 3), which can easily elude observation. The first nymphs of the large Neotropical cercopid genus *Ocoaxo* were not reported until 2020 and then proved to be root-feeders (Cid-Muñoz et al. 2020). Do epipygid nymphs make spittle? We do not know. In the absence of contrary observational evidence, the default likelihood is that they do.

Spittle Protects Nymphs Against Predators (Sometimes)

Because protection against predators and parasites is an important function of spittle, I have elected to include predation and parasitism of nymphs in this chapter. Predation on adults is covered in the following chapter (Chapter 5).

Spittle has been interpreted as a defense against predators from early on, either as a direct physical barrier or as a curtain hiding the nymphs. DeGeer (1773), for example, believed that spittle protected nymphs from spiders. On the other hand, the conspicuous white spittles clearly flag the presence of nymphs. One author has even suggested that the white froth serves as a kind of warning coloration, warding off predators (Yamazaki 2017). Other authors, more plausibly, have suggested that spittles by their conspicuousness may call the existence of nymphs to the attention of predators (Halkka et al. 1967; Whittaker 1970; Chen, Meyer-Rochow et al. 2018). Rather than protecting nymphs, spittle may advertise their presence. So, does spittle protect nymphs or serve as a signal that dinner is served? The evidence is mixed.

In general, spittle is a good but imperfect defense against predators. Guagliumi's (1955b) extensive table of predators and parasites of Central and South American spittlebugs indicates that a much greater variety of predators eat adults. Guagliumi's nymphal predation records, pertinent to this chapter, appear in Table 4.1. Corresponding adult predation records appear in the next chapter (Table 5.1). Predation on nymphs is sometimes quantitatively significant. Comparing field-caged nymphs of the important Brazilian pasture pest *Deois flavopicta* to controls exposed to the ordinary run of predators, Sujii et al. (2002) estimated that predation reduced nymphal survival by 20 to 47%. More generally, however, predation on nymphs in spittles appears to be light (Whittaker 1970). Whittaker (1971b), for example, measured twenty-day *Neophilaenus lineatus* nymph mortality at 25 to 38%, but, based on caged controls, estimated that at most 5% of mortality was due to predation. Karban (1989) found few instances of predation on *P. spumarius* nymphs living on sea daisies in California.

The following is a breakdown of observations of nymph predation (or the absence thereof) by predator category:

Ants

C. B. Williams, an early investigator of sugarcane pest spittlebugs (Chapter 6), reports two striking examples of the power of spittle to protect nymphs from ants, among the world's most rapacious arthropod predators.

> On one occasion in British Guiana the ants during the night gained access to my laboratory table, entered my breeding cages and killed all adult froghoppers and other insects (including Syrphid flies) that I had there, but the froghopper nymphs remained untouched. On another occasion in Panama I found a large army of hunting ants spread over a plot of grass which I knew to contain froghopper nymphs. After they had passed, I made a close search and found not an insect living except the froghopper nymphs in their protecting froth. (Williams 1921)

TABLE 4.1. Predators of spittlebug nymphs in portions of the Neotropics

CATEGORY	SPECIES	PREY	GEOGRAPHIC LOCALE
Beetle	Lampyridae, unidentified	*Aeneolamia varia*	Trinidad, Venezuela
Ants	*Monomorium carbonarium*	*Aeneolamia varia*	Trinidad
	Pseudomyrma elegans	*Aeneolamia flavilatera*	Guyana
	Solenopsis geminata	*Aeneolamia flavilatera*	Guyana
	..	*Aeneolamia varia*	Venezuela
	Solenopsis tenuis	*Aeneolamia flavilatera*	Guyana
Wasp	*Polybia rejecta*	*Deois morialis*	Brazil
Flies	*Salpingogaster nigra*	*Prosapia simulans*	Costa Rica
	..	*Zulia pubescens*	Trinidad, Guyana
	..	*Aeneolamia flavilatera*	Guyana
	..	*Aeneolamia lepidior*	Panama
	..	*Aeneolamia varia*	Guyana, Venezuela
	..	*Prosapia bicincta*	Cuba, Jamaica
	..	*Sphenorhina rubra*	Trinidad
	..	*Deois incompleta*	Brazil
Spider	Salticidae sp.	*Mahanarva tristis*	Trinidad
Scorpion	Scorpionidea sp.	*Mahanarva tristis*	Venezuela
Nematodes	*Mermis* sp.	*Mahanarva tristis*	Brazil
	..	*Deoisella picklesi*	Brazil
	..	*Aeneolamia varia*	Trinidad
	Nematoda sp.	*Prosapia bicincta*	Cuba
Lizards[a]	*Ameiva ameiva*	*Aeneolamia varia*	Venezuela
	Ameiva surinamensis	*Aeneolamia varia*	Trinidad
	..	*Aeneolamia flavilatera*	Guyana
	Cnemidophorus lemniscatus	*Aeneolamia varia*	Venezuela
Frogs[a]	*Atelopus cruciger*	*Aeneolamia varia*	Venezuela
	Bufo marinus (*Rhinella marina*)	*Aeneolamia varia*	Trinidad, Venezuela
	Eupemphix sp.	*Aeneolamia varia*	Tobago
	Hyla crepitans	*Aeneolamia varia*	Venezuela
	Hyla guogi	*Aeneolamia varia*	Trinidad, Venezuela
	Hyla punctatissima	*Aeneolamia varia*	Trinidad, Venezuela
	Hyla rubra	*Aeneolamia varia*	Trinidad, Venezuela
	Leptodactylus typhonius	*Aeneolamia varia*	Trinidad, Venezuela
	Phyllomedusa bicolor	*Aeneolamia varia*	Venezuela
Birds	*Agelaius icterocephalus*	*Aeneolamia varia*	Trinidad
	Icterus xanthornis	*Aeneolamia varia*	Trinidad
	Quisquqlis lugubris	*Aeneolamia varia*	Trinidad
	Geothlyypis aequinoctialis	*Aeneolamia varia*	Trinidad

Source: Excerpted from Guagliumi 1955b

Note: Double dots (**..**) represent repetition of the predator species listed above.

Note: Spittlebug species names have been updated to current usage.

[a] These records should be viewed with caution since in the source every one of these species is said to eat both adults and nymphs, which is unlikely.

I tested this observation on an army ant raiding column (genus *Exciton*) in Costa Rica, using tuna fish baits covered or not covered by spittlebug spittle. The ants rapidly made away with unprotected baits. They examined but passed up baits protected by spittle. Myers (1935), working on Marjó Island at the mouth of the Amazon, found that half or more of the spittles of *Mahanarva terrea* were in the soft soil of *Solenopsis* fire ant nests built around *Paspalum* grass clumps. Since ants are the major foraging predators in many ecosystems, especially the tropics, this is strong evidence for the protective power of spittle. Hoch et al. (2024) tested the reaction of black garden ants (*Lasius niger*) to spittle, with water as a control. The ants generally avoided contact with the spittle, but when they did contact the foam, they moved assertively to free themselves and groomed until foam-free. Several other lab and field experiments have demonstrated the deterrent effects of spittle on ant predation (Whitaker 1970; Nachappa, Guillebeau, Braman, and All 2006; Campo et al. 2011; Tonelli 2019; Balzani et al. 2023).

However, spittle protection of nymphs from ants is not complete (Sujii et al. 2002). Ants regularly attack nymphs temporarily unprotected outside spittle, such as when newly hatched, molting, or changing feeding places on a plant (Nachappa, Guillebeau, Braman, and All 2006). In Wisconsin, spittle provided some protection against *Formica montana*, but the ants could breach spittle defenses by placing soil and thatch on spittle to gain access to nymphs (Henderson et al. 1990). In the Llanos region of Colombia, Medina (1995) observed ant predation on first instar *Aeneolamia varia* nymphs in soil surface spittles in pasture grass. In an experiment in Brazil, ants carried off an extraordinary 749 of 750 newly hatched first instar nymphs before they could establish on *Urochloa decumbens* (Hewitt and Nilakhe 1986). Ants also successfully foraged for *Clastoptera* nymphs and adults in cacao trees in Mexico (Valenzuela-Gonzalez et al. 1994).

Predatory Wasps

Three species of the solitary wasp genus *Argogorytes* provision brooding burrows exclusively with paralyzed, living spittlebug nymphs. The Eurasian species *A. mystaceus* and *A. fargeii* load their burrows with *P. spumarius* and *Aphrophora* spp. nymphs (Callan 1980). The New Zealand species *A. carbonarius* provides its larvae with nymphs of two *Carystoterpa* species and *P. spumarius*. In each case, the female wasp grabs nymphs directly from spittles and lays eggs on the paralyzed, sequestered nymphs (Callan 1980; Evans and O'Neill 2007). Hatching wasp larvae devour the paralyzed nymphs from within, making these species both predators and parasitoids of their spittlebug nymph victims.

Spiders

Spiders commonly attack adult spittlebugs (Chapter 5) but do not prey heavily on nymphs (Sujii et al. 2002). They do occasionally eat nymphs (Plate 9d), as corroborated by radioactive labelling demonstrating predation on *P. spumarius* nymphs by wolf spiders and web spinners (Harper and Whittaker 1976).

Beetles, Predatory Bugs, Crickets, and Dragonflies

In Japan, both adults and larvae of the arboreal carabid beetle *Lebidia octoguttata* prey on nymphs of *Aphrophora* species, the larvae entering spittles to find nymphs, the adults pulling them out of spittle or ambushing them when they leave it (Hayashi 2023). In England, soldier beetles, rove beetles, and damsel bugs eat *P. spumarius* nymphs (Harper and Whittaker 1976). Cecil (1930) reports that the spined soldier bug (*Podisus maculiventris*) occasionally takes unprotected nymphs. Sahayaraj et al. (2021) found nymphs of the heteropteran bug *Antilochus coquebertii* (Pyrrhocoridae) feeding on *Poophilus costalis* nymphs in India. Along with lacewing larvae, heteropteran bugs were also the major predators of the pecan spittlebug *Clastoptera achatina* (Plate 5d) in Georgia (Tedders 1995) and of nymphal *P. spumarius* in Wisconsin (Williams 2004). In Sumatra, bush crickets (Tettigoniidae) have been implicated in heavy natural mortality of the machaerotid *Hindola fulva*, breaking into machaerotid tubes to feed on nymphs (Lomer et al. 1993). In China, dragonflies are reported to prey on *Callitettix versicolor* nymphs (Lei et al. 1992), an observation anomalous enough in light of dragonfly predation behavior to require confirmation.

Syrphid Flies

By far the single most voracious known predator of spittlebug nymphs is *Salpingogaster nigra* (Fig. 4.6), a syrphid fly whose larvae move from spittle to spittle devouring the inhabitants. In Trinidad, it is the major predator of *Aeneolamia varia*, estimated to reduce field populations by 17 to 40% (Pickles 1932, 1933), while a second species, *S. pygophora* attacks *Sphenorhina rubra* (Myers 1935). In Guyana *S. nigra* feeds on *Aeneolamia flavilatera* (Williams 1918). In Brazil it attacks *Mahanarva spectabilis*, apparently tracking nymphs by odor (Veríssimo et al. 2018). Brazilian *Salpingogaster* are also reported to attack *Cephisus siccifolius* and the sugarcane pests *Mahanarva posticata* and *Mahanarva liturata* (Knutson 1971). In Costa Rica, I reared an unidentified *Salpingogaster* that preys on *Clastoptera laenata* nymphs, and Kenji Nishida observed a syrphid, likely *Salpingogaster*, preying on nymphs of an undescribed *Clastoptera* species

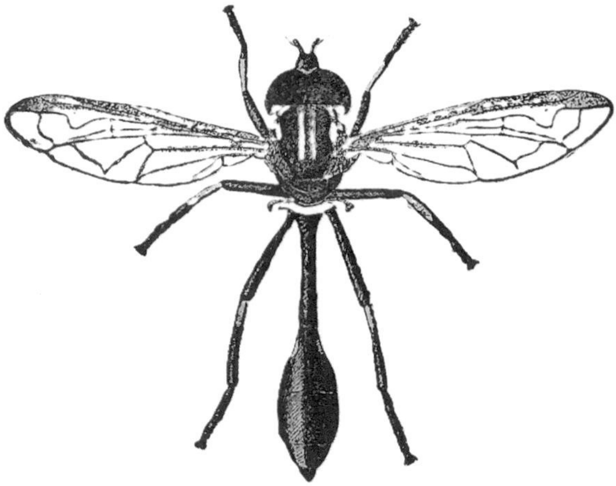

FIGURE 4.6. Adult *Salpingogaster nigra*, a Neotropical syrphid fly. The larvae are ferocious, specialized predators of spittlebug nymphs. From Williams (1921).

living on oaks. These examples demonstrate that *Salpingogaster* prey on a very wide range of spittlebugs (some feed on other Hemiptera, Knutson 1971).

Salpingogaster nigra are big eaters, each one wolfing down about 14 to 40 nymphs in the course of development (Guppy 1913; Páez et al. 1985). This has stimulated efforts to use them for biological control, but they have proven intractable to mass rearing (Pickles 1932, 1933; Veríssimo et al. 2018; and see Chapter 6). The genus *Salpingogaster* is endemic to the New World, but at least one African syrphid fly, *Asarkina ericetorum*, has larvae that prey on nymphs of the aphrophorid *Poophilus* nr. *costalis* in Nigeria (Musa 1975) and the cercopid *Locris rubra* in Uganda (Fewkes and Buxo 1964; Greathead 1970). There was discussion of importing this "voracious" predator from Uganda to Trinidad in the hope that it might control *A. varia* where the native *S. nigra* failed (Williams 1931, Fewkes and Buxo 1964). Fewkes demonstrated that *A. ericetorum* larvae will eat *A. varia* nymphs in the laboratory (Cock 1985), but I find no evidence control was ever attempted in the field. Larvae of an unidentified fly species attack spittlebug nymphs in the Philippines (Plate 9e).

Birds

Overall, birds appear to be infrequent predators of spittlebug nymphs. A study of prey contents in 80,000 Nearctic bird stomachs revealed no spittlebug nymphs, while adult spittlebug remains were found in the stomachs of forty-one species

(McAtee 1932). Nevertheless, birds have been recorded many times preying on nymphs but only on limited occasions as major food items (Whittaker 1970). In one spectacular example, "a great invasion" of long-tailed tits (*Aegithalos caudatus*) and other birds descended on willows in Hampshire, England, "phenomenally infested with *Aphrophora* spp." Three days later, "hardly a drop of the spume was to be found" (Haines, 1942). In Indiana, Nolan (1956) observed several instances in which Prairie warblers (*Setophaga discolor*) systematically preyed on nymphs in spittles (probably *P. spumarius*), as did red-winged blackbirds (*Agellaius phoenicius*) in the same state (Macklin 1958). Brindley (1934) reports similar behavior by house sparrows in England. Verbeek (1998) recorded 245 spittlebugs, most of them nymphs, in the nestling diet of Northwestern crows (*Corvus caurinus*) in British Columbia. The savannah sparrow of North America also extracts nymphs from spittles, a significant part of its diet (Wheelright and Rising 1993), and some black-capped chickadees in West Virginia concentrated on eating very small spittlebug nymphs (Cooper 1988). Finally, evidence culled from Super 8mm movie camera still photos captured California chestnut-backed chickadees bringing *Aphrophora* nymphs to nestlings, right down to residual spittle on their beaks (Kleintjes and Dahlsten 1994; Neff Kleintjes, personal comm. 2024). The nymphs accounted for 3.9% of the nestlings' diet.

In Europe, not without irony, cuckoos, once believed to be the source of spittle (Chapter 1), occasionally eat nymphs, as evidenced by stomach contents. Pheasant chicks foster reared by turkey hens eat and sometimes choke on nymphs (Gruner 1901). In Slovakia, Aphrophoridae nymphs make up a substantial portion of the diet of nestlings of the tit *Remiz pendulinus* (Krištín 1995), which is consistent with observations that *Aphrophora* nymphs are important to several other species of tits (Whitaker 1970). However, direct field experiments to detect higher survivorship of *P. spumarius* and *N. lineatus* nymphs in caged plots that excluded birds failed to demonstrate any difference from uncaged controls (Whittaker 1970). In Europe, Halkka and Kohila (1976) record only two bird species feeding on nymphs, the wood grouse (*Tetrao urogallus*) and ring-necked pheasant (*Phasianus colchicus*).

Reptiles and Amphibians

Whittaker (1965) records a single *Neophilaenus* second instar nymph among thirty sampled stomachs of *Rana temporaria* at a high elevation site in England. This is the only record I have encountered of any reptile or amphibian eating a nymph.

Mammals (Including Humans)

Three mammals, all primates, are confirmed predators of nymphs. In Madagascar, the lemur *Eulemur macaco* has been observed aggressively seeking nymphs in spittles of unidentified spittlebugs (Birkinshaw 1999 and Christopher R. Birkinshaw, letter to Vinton Thompson, January 20, 2000). In Costa Rica, white-faced capuchin monkeys have been observed reaching into large spittle masses in the forest canopy (Elizabeth K. Mallott, personal comm. 2017), and DNA barcoding has detected spittlebug DNA in their feces (Mallott et al. 2017). They were probably feasting on *Cephisus siccifolius*, whose communal spittles would provide ample nymphs for a nimble predator. Some of the fecal DNA matched *Mahanarva costaricensis* (Plate 4), which would more likely be encountered during ground foraging. Among adults, 12.5% of fecal samples included spittlebug DNA. Among juveniles the figure was only 4.0%, probably attributable to a less developed ability to catch agile prey (Mallot et al. 2017). In Gabon, humans consume *Ptyelus flavescens* (Plate 1c, 1d), both nymphs and adults, which are among seventy-five species of edible local insects (Detilleux et al. 2022). This species, the same one that David Livingstone experimented on in Angola (Chapter 3), is the only spittlebug reported to be subject to human predation.

Parasites of Spittlebug Nymphs Overcome Spittle Defenses

In addition to predators, spittlebug nymphs are victims to a select group of parasitic insects and nematode roundworms.

Carabunia Wasps

Wasps of the genus *Carabunia* (Encyrtidae) are the only well-characterized parasites of spittlebug nymphs. The best studied species is *Carabunia myersi*, which parasitizes nymphs of the genus *Clastoptera* and was first discovered in Cuba during a search for biological control agents against the sugarcane froghopper *Aeneolamia varia* (Myers 1930). Walking undeterred over the surface of spittle masses, female *C. myersi* insert their ovipositors into spittles and inject one to six eggs into each parasitized nymph. Infected nymphs have gigantic, swollen abdomens, 50% wider and longer than normal abdomens of the same instar. The wasp larvae grow to maturity inside nymphs, chewing great holes in the abdomen to emerge. Nymphs live for several more hours and may wander from

TABLE 4.2. *Carabunia* wasps attacking nymphal spittlebugs

CARABUNIA SPECIES	SPITTLEBUG ATTACKED	LOCALE	REFERENCE
C. gigantica[a]	*Ptyelus flavescens*	Uganda	Subba Rao 1973
C. longmarginalis[a]	Unidentified species	Malaysia	Subba Rao 1973
C. myersi	*Clastoptera flavidorsa*	Jamaica	Bennett & Hughes 1963
..	*Clastoptera undulata*	Cuba	Myers 1930
..	*Clastoptera* sp.	Cuba	Myers 1930
..	*Clastoptera globosa*	Mexico	Garcia J. 1982
..	*Clastoptera undulata*	Florida	Mead & Bennett 1987
..	*Clastoptera undulata*	Bermuda	Bennett & Hughes 1963
C. orientalis[a]	*Clovia puncta*	India	Ananthasubramanian & William 1989
..	*Ptyelus nebulas*	India	Subba Rao 1971
C. nr. *orientalis*[a]	*Hindola viridicans*	Sumatra	Lomer et al. 1993
Unidentified species	*Hindola viridicans*	Java	Lomer et al. 1993
Unidentified species	*Hindola fulva*	Sumatra	Lomer et al. 1993
Unidentified species	*Hindola sp.*	Sumatra	Lomer et al. 1993
C. waterstoni	*Clastoptera* sp.[b]	Puerto Rico	Subba Rao 1971
..	*Leocomia uprae*[c]	Puerto Rico	Wolcott 1948
C. zora[a]	*Clovia puncta*	India	Manickavasagam & Rameshkumar 2010
Unidentified species[d]	*Clastoptera* sp.	Haiti	Myers 1930
..	*Hindola fulva*	Indonesia	Lomer et al. 1993
..	*Clastoptera* sp.	El Salvador	Bennett & Hughes 1963

Note: Double dots (..) represent repetition of the *Carabunia* species listed above.
[a] Noyes 2010 has proposed moving all Old World Carabunia to the genus *Elijahia*.
[b] This is apparently *Clastoptera* sp. nr. *diminuata* studied by Bennett & Hughes 1963.
[c] Species I.D. from Ramos 1957.
[d] Known only from a single larva, which "appeared identical" to *C. myersi* larvae.

the spittle, like post-apocalyptic zombies, sometimes with protruding guts, until they die (Myers 1930; Bennett and Hughes 1963). *Carabunia waterstoni* was introduced into Bermuda for attempted biocontrol of *Clastoptera undulata* (Bennett and Hughes 1963; Cock 1985), and a *Carabunia* sp. is credited with controlling the same invasive species in Florida (Mead and Bennett 1987).

Several other *Carabunia* species have been reported to attack spittlebug nymphs, including other *Clastoptera* species in Cuba, Jamaica, Florida, and Mexico; Aphrophoridae in Africa and Asia; and Machaerotidae in Indonesia (Table 4.2). *Carabunia*'s parasitization of spittlebugs of three families spread over multiple continents suggests an ancient origin for the relationship. The taxonomy of *Carabunia* is unsettled. Noyes (2010) has proposed moving the Old World species to the genus *Elijahia*. Trjapitzin and Triapitsyn (2010) have proposed folding *Carabunia* into the genus *Neocladia*. In view of the *Carabunia*'s widespread familiarity as a taxonomic term and the special association of this taxon with Cercopoidea, I retain the older use here. There is no evidence that the other species included in the expanded concept of *Neocladia* have any special relationship with spittlebugs.

Strepsiptera

Strepsiptera are an odd order parasitizing a large range of insects. The females develop as endoparasites in the abdomens of their hosts, projecting enough of their bodies externally for the free-ranging males to inseminate them. For spittlebugs, the only well studied system involves *Halictophagus pontifex* parasitizing adults of seven Aphrophoridae species in Uganda, including four species of *Poophilus*, two of *Clovia*, and one of *Cordia* (Greathead 1970). In North America, a *Halictophagus* sp. has been recorded parasitizing the spittlebug genus *Philaronia* (Bohart 1946), and large numbers of strepsipterous larvae have been reported in unidentified spittlebug nymphs in Ontario, Canada (Weaver and King 1954). In Indonesia, an unidentified strepsipteran attacks machaerotid spittlebugs of the genus *Hindola* (Lomer et al. 1993). Bohart (1946) suggests that all adult infections begin in nymphs. If so, this indicates successful evasion of the spittle defense, the rationale for including Strepsiptera among nymphal parasites.

Nematodes

Nematodes kill spittlebug nymphs by penetrating the body cavity and releasing symbiotic bacteria, which proliferate, causing death by septicemia. The nematode dines on a combination of the killer bacteria and degraded host tissue (Vicente-Díez et al. 2021). Gruner (1901) found heavy nematode parasitization of spittlebug nymphs (species unspecified) in German pastures. Urich (1913) notes the presence of nematodes in *Aeneolamia varia* spittles, and Poinar and Linares (1985) report an entomopathogenic nematode attacking *A. varia* nymphs in Venezuela.

The last two decades have seen a surge of interest in the use of nematodes for biological control of root-feeding nymphs, particularly of sugarcane pests. They have been used most widely against *Aeneolamia* species. In a field study in Panama, Pérez Milián et al. (2018) report that *Heterorhabditis bacteriophora* killed up to 66% of early instar and 83% of late instar *Aeneolamia* nymphs and in some instances was more effective than entomopathogenic fungus controls. In Mexico, Grifaldo-Alcántara et al. (2019) report field mortality rates of 63 to

73% using native nematodes against *Aeneolamia* nymphs. In Colombia, Moreno et al. (2012) achieved 42 to 81% field mortality against fourth instar *A. varia* with *Heterorhabditis* nematodes. In Brazil, Leite et al. (2005) found 96 to 100% mortality in lab tests and 56% control in field tests of six nematode strains against *Mahanarva fimbriolata* in sugarcane, equivalent to results using chemicals and entomopathogenic fungi. Nematodes seem to be drawn to *M. fimbriolata* nymphs by the odor of spittlebug-damaged roots (Tonelli et al. 2016), suggesting they use plant volatiles to home in on feeding nymphs. Also in Brazil, Batista et al. (2014) achieved 100% nymphal mortality using several nematode species against *Mahanarva spectabilis* nymphs in greenhouse experiments.

Nematodes have also been targeted against *Philaenus spumarius* nymphs (Vicente-Díez et al. 2021). Spittle did not deter them, and three of the four species tested significantly decreased survival, with mortality rates of 53 to 90%. Cell-free supernatant extracts from the nematodes' bacterial symbionts led to mortality rates up to 64%. The results suggest that nematodes might be used effectively against *P. spumarius* and hint that lethal compounds released by their bacterial symbionts might be used to control *P. spumarius* as a *Xylella fastidiosa* vector (Chapter 9).

Fly Larvae That Live in Spittle and May or May Not Be Parasites of Spittlebug Nymphs

Fly larvae of three genera, *Cladochaeta*, *Leucophenga*, and *Paraleucophenga* (all Drosophilidae), live in spittles in close association with spittlebug nymphs (Ashburner 1981; Thompson and Mohd-Saleh 1995). Nothing is known of the *Paraleucophenga* association beyond its initial discovery among nymphs of the machaerotid *Hindola viridicans* in Sumatra (Lomer et al. 1993). *Leucophenga* is a global genus with several hundred species, mostly feeding on fungus, largely Old World, but with a few species in the Neotropics (Grimaldi and Richenbacher 2023). A Malagasy species lives in the spittles of *Ptyelus goudoti*, a Nigerian species in spittles of *Ptyelus grossus* and *Poophilus adustus*, a Ugandan species in spittles of *Ptyelus flavescens* (Decary 1928; Odhiambo 1958; Medler and Adenuga 1969; Ashburner 1981), and a South African species, perhaps the same, in *Ptyelus grossus* spittles (Plate 8c). Little is known of the ecology of these associations, but they are believed to be commensal, with no harm to the spittlebug nymphs (Medler and Adenuga 1969; Ashburner 1981).

Cladochaeta–spittlebug interactions are slightly better studied. Ten or so of 138 described *Cladochaeta* species are reported to live in *Clastoptera* spittle

FIGURE 4.7. Scanning electron micrograph of a third (last) instar larva of *Cladochaeta inversa*. Larvae of many *Cladochaeta* species live in spittles in close association with spittlebug nymphs. Reproduced from Grimaldi and Nguyen (1999) with permission of the American Museum of Natural History.

(Thompson and Mohd-Saleh 1995; Grimaldi and Richenbacher 2023). The *Cladochaeta* larvae (Fig. 4.7) often cling to nymphs (Fig. 4.8, Plate 8d). They pupate within the spittles. Pupae or empty puparia can often be found in dried spittle remnants or even directly adhering to late instar host nymphs (Grimaldi and Nguyen 1999). In addition to *Clastoptera*, one species inhabits large communal spittles of *Cephisus erythrocephala* in Ecuador (Tsacas and Couturier

FIGURE 4.8. Illustration of a *Cladochaeta inversa* larva piggybacking on a fifth instar *Clastoptera obtusa* nymph. From Baerg (1920).

1993), two live on *Sphodroscarta trivirgata* in Brazil (Carvalho-Filho et al. 2018; Pirani and Carvalho-Filho 2019), and another on an undetermined *Sphodroscarta* species in Ecuador (Grimaldi and Nguyen 1999). Fly larvae of an unidentified species, perhaps *Cladochaeta*, have also been reported in spittles of *Leocomia uprae* in Puerto Rico (Wolcott 1948; spittlebug identification from Ramos 1957).

Not all *Cladochaeta* have spittlebug hosts (Grimaldi and Nguyen 1999). One Brazilian species feeds on the contents of spider egg sacks (Carvalho-Filho et al.2018), and several have been reared from flowers, though it is impossible to rule out the possibility that some of the flowers may have harbored pupae in the remains of spittles.

The best studied North American species, *Cladochaeta inversa* and *C. sturtevanti*, appear to have negative consequences for their *Clastoptera* hosts, but it has not been conclusively demonstrated that they are direct parasites, actively consuming, for example, host haemolymph (Thompson and Mohd-Saleh 1995; Grimaldi and Nguyen 1999; Nguyen 2001; Thompson 2023a). Depending on species pair, there may be a range of *Cladochaeta*–spittlebug interactions. Since both *Cladochaeta* (Grimaldi and Nguyen 1999) and *Clastoptera* (Paladini et al. 2020) likely number in the hundreds of species, there is potential for the discovery of many more *Cladochaeta*–spittlebug interactions.

Chapter Overview and Looking Forward to Predation on Adults

Spittle production is the most characteristic feature of spittlebugs. With two possible exceptions, neither securely documented, it seems to be universal among the Cercopoidea. Despite spittle's central role in spittlebug biology, the biochemical composition and structure of spittle are still only partly understood. While spittle provides effective protection against many predators and parasites, this protection is incomplete, and a number of specialized or occasional enemies eat or parasitize spittlebug nymphs. Adult spittlebugs are also subject to predation, minus the protection of spittle. Their predators and their defenses against them form the subject of the next chapter. In contrast to nymphs in gooey spittle, get ready for a lot of jumping and a dazzling array of color variation.

5

EVADING PREDATORS

How to Keep from Getting Caught

In the preceding chapter we learned that spittle is a good, though not foolproof, defense for nymphs against predators. When spittlebugs become adults they no longer have spittle as a defense. This chapter examines the roles of jumping, reflex bleeding, warning coloration, camouflage, and polymorphism as defenses adults have evolved against birds and other predators.

Spittlebugs: World Champion Jumpers

On 31 July 2003, the world woke up to press reports declaring that the meadow spittlebug, *Philaenus spumarius*, had dethroned the flea as the world's champion jumper. "Leapin' better than lizards, jumpin' beyond Jehoshaphat, and more than hopping mad, the lowly spittlebug now ranks as nature's best jumper" declared an *Associated Press* wire report, adding that "If a human could leap with equal facility, he or she might clear Manhattan's 625-foot-tall Waldorf-Astoria hotel in a single bound" (Cooke 2003). A few days later the Tuesday science section of the *New York Times* picked up the story with the observation that it would be like 7-foot-1-inch (2.16 m) basketball superstar Shaquille O'Neal leaping 500 feet above the basket (Fountain 2003). In a memorable blooper, the same article included a cartoon of a spittlebug nymph leaping high over a flea. Spittlebug nymphs never jump, only adults. Even the most august periodicals occasionally get things wrong.

The fuss arose from a one-page publication in *Nature* by Malcolm Burrows of Cambridge University (Burrows 2003). Using ultrafast photography, he demonstrated that *P. spumarius* could jump with extraordinary speed, taking off in less than one thousandth of a second, and reaching heights of 700 mm (about 2 feet), or 115 body lengths. Relative to size, they outleaped fleas and generated far more jumping force, 414 times their body weight versus 135 in fleas and a paltry 2 to 3 times in human jumpers. They accomplished these feats by virtue of a unique jumping mechanism.

A Crossbow-like Firing Mechanism Makes Spittlebug Super Jumps Possible

Despite the alternative name froghopper, spittlebugs do not hop like frogs at all. Instead, they catapult into the air abruptly, like a projectile suddenly released from a crossbow. They perform this trick with the help of massive thoracic muscles, a full 11% of their body weight, that contract slowly while locking the hind legs in a "firing" position. The energy from the slow contraction is stored by bending bow-like laminated struts on either side of the thorax (Burrows et al. 2008; Siwanowicz and Burrows 2017), and it is triggered suddenly to forcefully extend the hind legs, thrusting them against the substrate on which the insect is sitting. The hind leg tibial and tarsal spines (Figs. 1.4, 1.7) pierce the substrate, providing traction for takeoff (Goetzke et al. 2019). The struts, technically metathoracic pleural arches, are composed of the rubber-like protein resilin laminated with chitin, the basic material of insect exoskeletons. Resilin provides flexibility plus resiliency after firing. Stressed deformation of the laminated chitin provides energy storage. Like a crossbow, energy is stored until triggered. Most sitting spittlebugs are ready and loaded to escape by launching away.

Upon release, a spittlebug takes off like a bullet, startling potential predators and catapulting quickly out of reach. Though spittlebugs can and do fly, sometimes long distances (Chapter 9), they do not open their wings before or during takeoff, and only occasionally after becoming airborne (Burrows 2006a). The ability to jump extends to mating pairs in copula. They jump ensemble, without separating (Xie 1994 for *Makiptyelus dimorphus*; my observations in multiple species). This has obvious survival benefits. It also means that coupling must leave the legs free in at least one sex. I do not know whether the jump is triggered in the male or the female, but it is unlikely to be both, since that would require difficult coordination. Sherry et al. (2020) suggest that propulsive jumpers like spittlebugs may benefit from substantial body mass so that air drag decreases their velocity less quickly (Vogel 2005). This may account, in part, for the comparatively large size of spittlebugs, discussed in Chapter 3.

The hind leg locking mechanism is formed by bilateral protrusions from the coxae (Fig. 1.6), which engage corresponding protrusions of the femurs to form a "mechanical latch" held in place by Velcro-like surfaces (Burrows 2006b). These interlocking protrusions are absent from the fore and middle legs, which are not involved in jumping, and from nymphs, which do not jump. Release of the "cocked" hind legs is extraordinarily quick and produces a measurable click. To keep the bugs from veering to the side, firing is tightly coordinated between the legs, synchronized within a remarkable 32 millionths of a second. Spittlebugs also exercise some control over jumping direction through initial orientation of the hind tibiae in contact with the surface (Sutton and Burrows 2010). When the hind legs are locked in firing position, spittlebugs walk on the four anterior legs, dragging their hind legs and giving the bugs a characteristic gait. The Burrows studies covered six spittlebug genera in three families: *Aphrophora*, *Neophilaenus*, *Philaenus*, *Lepyronia*, *Cercopis*, and *Clastoptera*. They are all great jumpers, but proportional to size, *P. spumarius* takes the prize.

Fossils indicate that this unique spittlebug jumping mechanism dates back to the Jurassic Procercopidae (Wang et al. 2010), the earliest spittlebugs and the apparent ancestors of modern forms (Chapter 2). If so, propulsive jumping has been a feature adaptation over the whole 200 million years of spittlebug evolution. Newly emerging adults trade the protection of spittle for the ability to jump (Gruner 1901). Although they can walk and run, including backwards (Ulrich 1913; Doering 1931; my observation for *Prosapia ignipectus*), as well as fly and even play dead when caught (Xu 1986; my observation for some *Aphrophora*), their extraordinary jumping ability becomes their primary defense against predators.

Many Spittlebugs Also Fend off Predators with Reflex Bleeding

Some adult spittlebugs have an additional defense mechanism, reflex bleeding, familiar to many people in the form of the orange odiferous liquid excreted by ladybird beetles (Coccinellidae). When molested by predators, these spittlebugs exude smelly liquid hemolymph, the insect equivalent of blood, from their tarsal pads (Plate 7a, 7b, 7g). This was first reported in the sugarcane pest *Aeneolamia varia* in Trinidad (Chapter 6). Urich (1913) observed that most insectivorous birds "do not appear to like adult froghoppers . . . no doubt due to the repellant secretion which the adults emit from the ends of their tarsi . . . a yellow repellant liquid" with "a peculiar odor." Kershaw (1914) noted that the "bright, clear oil-yellow hemolymph" is limited to adults. Nymphal hemolymph is nearly colorless and odorless, and nymphs do not reflex bleed.

The only substantial work on reflex bleeding in spittlebugs comes from Daniel Peck (2000a). He found that it is limited to the family Cercopidae, likely universal in New World Cercopidae, and found in at least some Old World Cercopidae. It is not present in the Clastopteridae or Aphrophoridae, nor, based on my observation of one species each, in the Epipygidae or Machaerotidae. The published evidence for reflex bleeding in the Old World Cercopidae is scanty but includes the genera *Locris* (Greathead 1970) and *Cercopis* (refs. in Peck 2000a). I have observed reflex bleeding in *Homalostethus*, *Mioscarta* (Plate 9b), and *Serapita* (Plate 7g, 7f), though in these cases associated with clear, not yellow, haemolymph. This is consistent with observations that the haemolymph emitted by *Cercopis vulnerata* is also clear (Körner 2006), suggesting that the unknown substance that makes New World cercopid haemolymph yellow is not directly related to the characteristic reflex bleeding smell. *Cercopis* is at the root of one of two Old World Cercopidae major lineages, and *Serapita* occurs in the other (Crispolon et al. 2023). This supports Peck's suggestion that reflex bleeding is an ancestral trait shared by the whole Cercopidae lineage (a synapomorphy in technical terms).

Peck studied reflex bleeding in detail in four Costa Rican Cercopidae species. They bled when molested in lab and field settings and, following a bleeding episode, recovered the ability to bleed again within six hours, provided they were permitted to feed in between. Despite the color and odor, the bleeding was not clearly a direct deterrent to predators, either mechanically or chemically. The evidence was mixed. In experiments, fire ants, rove beetles, funnel spiders, and an anole lizard accepted adult *Prosapia* as prey, but chickens and army ants (*Exciton burchellii*) rejected them. In a subsequent lab experiment, Nachappa, Guillebeau, Braman, and All (2006) found that reflex bleeding failed to deter a range of arthropod predators from preying on *Prosapia bicincta*. They did observe extensive grooming of mouthparts and antennae of predators that came in contact with haemolymph, suggesting some negative effect. In an unpublished experiment at Las Cruces in Costa Rica, I presented large cane toads (*Rhinella marina*) with live specimens of the big spittlebug *Mahanarva insignita* and cryptic moth controls. The toads, legendarily voracious predators, ate fewer spittlebugs and occasionally regurgitated them, suggesting they found them at least partially distasteful. They ate the moths without hesitation or aftereffect.

Human taste testers have not had a negative response. Williams (1921) tasted *A. varia* adults and reported "that the taste resembled that of grass and was not unpleasant." Peck (2000a) makes a similar observation. I concur.

In 2024 I held a live specimen of the fairly large Mexican species *Prosapia ignifera* under my tongue for a couple of minutes. The taste was similar to the smell, markedly present but not repulsive. It lingered for at least half an hour, affecting the taste of my coffee. Thus, by existing accounts, the bugs do not elicit a strong, negative, get-this-thing-out-of-my-mouth reaction. The specimen in my mouth emerged from its ordeal with no evident ill effects, indicating that in at least some circumstances reflex bleeding spittlebugs can survive temporary ingestion.

Despite the absence of evidence for strong deterrence, spittlebug reflex bleeding is highly correlated with conspicuous coloration in the Cercopidae (Peck 2000a; Paladini, Takiya et al. 2018). Many Cercopidae, both New World and Old, display bold color patterns, mixing red, orange, yellow, and black, often with stripes and spots. In aggregate, these striking patterns scream warning coloration. The impression of warning coloration is reinforced by the tendency of some conspicuous spittlebugs to aggregate as adults (Plate 4c) (Peck 2000b; Thompson 2013), a recurring phenomenon among *aposematic* (warningly colored) insects (Ruxton and Sherratt 2006).

Warning Coloration Advertises Difficulty of Capture: Reflex Bleeding Can Reinforce the Message

Warning coloration evolves in well-defended prey as a signal to predators that pursuit is likely to be futile or at least not worth the effort (Ruxton, Sherratt, and Speed 2004). A predator that has a frustrating encounter with its prey is more likely to recognize and recall the encounter if the prey has a conspicuous pattern. As a result, well-defended prey benefit from a shared, memorable color pattern or other shared features easily recognized from a distance. Natural selection will favor convergence on a memorable, conspicuous pattern.

How do we reconcile the association of reflex bleeding with strong warning coloration when, as a deterrent, it appears to be weak? The answer, I suggest, is that it serves as an auxiliary and reinforcement to spittlebugs' prime defense, their uniquely effective capacity to leap to safety. Warning coloration is not limited to defense by distasteful chemicals, stings, or other mechanical defenses. It can also be based on effective escape ability. Predator learning to avoid escape artist prey can be faster than learning to avoid distasteful prey (Gibson 1974, 1980; Ruxton, Speed, and Sherratt 2004; Páez et al. 2021). Spittlebugs may reinforce this learning by sitting still until they suddenly catapult away, providing extra time for a predator to secure an image and reinforce memory.

Reflex bleeding, though not a strong independent deterrent, reinforces the primary jumping defense with a tactile fluid and odor defense that is applied at close range. Multimodal defenses of this kind are relatively common (Rowe and Halpin 2013). Rothchild (1961) suggested that *Cercopis vulnerata* (Plate 11g), a striking red and black European spittlebug with smelly reflex bleeding, fits this pattern. Peck (2000a) and Nachappa, Guillebeau, Braman, and All (2006) also interpret reflex bleeding as complementing but subordinate to the primary mechanism of escape by jumping. As a caution against overgeneralization, I note that reflex bleeding may play very different roles in defense against vertebrate predators with high visual acuity and good memories versus invertebrate predators that react on a smaller scale to the immediate physical consequences of haemolymph release (Plate 9c).

The Chemical Nature of the Smell Associated with Reflex Bleeding

The odor associated with reflex bleeding in Cercopidae is quite distinctive and permits quick pinch tests for reflex bleeding by smell alone. An agronomist in Trinidad told me in 2003 that he could gauge the level of *A. varia* infestation in a sugarcane field simply by the intensity of the smell. Despite its olfactory prominence, no one has characterized the chemical basis of the odor. A single, isolated report identified the presence of 2-methoxy-3-*sec*butylpyrazine in a whole-body extract of *Cercopis vulnerata* (Moore et al. 1990). Pyrazines are hypothesized to serve a role as alerting odors in a wide variety of insects. Whether this chemical plays a role in the odor associated with reflex bleeding remains to be determined.

Mimicry Rings Bring Safety in Numbers for Multiple Warning Colored Species

Just as warning coloration provides selective pressure for individuals of the same species to all look alike, it also leads to selection for mutual resemblance of co-occurring, well-defended species, resulting in multiple local species sharing the same warning pattern. This phenomenon is called Mullerian mimicry, named for Fritz Müller who first described it in Neotropical butterflies (Ruxton, Sherratt, and Speed 2004). Groups of co-occurring species that share similar color patterns are called mimicry rings. One putative case involves North American species of the genera *Prosapia* and *Aeneolamia* (Thompson and Carvalho 2016). In an unusual case in Peru, an aphrophorid, *Sphodroscarta bimaculata*, seems to mimic a cercopid, *Zuata ephippiata* (Hamilton and Thompson 2007). Though billed as *Batesian mimicry*, in which an undefended species mimics a defended species to gain protection, this case could also be interpreted as *Mullerian mimicry*, in which both species share a pattern that advertises good jumping defenses. *Cercopis vulnerata* (Plate 11g) and *Cercopis intermedia* have been implicated in Mullerian mimicry rings with red and black insects of several other groups in Europe and Lebanon, respectively (Sbordoni et al. 1985; Evans et al. 1985).

Spittlebug Warning Coloration is Old and Apparently Evolutionarily Volatile

Warning coloration in spittlebugs appears to be volatile over evolutionary time. Paladini, Takiya et al. (2018) present evidence that it has been gained or lost in multiple lineages of New World Cercopidae. However, warning coloration is also inherently contextual. Paladini, Takiya et al. scored it as absent in species with uniformly brown or black dorsal coloration, which is an oversimplification since some black or brown species are quite conspicuous in nature. For example, they categorize *Zulia vilior*, the most common spittlebug in Costa Rica, as cryptic, but it stands out against its customary green leaf backgrounds (Plate 10f). It also reveals a conspicuously orange colored abdomen when flying, as do many other cercopid spittlebugs. Thus, while there is good evidence that warning coloration has transmuted over evolutionary time in a variety of cercopid lineages, assertions of absolute loss or gain should be interpreted with caution.

There is direct fossil evidence that warning coloration originated early in the Cercopidae, with fossils from the Eocene (approximately 40 mya) exhibiting conspicuous wing patterns (Fig. 2.8; Wilson 1977; Xu et al. 2022). This is consistent with the hypothesis that reflex bleeding evolved in the earliest Cercopidae. In contrast, Eocene Aphrophoridae seem to have had subdued color patterns (Swedo et al. 2019) like the better part of contemporary forms. The few wing patterns known from the Jurassic and Cretaceous Procercopidae and Sinoalidae are subdued and consistent with cryptic coloration (Fu et al. 2018a, 2018b; Chen, Wang, Zhang et al. 2019).

Some Spittlebugs Without Reflex Bleeding Have Warning Coloration

Some Aphrophoridae do have bold colors that appear to be warning coloration. For example, *P. spumarius* has highly conspicuous black and white minority color forms (Fig. 5.1, Plate 11b) interpreted as warning patterns (Thompson

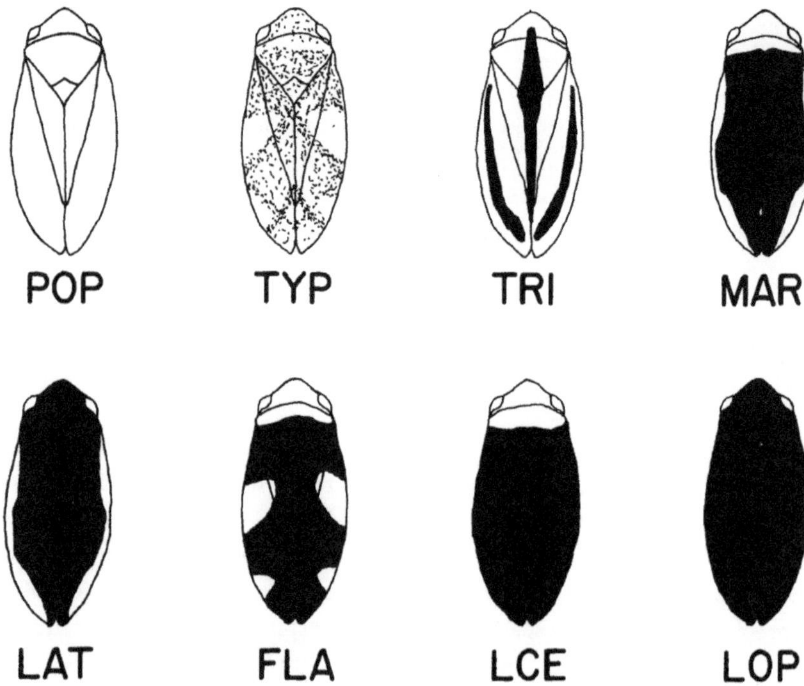

FIGURE 5.1. Eight common genetically determined color forms in the meadow spittlebug, *Philaenus spumarius*. The three-letter abbreviations stand for populi (POP), typicus (TYP), trilineatus (TRI), marginellus (MAR), lateralis (LAT), flavicollis (FLA), leucocephalus (LEU), and leucophthalmus (LOP). The forms POP, TYP, and TRI are relatively cryptic, in brownish shades of cream to dark brown. MAR, LAT, FLA, LCE, and LOP are melanic forms characterized by bold white and black patterns. TYP is representative of the cryptic "aphrophorid ground plan." From Thompson 1984b, reproduced with permission of the American Midland Naturalist, University of Notre Dame. For less stylized drawings of these color forms, see Fig. 8.1. For color photos of TYP, MAR, and TRI, see Plate 11a–c.

1973, 1988). In tropical Asia and Oceania, many species of the aphrophorid genera *Clovia*, *Liorhina*, *Perinoia*, and *Lallemandana* (Fig. 5.2, Plate 13c, 13e) have conspicuous patterns of yellow, black, and sometimes red. In the absence of evidence for reflex bleeding, it appears that all of these species may have evolved warning coloration solely to advertise escape by jumping. The same may be true for several patterned species of the Neotropical aphrophorid genus *Sphodroscarta*. There are also many examples of conspicuous patterns in species of the New World genus *Clastoptera* (Plate 5d, 5e). In a particularly striking

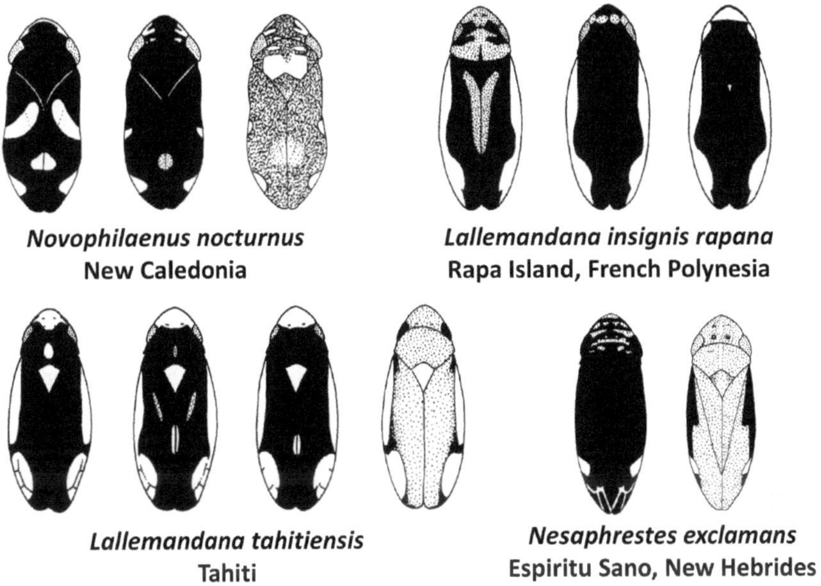

FIGURE 5.2. Color polymorphisms for bold and cryptic patterns in four species of Pacific islands Aphrophoridae: *Novophilaenus nocturnus*, *Lallemandana insignis*, *Lallemandana tahitiensis*, and *Nesaphrestes exclamans*. In each case there are multiple color forms, one of which appears to be cryptic or semicryptic. In life, the black areas are black or dark brown, the white areas yellow or white (Plate 13e). Compiled from Hamilton 1980, 1981a, 1981b. Reproduced with permission of the Bishop Museum, Honolulu.

example, in 2024 I collected a uniformly bright red *Clastoptera*, species unidentified, at the Las Cruces Biological Station in Southern Costa Rica. Reflex bleeding clearly tips spittlebug evolution towards warning coloration but is not a necessary condition. This is consistent with observations in butterflies, in which escape warning coloration appears to occur in both palatable and unpalatable species (Pinheiro et al. 2016)

Cryptic Coloration: Another Important Form of Spittlebug Defense

If you are potential prey, it is good to be passed over if seen, but even better not to be seen at all. Most temperate and many tropical climate Aphrophoridae have *cryptic coloration*, camouflage that helps conceal them from visual predators.

Many *Aphrophora* species, for example, exhibit variations on a characteristic double chevron dorsal pattern composed of muted shades of brown, black, tan, and off-white that combine to break up its visual image (Fig. 1.9l, Plates 11a , 11f, 13a) and are particularly effective in conifer hosts (Plate 13f). This cryptic pattern, which I call the aphrophorid ground plan, recurs across a wide range of other Aphrophoridae, including species in the genera *Ariptyelus, Cephisus, Clovia, Jembra, Lepyronia, Mesoptyelus, Neophilaenus, Peuceptyelus, Philaenus, Plinia, Poophilus, Pseudaphrophora, Ptyelus*, and *Sinophora*. In *P. spumarius*, which has multiple color forms, the ground plan is expressed in the most frequent typical form (Figs. 5.1, 8.1, Plate 11a) but is also faintly visible immediately post-emergence in adults of all color forms, disappearing in the other forms as individuals assume adult coloration (Halkka and Halkka 1990). This plus the ground plan's occurrence in multiple phyletic lines suggests that it is an old ancestral form now overlain in many lines by patterns evolved later in evolution.

Ptyelus flavescens (Plate 1c), the species David Livingstone observed in the African interior (Chapter 3), exhibits a different approach to camouflage called protective resemblance. Multiple bright yellow adults cluster on a branch to give the appearance of a multiple-flower inflorescence (Hinde 1906). They are visually conspicuous but disguised in non-prey form.

Color Polymorphism

In many spittlebugs, local populations exhibit multiple color forms. This is called color polymorphism. It is widespread among the Aphrophoridae and Cercopidae and also occurs among the *Clastoptera*.

Polymorphism and Apostatic Selection in *Philaenus spumarius*

Philaenus spumarius is by far the best-studied example of color polymorphism in spittlebugs. It has eight common color forms (Fig. 5.1) determined by seven genes (*alleles*) at a single Mendelian genetic locus (Halkka et al. 1973; Stewart and Lees 1988; Halkka and Halkka 1990). The color form frequencies vary on a continental scale, but in most localities they are stable over time. In a reciprocal transfer experiment between small Baltic islands in Finland, artificially altered frequencies returned to preexisting equilibria in a handful of years (Halkka and Halkka 1990). Continent-wide patterns combined with local stability and resistance to perturbation strongly suggest that *P. spumarius* color form frequencies are a product of selection.

What maintains widespread, stable color polymorphism when selection driven by escape ability should push towards uniformity of appearance? The generally accepted answer, first proposed by Owen (1963) for *P. spumarius*, is *apostatic selection*, a fancy term for selection in favor of rarer color forms (Halkka et al. 1970; Thompson 1973, 1984a; Halkka and Halkka 1990). The basic idea is that predators develop search images for the most common form or forms, leading them to overlook rarer forms, even when such forms are more conspicuous. This can lead to balanced polymorphism with stable frequencies over time, which is consistent with many observations on *P. spumarius* populations (Owen and Wiegert 1962; Hutchinson 1963; Stewart and Lees 1996; Thompson and Moscovitch 2022). It can even lead, counterintuitively, to a stable balance including forms with warning coloration (Thompson 1984a), which appears to be the case in *P. spumarius* (Thompson 1973), where striking black and white color forms regularly co-occur with more cryptic forms, and probably in some of the colorful Oceanic region Aphrophoridae, where polymorphic bold color forms co-occur with apparently cryptic forms (Fig. 5.2) (Hamilton 1980, 1981a, 1981b). Direct evidence for apostatic selection in nature has been elusive, but Harper and Whittaker (1976) present evidence that, over an eight-year period, parasitization of the most common *P. spumarius* color form at one site in England by a big-headed fly was positively correlated with its frequency. This is perhaps the single piece of direct evidence for frequency-dependent selection in spittlebugs.

Color Polymorphism Is Common Among the Cercopidae, Including Many Pests

Color polymorphism is common among the Cercopidae (Paladini et al. 2015; Paladini, Domahovski et al. 2018; Thompson and Carvalho 2016), including many prominent pest species. Among eight pasture pest species in Costa Rica, three—*Prosapia plagiata* (Plate 10c, 10d), *Prosapia simulans*, and *Zulia vilior*—are polymorphic (Washbon 1968; Clark et al. 1976; Thompson and Leon 2005). In Venezuela, the pasture and sugarcane pest *Aeneolamia varia* exhibits a cornucopia of wing color variation, with twenty named subspecies based partly on this variation (Guagliumi 1955a). In Brazil, the important pasture pests *Notozulia entreriana*, *Deois flavopicta*, and *Deois schach* all exhibit striking wing pattern polymorphisms (see color illustrations in Tolotti et al. 2018; Paladini, Domahovski et al. 2018), with a total of fifteen, seventeen, and nine recorded wing patterns, respectively (Fig. 5.3) (Sá 1981).

South American pasture and sugarcane pests of the genus *Mahanarva* are given to especially extravagant displays of colorful polymorphism. The important pasture pest *Mahanarva spectabilis* has four common coexisting color

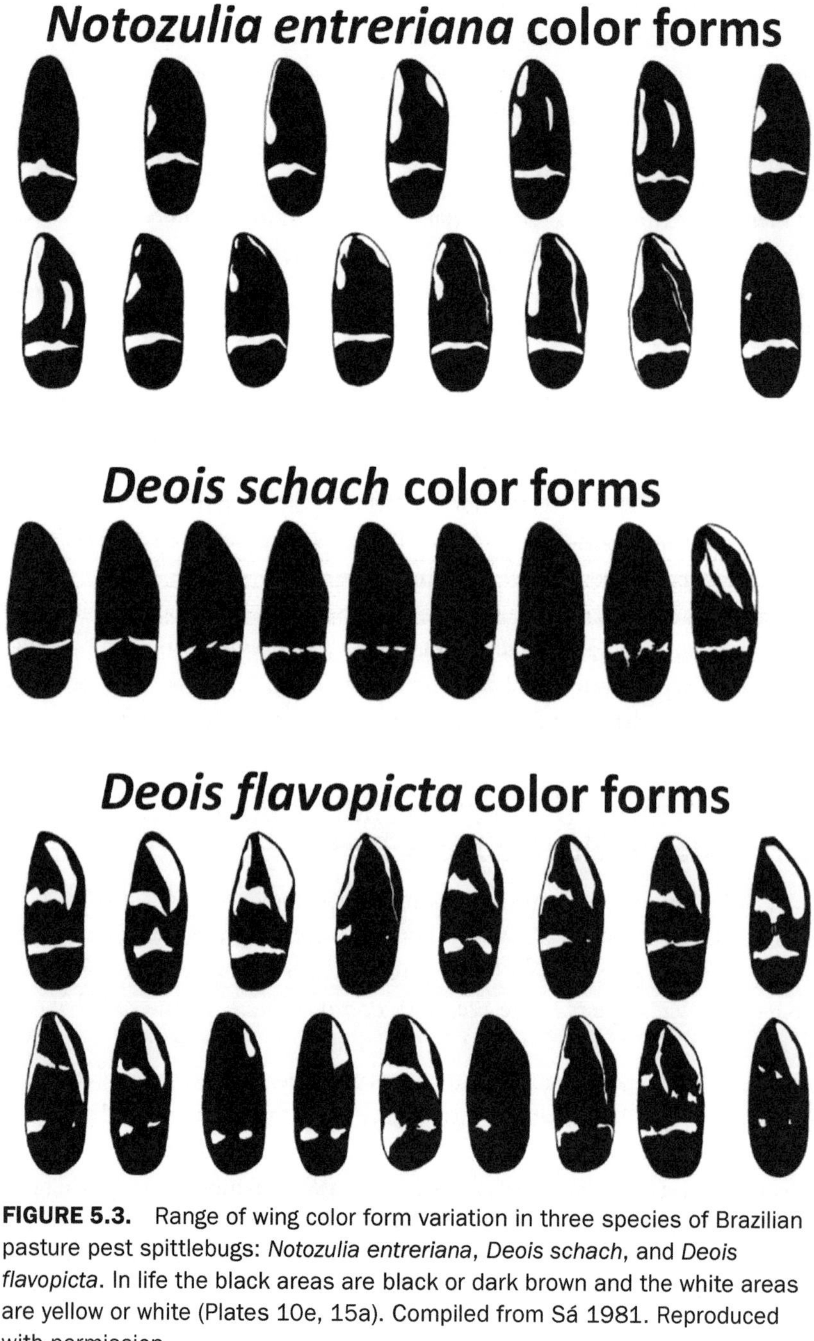

FIGURE 5.3. Range of wing color form variation in three species of Brazilian pasture pest spittlebugs: *Notozulia entreriana*, *Deois schach*, and *Deois flavopicta*. In life the black areas are black or dark brown and the white areas are yellow or white (Plates 10e, 15a). Compiled from Sá 1981. Reproduced with permission.

forms: black, red, red striped, and yellow striped (Borges et al. 2020; Plate 10a). Brazil's most important sugarcane pests, *Mahanarva fimbriolata* and *Mahanarva posticata*, are also polymorphic (Tolotti et al. 2018). *Mahanarva tristis*, an occasional pest of pastures, exhibits multiple color forms (Paladini, Domahovski et al. 2018; Schobel et al. 2023), as does *Mahanarva bipars*, a sugarcane pest in Colombia (Peck et al. 2004).

Polymorphic Warning Coloration Poses an Evolutionary Conundrum

The widespread recurrence of striking color polymorphisms in spittlebugs, many forms of which are clearly aposematic (warning) in nature, raises an evolutionary conundrum. As noted previously, selection for warning coloration should, in theory, lead to *monomorphism*, uniformity for a single color form (Ruxton, Sherratt, and Speed 2004). Instead, many insects in many taxonomic groups exhibit an exuberance of polymorphic warning color forms. Such cases are, as one commentator laments, "almost embarrassingly frequent" (Mallet 2010). What might be at work here? In addition to counterbalancing apostatic selection, a variety of possible explanations, summarized in Thompson and Carvalho 2016, have been offered. Most apt for spittlebugs, it appears that relatively weak defenses or defenses that vary in effectiveness against different types of predators might be especially conducive to polymorphism for aposematic forms. Going forward, spittlebugs offer excellent opportunities to test hypotheses regarding the enigmatic phenomenon of polymorphic warning coloration.

Multiple Factors Influence Color Form Frequencies in *Philaenus spumarius*

For a wide range of organisms there is considerable evidence that polymorphic color forms are correlated with other biological characteristics (McKinnon and Pierotti 2010). Factors other than selection for warning coloration, camouflage, and rarity certainly influence *P. spumarius* color form frequency. *Thermal melanism*, selection for darker forms in colder climates, is best documented (Thompson 1984b, 1988; Berry and Willmer 1986; Thompson and Moscovitch 2022). In general, cooler high latitude and high elevation climates have a higher proportion of dark forms (Halkka and Halkka 1990), though populations in Asian Russia depart from this pattern (Whittaker 1972).

Extraordinarily high frequencies of dark forms (greater than 90%) associated with particulate air pollution from a coking plant in Wales have also suggested a form of industrial melanism, but the causal mechanism is unclear (Stewart and Lees 1996). Following subsequent closure of the plant, dark form frequencies have fallen to about 50% in thirty-three generations (Alan J. A. Stewart, personal comm. 2024). Some other polluted urban sites in Britain also showed elevated frequencies of dark color forms, though not to the same extreme (Stewart and Lees 1996), while a Chicago area survey showed no effect of urban pollution (Thompson and Halkka 1973). There is also a paper suggesting color form frequency differences between areas with greater and lesser amounts of Chernobyl radioactive contamination five years after the nuclear reactor disaster (Klimaszewski 1993).

Continental-scale geographic variation in color form frequencies is thought to reflect variation in climate (Halkka and Halkka 1990; Stewart and Lees 1996), but, with the exception of thermal melanism, the possible selective causes are obscure. Studies of microgeographic variation have likewise been inconclusive and, despite hints to the contrary (Halkka and Halkka 1990), there has been no convincing, consistent correlation of color forms with host plants. One set of observations suggests that the lined color form TRI (trilineatus; Figs.5.1, 8.1, Plate 11c) might have advantages in egg laying and longevity (Silva et al. 2015), but the measured differences are so large that they would require strong, unknown compensating disadvantages to prevent that form from taking over completely instead of maintaining an observed frequency of about 6%.

There is speculation that color form alleles are closely linked to other genes that make them markers for ecologically differentiated biotypes influenced by multiniche selection (Halkka and Halkka 1990). Molecular analysis offers the most promising path to test this hypothesis. Rodrigues et al. (2016) have reported associations between *P. spumarius* color forms and single base pair DNA variations, which could contribute to noncolor-related biological differences among color forms. So far, however, no one has pinned down the Mendelian genetic locus that determines color form or elucidated the nature of the genes governing color pattern. This is a work in progress, which is made more difficult by the fact that *P. spumarius* has one of the largest known insect genomes, about 5.3 billion base pairs (Rodrigues et al. 2016). The upshot is that 60-plus years into the work on *P. spumarius* polymorphism initiated by Halkka and his colleagues in Finland, we are still searching for sources of variation in color form frequencies, both molecular and ecological. As in the case of polymorphic *Cepaea* snails, *P. spumarius* polymorphism is presently "a problem with too many solutions" (Jones et al. 1977), leaving us to sort among them on the basis of insufficient evidence.

Color Polymorphism in *Philaenus spumarius* and Its Relatives Is Ancient

Though we do not know what determines the *P. spumarius* color form frequencies, we do have considerable insight into the polymorphism's evolutionary antiquity. Halkka and Lallukka (1969) noted early on that *P. spumarius* shares most of its color forms with two Mediterranean species, *Philaenus signatus* and *Philaenus tesselatus*, the latter considered by them to be a *P. spumarius* subspecies but now recognized as a closely related sister species (Drosopoulos and Quartau 2002; Seabra et al. 2021). Drosopoulos and his colleagues later demonstrated that this shared polymorphism extends to several other Mediterranean *Philaenus* species—eight in all—and hypothesized that the polymorphism predated the origin of *P. spumarius* (Drosopoulos 2003; Drosopoulos et al. 2010). It apparently even predates the origin of the genus *Philaenus*, since parallel forms occur in the related genus *Mesoptyelus* (Tanyeri et al. 2023). This plus the fact that *P. spumarius* diverged from other *Philaenus* at least 3.7 mya (Maryańska-Nadachowska et al. 2010) suggests that the polymorphism dates back at least several million years.

Some Spittlebugs Deflect Predators with Fake Eyes Spots and a Few Even Bite

Jumping insects have another card up their sleeve when it comes to escape, diversionary eyespots (Dietrich et al. 2023), an anti-predator strategy that falls in a category called deflection (Humphries and Ruxton 2018). In some spittlebugs, posterior portions of their bodies have evolved markings that resemble eyes, making it look as if the rear of the insect is the head. Spittlebugs jump forward and up (at a 28° angle in *P. spumarius* [Burrows 2006a]). Based on the perceived position of the head, predators expect them to jump in one direction. Then they jump in the other, giving the spittlebug another advantage in escape. Fake eyespots are common in the genus *Clastoptera* (e.g., *Clastoptera mineralis* [Thompson 2023a] and *C. cimicoides* [Plate 5e]). They are also evident in the Neotropical cercopid *Microsargane vittata*. In addition to eye spots, some *Clastoptera* have the ability to bite humans. Doering (1931) reports that *C. xanthocephala* bites aggressively and viciously. William Eberhard confirms being bitten by an unidentified Central American *Clastoptera* species (a personal communication through Paul Hanson). How the spittlebugs accomplish this feat in the absence of biting mouthparts remains to be determined.

Adult Spittlebugs Have Diverse Predators but the Level of Successful Predation Is Generally Low

Lots of predators and parasites attack adult spittlebugs. This has generated an enormous and dispersed literature of small reports and observations, most of it incidental to other work. These observations could fill a book in themselves. Here I present an overview of the most important predators, with the caveat that there are many more reports in the literature than those cited. Just as spittle is a good but not foolproof defense against predators and parasites, the adult defenses of jumping, camouflage, and warning coloration are also good but imperfect.

Birds: Important Visual Predators but Modest in Actual Predation

Birds are thought to be the most important vertebrate predators of spittlebugs (Halkka and Lallukka 1969; Halkka and Halkka 1990) and most likely to shape selection for crypsis, warning coloration, and apostatic polymorphism. Nevertheless, and in apparent partial contradiction, spittlebug adults are generally rare in bird diets. In his 170-page monograph on the Trinidad sugarcane pest *A. varia*, Williams (1921) devotes almost eight full pages to its local avian predators. He found seventeen species with spittlebugs in their stomachs. But despite large numbers of bugs in the cane fields, the number of individuals consumed was always small. Even the scissors-tail flycatcher (*Milvulus tyrannus*), "one of the most abundant birds in the cane fields and . . . without doubt the most important bird enemy of the froghopper," revealed only seven specimens consumed. He concluded that birds are not important in controlling pest populations, echoing the earlier opinion of Urich (1913), noted previously.

Table 5.1 summarizes Neotropical vertebrate predators of adult spittlebugs tabulated by Guagliumi (1955b) (see Table 4.1 for predators of nymphs). Halkka and Kohila (1976) present a similar list summarizing stomach contents studies for European birds. They identify ten species that had eaten *P. spumarius*. With the exception of rooks, *Corvus frugilegus*, which accounted for most of the specimens, predation seemed to be sporadic and low in intensity. Harper and Whittaker (1976) found no evidence for bird predation on adult spittlebugs at an English site. As a counterpoint, studies of vesper, field, and chipping sparrows by Francis Evans in an old field in Michigan indicated that they might have accounted for almost 10% of *P. spumarius* adult mortality (Wiegert 1964a). Evans never published the underlying data but shared a portion with me (Table 5.2). During the period of adult emergence, 54 to 90% of stomachs examined had

TABLE 5.1. Invertebrate predators of adult spittlebugs in portions of the neotropics

CATEGORY	SPECIES	PREY	GEOGRAPHIC LOCALE
Ants	*Anochetus inermis*	*Aeneolamia varia*	Guyana
	Ectatomma ruidum	*Aeneolamia lepidior*	Panama
	Monomorium carbonarium	*Aeneolamia lepidior*	Panama
	..	*Aeneolamia varia*	Trinidad
	Pseudomyrma elegans	*Aeneolamia flavilatera*	Guyana
	Solenopsis geminata	*Aeneolamia flavilatera*	Guyana
	..	*Aeneolamia varia*	Venezuela
	Solenopsis tenuis	*Aeneolamia flavilatera*	Guyana
Flies	Phoridae sp.	*Aeneolamia lepidior*	Panama
	..	*Aeneolamia varia*	Trinidad
	Tachinidae sp.	*Zulia pubescens*	Guyana
Assassin bugs	*Castolus plagiaticollis*	*Aeneolamia contigua*	Mexico
	..	*Aeneolamia varia*	Trinidad[a]
	Heza peramata	*Aeneolamia flavilatera*	Guyana
	Repipta taurus	*Aeneolamia lepidior*	Panama
	Zelus longipes	*Aeneolamia varia*	Venezuela
	Zelus rubidus	*Aeneolamia lepidior*	Panama
	Zelus mimus	*Aeneolamia flavilatera*	Guyana
Plant bug	Pentatomidae sp.	*Aeneolamia varia*	Venezuela
Dragonflies	Odonata sp.	*Aeneolamia flavilatera*	Guyana
	..	*Aeneolamia varia*	Venezuela
Mantis	Mantidae sp.	*Aeneolamia varia*	Venezuela
Katydids	*Phlugis teres*	*Aeneolamia varia*	Venezuela
	Phlugis mantispa	*Aeneolamia flavilatera*	Guyana
	..	*Aeneolamia varia*	Trinidad
Bush crickets	*Conocephalus fasciatum*	*Aeneolamia varia*	Trinidad
	Conocephalus propinquum	*Aeneolamia flavilatera*	Guyana
	Conocephalus saltator	*Aeneolamia varia*	Trinidad
	Conocephalus versicolor	*Aeneolamia varia*	Trinidad
Jumping spiders	*Scopocira* sp.	*Aeneolamia lepidior*	Panama
	Salticidae sp.	*Aeneolamia varia*	Trinidad, Tobago
	..	*Aeneolamia varia*	Venezuela
	..	*Aeneolamia flavilatera*	Guyana
	..	*Mahanarva tristis*	Trinidad
	..	*Zulia pubescens*	Trinidad
Scorpions	Scorpionidea sp.	*Aeneolamia varia*	Trinidad
	..	*Aeneolamia varia*	Venezuela
	..	*Mahanarva tristis*	Venezuela
Mite	*Rhyncholophus* sp.	*Aeneolamia varia*	Venezuela
Nematodes	*Mermis* sp.	*Aeneolamia varia*	Trinidad
	Nematoda sp.	*Prosapia bicincta*	Cuba
Lizards	*Ameiva ameiva*	*Aeneolamia varia*	Venezuela
	Ameiva surinamensis	*Aeneolamia flavilatera*	Guyana
	..	*Aeneolamia varia*	Trinidad
	Cnemidophorus lemniscatus	*Aeneolamia varia*	Venezuela

Source: Data from Guagliumi 1955b. Spittlebug nomenclature updated to current usage.
Note: Double dots (..) represent repetition of the predator species listed above.
[a] The assassin bug *C. plagiaticollis* was purposely introduced from Mexico to Trinidad in 1911 in an attempt at biological control of *A. varia* (Williams 1921).

TABLE 5.2. Stomach contents evidence for heavy predation on *Philaenus spumarius* by three species of sparrows

SPARROW SPECIES AND LIFE STAGE	NUMBER OF STOMACHS EXAMINED	POSITIVE FOR *P. SPUMARIUS*	PERCENT POSITIVE FOR *P. S.*	NUMBER WITH *P. S.* COUNTS	COUNT RANGE	MEAN COUNT	NUMBER WITH *P. S.* PERCENTAGE	PERCENT RANGE	MEAN PERCENT
Field sparrow, nestlings	11	6	54.5%	2	3–10	6.5	3	10–70%	37%
Field sparrow, adults	21	19	90.0%	11	1–8	4.0	13	25–75%	43%
Vesper sparrow, adults	20	8	40.0%	3	1–1	1.0	6	10–30%	19%
Chipping sparrow, adults	41	26	63.4%	17	1–15	3.7	13	5–80%	28%

Source: Data provided to the author circa 1972 by Francis C. Evans. Bird specimens were collected in or near the Old Field site (Wiegert 1964a) in the Edwin S. George Reserve in Livingston County, Michigan by F. C. Evans and P. S. Humphrey. Includes data for June, July and August. With the exception of eight field sparrow nestlings collected in 1953–1955, all bird specimens were collected in 1950 or 1951. All spittlebugs ingested were adults. There was no evidence of consumption of nymphs, or of adults before June. Counts refer to the number of *P. spumarius* individuals observed in stomachs, percentages to the estimated portion of the contents of individual stomachs attributed to *P. spumarius*. Mean counts and percentages exclude stomachs with no *P. spumarius*.

P. spumarius remains. Omitting stomachs with none, the mean number of specimens per stomach averaged 1.0 to 6.5, and in cases in which it could be estimated, the proportion of total stomach contents attributable to *P. spumarius* ranged from 28 to 43%. At this locality, sparrows were a major predator, and *P. spumarius* was a major food source.

Vesper sparrows have also been shown to eat the sage spittlebug *Clastoptera brunnea* in Utah (Knowlton and Nye 1948). In the southeastern USA, the pasture pest *Prosapia bicincta* is a common prey item for red-winged blackbirds (present in 13% of stomachs examined) and meadowlarks (present in 44%) (Genung and Green 1974; Genung et al. 1976). In Michigan, adults of the pine spittlebug *Aphrophora cribrata* are a major food source for the endangered Kirtland's warbler (Deloria-Sheffield et al. 2001; Christie Deloria-Sheffield, personal comm. 2003), while in Oregon, spittlebugs were found in 23% of fecal samples from Wilson's warblers (Hagar et al. 2007). Spittlebugs made up 6.5% of the willow flycatcher's prey in Wisconsin (McCabe 1991) and five specimens were recorded from forty-eight stomachs of juvenile ruffed grouse on Vancouver Island, Canada, a very small portion of their diet, which was largely vegetarian (King 1969).

On the more general question of the extent to which North American birds eat spittlebugs, McAtee (1932) reports that forty-one species of birds eat adults,

TABLE 5.3. US Fish and Wildlife Service records of spittlebugs in bird stomach contents

PREY	AVIAN PREDATOR	COMMON NAME	NUMBER OF STOMACHS WITH SPITTLEBUGS	NUMBER OF SPITTLEBUG SPECIMENS
Philaenus spumarius				
	Passer domesticus	House sparrow	8	16
	Sternus vulgaris	Starling	7	20
	Acridotheres cristatellus	Crested myna	5	5
	Bartramia longicauda	Upland sandpiper	1	1
	Bonasa umbellus	Ruffed grouse	1	1
	Junco hyemalis	Dark-eyed junco	1	1
	Pluvialis squatarola	Grey plover	1	1
Other Cercopoidea				
	Sitta pygmaea	Pygmy nuthatch	18	18
	Acridotheres cristatellus	Crested myna	10	19
	Passer domesticus	House sparrow	5	5
	Parus atricapillus	Black-capped chickadee	5	5
	Loxia curvirostra	Red crossbill	4	4
	Melospiza melodia	Song sparrow	3	7
	Pica hudsonia	Black-billed magpie	3	3
	Dendroica discolor	Prairie warbler	2	7
	Icturus spurius	Orchard Oriole	2	3
	Eremophila alpestris	Horned lark	2	2
	Petrochelidon pyrrhonota	Cliff swallow	2	2
	Polioptila caerulea	Blue-gray gnatcatcher	2	2
	Troglodytes aedon	House wren	2	2
	Vireo gilvus	Warbling vireo	2	2

Source: Mitchell (1972). Compiled in 1972. Bird nomenclature updated.

with no records for nymphs. A single chimney swift had eaten one hundred specimens of *Clastoptera*, but, overall, the presence of spittlebugs in North American bird diets seems to be occasional, sporadic, and generally low. For example, 1,951 stomachs taken from eight species of vireos revealed only seven specimens of spittlebugs, including three *Aphrophora* (Chapin 1925). Likewise, studies of 2,030 swallow stomachs (Beal 1918) and 410 cattle egret stomachs (Fogarty and Hetrick 1973) revealed a single spittlebug each. In response to my inquiry, R. T. Michell of the U.S. Fish and Wildlife Service provided a summary of agency records of spittlebug prey as of 1972 (Table 5.3). He commented that "our food habits files show that *Philaenus spumarius* is not a common food item" and goes on to say that "the low frequency of Cercopidae [Cercopoidea] in the stomachs surprises me in view of their natural abundance and wide distribution" (Mitchell 1972).

Aside from the studies on sugarcane and pasture pests in the Neotropics, observations on spittlebugs and bird predation outside Europe and the United States are sparse. In one report from Algeria, unidentified spittlebugs accounted for 10 of 277 specimens recovered from fecal sacs of the short-toed tree creeper (Boulaouad et al. 2018). In Madagascar, again based on fecal samples, spittlebugs accounted for a minor part of the diet of three dry forest understory birds (Faliarivola et al. 2020). In Mexico, a study of 72 stomachs from 15 combined species of hummingbirds found that 5 of 103 arthropods consumed were spittlebugs, the second most common category after spiders (Partida-Lara et al. 2022).

An intriguing set of studies in Costa Rica suggests that some insectivorous birds specialize on spittlebugs and related jumping planthoppers. Sherry (1984) studied the diets and behaviors of sixteen species of flycatchers in the Caribbean lowlands, recording the contents of 126 stomachs with a total of 2,834 prey items. Thirteen of the sixteen species had eaten spittlebug prey, totaling fifty-two specimens in forty-three stomachs. Just over a third of the stomachs yielded spittlebugs, which accounted for 1.8% of all prey taken. Clearly there was widespread predation on spittlebugs by numerous species but overall low numbers of items ingested. Two species, the ruddy-tailed flycatcher (*Terenotriccus erythrurus*) and the black-capped pygmy tyrant (*Myiornis atricapillus*), accounted for a large proportion of the spittlebugs taken, twenty out of fifty-two (38%). The ruddy-tailed flycatcher is a narrow specialist, with Auchenorrhyncha accounting for 94% of its diet (Sherry 1984; Sherry et al. 2020). After flushing jumping prey, it pursues them aerobatically through the vegetation, while the black-capped pygmy tyrant has a broader diet and strikes upward, gleaning vegetation. Sherry et al. (2020) suggest that predators like these may drive selection for strong jumping defense behaviors.

Bats: Alternative Flying Vertebrate Predators

Spittlebugs appear in the diets of bats fairly often but usually as a very small portion of captured prey. In an exception, Griffith and Gates (1985) found that "Cercopidae" (likely Aphrophoridae or *Clastoptera*) were found in 7.1 to 33.3% of individuals of four cave-nesting bat species in Maryland. In another exception, spittlebugs formed a sizeable portion of the diet of two out of five insectivorous bat species studied in Idaho, comprising 28.6% of the specimens eaten in one case and 12.5% in the other (Lacki et al. 2007). As a harbinger of more detailed information to come from new techniques, Galey (2020) has identified

DNA from *Clastoptera obtusa*, *Aphrophora saratogensis*, and *Aphrophora cribrata* in the feces of two species of Minnesota bats, a technique sure to be replicated more widely.

Reptiles, Amphibians, and Fish

Guagliumi (Table 5.1) lists nine species of frogs and three of lizards that eat spittlebugs in northern South America. In Utah, Knowlton (1942) reports eight *Clastoptera delicata* among the stomach contents of fifty-nine swift lizards. In Bermuda, 10% of stomachs of the lizard *Anolis roqu*et included *Clastoptera* (Wingate 1965), likely *C. undulata*, an introduced species that badly damaged local *Casuarina* plantings (Bennett and Hughes 1963). Collier et al. (1998) found five spittlebugs in the stomachs of leopard frogs in Ohio, which, given the frog's aquatic environment, must have been taken on pond edges or from individuals fallen into ponds. Likewise, Blackith and Speight (cited in Harper and Whittaker 1976) recorded *P. spumarius* in frog stomachs in Ireland. Newly metamorphosed individuals of two British Columbian amphibians, the red-legged frog and the western spotted frog, may hold the vertebrate record for spittlebug predation. Over half (52.6%) of the red-legged frog stomachs held spittlebugs, which accounted for 22.9% of all food items. For western spotted frogs living in the same habitat, the equivalent numbers were 27.8% and 14.7% (Licht 1971).

In addition to lizards and frogs, there is at least one report of a salamander in Oregon eating spittlebugs (Whitaker et al. 1986). They were not a large portion of the diet but were eaten year around. There is even a report of a small native Indianna fish eating spittlebugs, which comprised about 0.5% of its diet (Clem and Whitaker 1995).

Spiders: The Most Important Invertebrate Predators

While spiders are not important predators of spittlebug nymphs (Chapter 4), they are probably the most important invertebrate predators of adults and account for a lot of adult mortality (Weigert 1964a; Sujii et al. 2002). Over years of collecting, I have been struck by the large numbers of spiders present in sweep net samples from meadows and pastures with large populations of spittlebugs. Spiders are generally the most abundant predators in the herb layer, where spittlebugs abound.

Jumping spiders appear to be common predators of sugarcane and pasture pests in the Neotropics (Table 5.1) and have been observed preying on *P. spumarius* in Michigan (Weigert 1964a) and *Mioscarta obscuripennis* in the

Philippines (Plate 9b). Orb-weaving spiders have been reported to snare a few *P. spumarius* and *N. lineatus* in Europe (Kajak 1965; Harper and Whittaker 1976) as well as *Aphrophora maritima* in large numbers in Japan (Murakami 1983). *Prosapia bicincta* is prey to orb-weaving spiders in Georgia (Fagan and Kuitert 1969) and has been photographed deploying reflex bleeding against a funnel-web spider (Plate 9c) (Thompson and Carvalho 2016). A lynx spider has been observed preying on *Clastoptera xanthocephala* in Florida (Altieri and Whitcomb 1979). Sujii et al. (2002) report that spiders are the most abundant potential predators of adult *Deois flavopicta* in Brazil but include only two direct observations of predation, both by orb-weavers. Harper and Whittaker (1976) report predation of *P. spumarius* by opiliones (harvestmen or daddy longlegs, collateral relatives of spiders) in England. In addition to these field observations, there have been several lab experiments exposing spittlebugs to spider predation (Howell and Pienkowski 1971; Harper and Whittaker 1976; Nachappa, Guillebeau, Braman, and All 2006; Benhadi-Marín et al. 2020; López-Mercadal et al. 2024).

Vertebrate predators conveniently preserve remnants of prey in the form of fecal or stomach contents. With the exception of web-makers, spiders do not. This probably accounts for both the relative paucity of spider prey records and the high proportion of web-makers among the observations we have. Advances in technology promise to remedy this bias. Rodrigues et al. (2022) show that it is possible to detect spittlebug DNA in spiders for several days after feeding and have used this technology to demonstrate that 20% of fifty field-collected Portuguese lynx spiders had ingested measurable *P. spumarius* DNA.

Predatory Insects: A Diverse Set of Enemies, Including a Group of Spittlebug Specialist Wasps

Several species of ant prey on adult spittlebugs of the Neotropical pest genus *Aeneolamia* (Table 5.1), and one species preys on adults and nymphs of *Clastoptera* in Mexico (Valenzuela-G. et al. 1994), but, overall, ants do not appear to be major predators on adult spittlebugs and probably take their greatest toll on eggs (discussed in this chapter in the section on Predators and Parasitoids of Eggs). Predatory bugs appear to be more important hunters of adults. The assassin bug *Castolus plagiaticollis* (Fig. 5.4) was purposely introduced to Trinidad from Mexico in the early twentieth century in an attempt to control sugarcane spittlebugs (Williams 1921). The attempt failed, but hope in predatory bugs lives on in the form of current studies of the invasive assassin bug *Zelus renardii* (Fig. 5.5) as a potential control for *P. spumarius* in Italy (Lahbib et al. 2022). A relative, *Zelus bilobus*, preys on *P. bicincta* in Florida (Fagan and Kuitert 1969).

FIGURE 5.4. The assassin bug, *Castolus plagiaticollis*, purposely introduced to Trinidad from Mexico in the early twentieth century in an attempt to control sugarcane spittlebugs. The individual illustrated has captured an adult cane spittlebug, *Aeneolamia varia* (Plate 11b). The attempt at control failed. From Urich 1913.

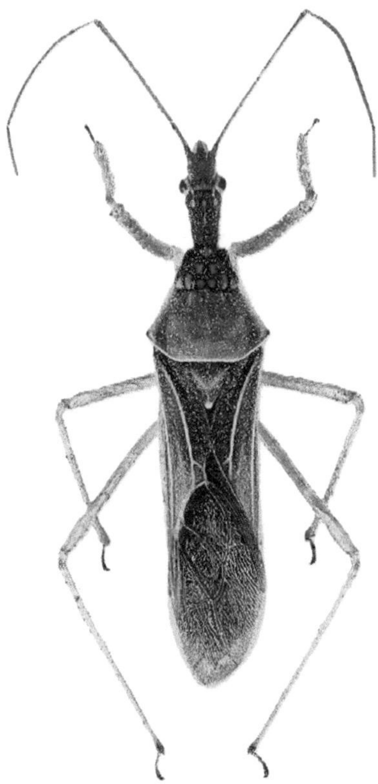

FIGURE 5.5. The assassin bug *Zelus renardii*, invasive in Europe, has been proposed as a potential control agent for *P. spumarius* in Italy. Body length 11.9 mm. From Kment and van der Heyden 2022. Reproduced under Creative Commons Attribution 4.0 International License (https://creativecommons.org/licenses/by/4.0/).

In Brazil an assassin bug has been reported preying on the sugarcane pest *Mahanarva fimbriolata* (Trevisan and Meneguetti 2012), and in Wisconsin they have attacked *Clastoptera proteus* (Hagerty and McPherson 2000). Several other kinds of bugs and beetles prey on spittlebugs in the field or in lab experiments (Table 5.4) (Harper and Whittaker 1976; Nachappa, Guillebeau, Braman, and All 2006), but none seem to be spittlebug specialists.

Robber flies (Asilidae, Plate 9a) have been reported to prey on at least twelve genera and eighteen species of spittlebugs (Dennis et al. 2010), and the pasture pest *Deois flavopicta* constituted 92.4% of their prey in a Brazilian study (Bueno 1986). The colony-forming wasp *Vespula vidua* has been observed to prey on substantial numbers of adult *Aphrophora cribrata* but seems to prefer treehoppers (Greene 1983). In China, dragonflies and damselflies are said to prey on egg-laying or mating adults of the rice and maize pest *Callitettix versicolor* (Li et al 2001). Because neither case involves the flying prey that are typical targets of these predators, this observation requires confirmation, but it is corroborated by Guagliumi's records in Table 5.1.

One group of solitary wasps does specialize on spittlebugs and related insects. Females of several species of the genus *Gorytes* provision their nests with paralyzed adult spittlebugs (Evans and O'Neill 2007). In North America, *G. atricornis* preys on spittlebugs and treehoppers. In Eurasia three species prey on *P. spumarius*. In Japan, *G. tricinctus* provisions burrows with dozens of *Aphrophora maritima*, the same species that falls victim in numbers to orb-web spiders. An African species, *Pseudoplius natalensis*, stocks its burrows with fewer but much larger spittlebugs, *Ptyelus grossus* (Plate 8a). Two European wasps of the genus *Bembecinus* also prey on adult spittlebugs, but as one of several Auchenorrhyncha prey families (Zolda and Holzinger 2002). Other solitary wasps specialize on spittlebug nymphs (Chapter 4).

Without doubt, the all-time award for heroic nest provisioning with spittlebugs goes to the Madagascan wasp *Pseudoplisus ranosahae*, as described by Callan (1991):

> The source of the prey was several large trees of the introduced black wattle *Acacia mollissima* Willdenow (Leguminosae) near the nest sites. The trees supported an immense population of froghoppers, *Ptyelus goudoti* (Bennett). This large, spectacular froghopper is endemic to Madagascar. Adults are 25-30mm long, yellowish with highly variable black maculations. Colour variants range from immaculate to all black. With the first rains nymphs appear in enormous numbers on leguminous and rosaceous trees. They are notorious for the discharge of large amounts of clear fluid, which is said to fall like fine rain, and

> give rise to prodigious adult populations . . . Wasps were seen to capture and sting adult froghoppers, and prey-laden females flew slowly down from the trees . . . As the prey was much larger than the wasp, it projected behind it. Females had some difficulty in flying with such a large burden, and prey-laden wasps, after alighting on the ground, straddled the prey and crawled laboriously with it to the nest . . . Sometimes the prey was dropped in flight, but was then picked up again, and . . . occasionally abandoned, and the ground at the base of the bank was littered with discarded froghoppers.

Predators and Parasitoids of Eggs

The egg stage is particularly vulnerable in spittlebugs; they produce no spittle and lack the ability to escape predators. This is especially true for many Neotropical grass and sugarcane pest spittlebugs, which lay loose, long-lived, dry season diapause (resting stage) eggs in the soil and soil litter. These eggs are highly vulnerable to predators and parasitoids (Table 5.4). In Brazilian Cerrado pastures, experimental dry season egg mortality was 73% in *N. entreriana* (Plate 10e) (Hewitt 1986) and 60% in *D. flavopicta* (Plate 15a) (Sujii et al. 2002), with ants inferred to make the greatest contribution to predation (Hewitt and Nilakhe 1986; Sujii et al. 2002). In Colombian Llanos pastures, ants scavenged more than 90% of *A. varia* (Plate 15b) eggs placed near their nests in the dry season (Medina 1995). *Prosapia bicincta* (Plate 10e) eggs are also consumed by a variety of ants, bugs, and beetles in lab experiments (Nachappa, Guillebeau, Braman, and All 2006), and thrips have been reported as egg predators of *Aeneolamia* in the Neotropics (Williams 1921) and *Clastoptera achatina* (Plate 5d) in Georgia (Tedders 1995). Comparable overwintering mortality of eggs in temperate climates, less likely to be due to predation, seems to be smaller but still high, 25 to 31% in *Neophilaenus lineatus* (Plate 11d) in England (Whittaker 1965, 1971a). In China, ants prey on the eggs of C. *versicolor* (Plate 12a) (Lei et al. 1992).

Spittlebugs are also subject to a wide range of tiny egg parasitoid wasps of the superfamily Chalcidoidea. They develop inside the eggs, often several to a single egg, consuming the contents and destroying the host embryos. This makes them parasitoids as opposed to parasites, which leave the host intact. Tables 5.5 and 5.6 provide a summary of known instances. Study of these parasitoids dates to early twentieth century efforts to control the sugarcane pest *A. varia* (Plate 15b) (Williams 1921 and see Chapter 6). Despite concerted exploration of diverse areas of the Neotropics, no species found proved to be effective for biological control (Myers 1931, 1935). The most promising, *Anagrus*

TABLE 5.4. Parasitoids and predators of spittlebug eggs in portions of the Neotropics

CATEGORY	SPECIES	PREY	GEOGRAPHIC LOCALE
Parasitoid wasps: Mymaridae			
	Anagrus flaveolus	*Aeneolamia varia*	Trinidad, Venezuela
	Anagrus urichi	*Aeneolamia varia*	Trinidad
	Anagrus sp.	*Aeneolamia flavilatera*	Guyana
	..	*Aeneolamia lepidior*	Panama
	..	*Aeneolamia varia*	Trinidad
	..	*Prosapia simulans*	Costa Rica
	..	*Prosapia* sp.?	Panama
Parasitoid wasps: Trichogrammatidae			
	Abellia tomaspis	*Aeneolamia varia*	Trinidad
	Lathomeris sp.	*Aeneolamia varia*	Trinidad
	Oligosita sp.	*Aeneolamia flavilatera*	Guyana
	..	*Aeneolamia lepidior*	Panama
	..	*Aeneolamia varia*	Trinidad, Tobago
	..	*Aeneolamia varia*	Venezuela
	..	*Prosapia simulans*	Costa Rica
	..	*Zulia pubescens*	Trinidad
Parasitoid wasps: Aphelinidae			
	Centrodora perkinsi	*Aeneolamia varia*	Guyana
	Centrodora tomaspidis	*Aeneolamia varia*	Trinidad
	Centrodora sp.	*Aeneolamia varia*	Tobago
Ants			
	Unidentified sp.	*Aeneolamia varia*	Venezuela
Thrips			
	[H]*aplothrips* sp.	*Aeneolamia lepidior*	Panama
	..	*Aeneolamia varia*	Trinidad
	..	*Aeneolamia flavilatera*	Guyana

Source: Data from Guagliumi 1955b.
Note: Double dots (..) represent repetition of the parasitoid/predator species listed above.
Note: Spittlebug nomenclature updated to current usage.

TABLE 5.5. Parasitoid wasps associated with spittlebugs of the Family Cercopidae

	PARASITOID WASP FAMILY AND SPECIES									
	APHELINIDAE		ENCYRTIDAE	MYMARIDAE			TRICHOGRAMMATIDAE			
Species	*Centrodora perkinsi*	*Centrodora tomaspidis*	*Syrphophagus* sp.	*Anagrus flaveolus*	*Anagrus urichi*	*Acmopolynema hervali*	*Lathromeris* sp.	*Oligosita giraulti*	*Paracentrobia tomaspidis*	*Uscanoidea aliciae*
Aeneolamia flavilatera	X			X	X			X		
Aeneolamia lepidior				X				X		
Aeneolamia reducta				X						
Aeneolamia selecta		X				X				
Aeneolamia varia	X	X	X	X	X		X	X	X	
Deois flavopicta					X					
Deois terrea		X								
Kanaima fluvialis						X				
Mahanarva fimbriolata					X	X				
Mahanarva liturata						X				
Mahanarva posticata		X				X		X		
Mahanarva rubicunda						X		X		X
Zulia pubescens				X						

Source: Data from the Universal Chalcidoidea Database (Noyes 2019).
Note: Spittlebug nomenclature has been updated.

TABLE 5.6. Parasitoid wasps associated with the spittlebug families Aphrophoridae, Clastopteridae, and Machaerotidae

SPITTLEBUG FAMILY AND SPECIES	PARASITOID WASP FAMILY AND SPECIES							
	APHELINIDAE			ENCYRTIDAE			MYMARIDAE	
Aphrophoridae	*Centrodora salicis*	*Centrodora cercopiphagus*	*Centrodora* sp.	*Neocladia gigantica*	*Neocladia myersi*	*Neocladia orientalis*	*Ooctonus aphrophorae*	*Ooctonus vulgatus*
Aphrophora salicina	X							
Aphrophora saratogensis		X					X	
Clovia puncta						X		X
Epicranion sp.					X			
Philaenus spumarius			X					X
Ptyelus flavescens				X				
Ptyelus nebulas						X		
	Encyrtidae			**Signiphoridae**	**Trichogrammatidae**			
Clastopteridae	*Neocladia myersi*	*Neocladia waterstoni*	*Ooencyrtus sorkula*	*Signiphora tridentata*	*Uscanoidea nigriventris*			
Clastoptera flavidorsa	X	X			X			
Clastoptera globosa	X							
Clastoptera undulata	X	X			X			
Clastoptera sp.			X	X				
	Aphelinidae	**Eupelmidae**	**Mymaridae**					
Machaerotidae	*Centrodora* sp.	*Eupelmus tachardiae*	*Acmopolynema sp.*	*Himopolynema* sp.				
Hindola fulva	X							
Hindola patrielis	X							
Hindola striata	X		X	X				
Machaerota planitiae		X						

Source: Universal Chalcidoidea Database (Noyes 2019). Observation of *Centrodora* sp. associated with *P. spumarius* from Weaver and King (1954).
Note: Spittlebug nomenclature has been updated.

urichi, rarely achieved infection levels above 10% in the field, and it was ultimately deemed economically unimportant (Pickles 1932, 1933). There have been occasional revivals of interest in parasitoid wasps in Brazil, including two attempts to establish *Acmopolynema hevali* for biological control of *Mahanarva* pests of sugarcane in Pernambuco (Boas and Andrade 1990) and proposals that *A. urichi* and an *Anagrus* species might be effective against the pasture pests *N. entreriana* (Valério and Oliveira 2005) and *D. flavopicta* (Pires et al. 1993), respectively.

"*Muy Pequeñas Avispas*!"

I do not work with parasitic wasps and have only reared them on one occasion, accidentally, while studying flies that grow in spittle (Thompson 2023a). However, in summer 1997 my host in Nicaragua, Jean Michel Maes, tasked me with carrying a few thousand of these tiny creatures in vials of alcohol back to Costa Rica for help with identification by my host there, hymenopterist Paul Hanson. A Costa Rican customs inspector rapidly found the several dozen plastic vials of clear liquid, no doubt taking them for a drug shipment. They had no visible nonliquid contents, but I tried to explain in my rudimentary Spanish that they held *muy pequeñas avispas*, very small wasps. He was unmoved and the Managua-San José bus, with all its passengers, was parked on indefinite hold while they tried to decide what to do with me. After an interval, a fellow pulled up on a motor bike, an agricultural inspector it turned out. Did I know Humberto Lezama, the entomologist at the university ag school? Yes, I did. "Great. Show me some of the wasps. They look interesting." At this point a crestfallen customs agent gave up on the prospective bust and the driver pulled me toward the bus in a hurry to finally move on. The *pequeñas avispas* got to their destination, I avoided detention, and the San José bus was only a bit late that day.

Outside the New World, it has been suggested that an *Acmopolynema* parasitoid wasp might be useful for controlling species of the tube spittlebug genus *Hindola* in Indonesia, where they transmit a xylem-dwelling bacterium responsible for Sumatra disease of cloves (Balfas et al. 1990; Lomer et al. 1993; and see Chapter 9). Finally, there is renewed interest in *Ooctonus vulgatus* (Fig. 5.6) as a possible biological control for *P. spumarius*. This wasp was reported at infestation levels of less than 10% in North America (Weaver and King 1954) but infects up to 69% of eggs in Corsica (Mesmin et al. 2020).

FIGURE 5.6. The parasitic wasp *Ooctonus vulgatus*, a natural enemy and possible biological control agent for *P. spumarius*. Scale bar = 0.2 mm. From Mesmin et al. 2020. Reproduced in black and white under Creative Commons Attribution 4.0 International License (https://creativecommons.org/licenses/by/4.0/).

Parasitoids also affect spittlebugs indirectly. In California, wasps of the genus *Pachyneuron* hyperparasitize *Cladochaeta* fly larvae living in spittles with *Clastoptera mineralis* (Thompson 2023a), while in Costa Rica, the wasp *Horsimenus persimilis* hyperparasitizes *Cladochaeta* living in association with *Clastoptera* living on *Casuarina* (Hansson 2009). Counting the tree hosts, each of these examples represents three successive layers of parasitism.

Big-headed Flies

Big-headed flies (Pipunculidae), named for enormous eyes that cover almost their entire head (Fig. 5.7), are specialist parasitoids of Auchenorrhyncha. Most species parasitize nymphs of leafhoppers, treehoppers, or planthoppers, but species of the genus *Verrallia* specialize on adult spittlebugs (Coe 1966). They inject an egg into the host, where it hatches and grows into a larva that eventually fills

FIGURE 5.7. The big-headed fly *Verrallia aucta*, a major parasite of *P. spumarius* adults in parts of Europe. Photograph by Dick Belgers through Wikimedia Commons. Reproduced in black and white under Creative Commons Attribution 3.0 Unported Deed (https://creativecommons.org/licenses/by/3.0/).

the entire abdomen. When ready to pupate, it exits through a hole in the body wall, killing the host, though in the case of spittlebugs not until late in the season. Spittlebug victims become walking dead, reproductively doomed and devoured from within, reminiscent of the film *Alien*. Coe (1966) speculates that

> there are two probable reasons why the adults only are parasitized; firstly, the soft bodied nymph of the hopper in its spittle-like secretion would most likely not have the endurance to attain maturity with the developing pipunculid larva in its abdomen, and secondly, the spume itself apparently deters the fly from ovipositing on or in the nymph.

He adds, "I have watched females of *Verrallia* seeking their Cercopid hosts, and without exception when spume was encountered the fly would veer away from it." Whittaker (1969) suggests that the flies attack teneral adults shortly after their emergence from spittle.

Verrallia aucta, a common European big-headed fly, specializes in parasitizing *P. spumarius* and *Neophilaenus lineatus* (Whittaker 1969, 1973). The following delightful description of *V. aucta* in action was recorded by Francis Jenkinson while he had a day job as head librarian of Cambridge University. It is a striking record of jumping as a good but not foolproof defense:

> Noticing *V. aucta* to be common in my garden on the morning of July 5th, I determined to watch them as they busily beat over every inch of the herbage . . . Frog-hoppers being as scarce as *Verrallia* was common (perhaps these facts are not entirely unconnected with each other),

> I occasionally caught one and put it in the *Verrallia's* path . . . As soon as a *Verrallia* saw a frog-hopper it poised itself in the air (like a kestrel hovering, but with a certain intensity perceptible in its motionlessness), and if the position of its victim was favourable, it pounced upon it immediately. Then the frog-hopper hopped; in some cases the fly lost it; in some cases the fly re-appeared instantly from the place to which the frog-hopper hopped. In one case I saw the frog-hopper land with the fly still on its back . . . (Jenkinson 1903)

In contrast to Whittaker's suggestion that *Verrallia* attack teneral adults, which would seem to rule out discrimination based on color form, this observation suggests that they attack fully colored adults, behavior that would be compatible with the possibility of selection on color polymorphism (Harper and Whittaker 1976).

Parasitism rates of *P. spumarius* by *V. aucta* are high in England (34–40%) and Scotland (17–57%), but lower in Northern Italy (0–17%) (Whittaker 1969; Molinatto et al. 2020; McCaw et al. 2024). *Verrallia aucta* also parasitizes *Aphrophora alni* (Plate 11f), at least at low levels (McCaw et al. 2024). A second big-headed fly, *V. setosa*, specializes on *Neophilaenus campestris* (Whittaker 1969), a species also attacked at low level by *V. aucta* (Molinatto et al. 2020). A North American species, *Verrallia virginica*, commonly parasitizes the pine pest *Aphrophora saratogensis* (Wilson 1987). Present interest in *V. aucta* centers on its potential as a biocontrol agent for *P. spumarius* (Molinatto et al. 2020; McCaw et al. 2024; and see Chapter 9). There is precedent for the use of big-eyed flies in biocontrol, including against the sugarcane planthopper *Perkinsiella saccharicida* (refs. in Coe 1966), but, given that *V. aucta* is already abundant in areas where *P. spumarius* achieves substantial population levels in Europe, its utility would appear to be minimal, barring successful mass rearing.

The Relationship Between Predation and Selection on Color Forms

There is a persistent, unresolved tension in studies on the relationship between visual selection by predators and the maintenance of warning coloration, cryptic coloration, and polymorphism in spittlebugs. On the one hand, most observers believe that selective visual predation, particularly by birds, is the most likely driving force for these phenomena. Toward the end of their long series of studies, Halkka and his colleagues (as summarized by Halkka and Halkka 1990) continued to include visual selection as the first of three selective factors governing the frequencies of color forms in *P. spumarius* along with climatic selection and "multiniche selection." On the other hand, evidence for substantial predation, especially selective predation, is sparse (Halkka and Kohila 1976; Harper and Whittaker 1976).

As demonstrated in this chapter, bird predation on spittlebugs is sometimes substantial, and laboratory experiments demonstrate that birds can exercise effective selection on color forms linked to escape ability (Gibson 1974, 1980; Páez et al. 2021). Direct demonstration that this happens in nature is elusive. Similar limitations apply to evidence for invertebrate predation. Although spiders, for example, do not compare to birds in visual acuity, they do respond to differences in prey color (Powell et al. 2019) and pattern (Gawel et al. 2023) and may exert significant selective influence on the evolution of spittlebug color patterns. Though earlier evidence suggested that big-headed fly parasitization of *P. spumarius* is indiscriminate with respect to color form (Whittaker1971a), Harper and Whittaker (1976) present evidence that these flies may maintain polymorphic balance in an English population by attacking the most common color form more frequently the more common it becomes. However, as they note, big-headed fly parasitization is absent from places like North America and cannot be a universal explanation for *P. spumarius* polymorphism. This is an area of spittlebug biology begging for further study.

Predation in Overview

Hamilton (1982) argues that the relatively small number of eggs laid by spittlebugs testifies to their relative immunity to predators, parasites, and disease. This is consistent with the evidence reviewed in this and the preceding chapter. Overall, it appears that predation on spittlebugs is consequential but, compared with many other insects, not particularly heavy, suggesting that spittlebug defense mechanisms, particularly spittle and their spectacular jumping ability, are effective. Most mortality appears to occur in the egg stage and among early instar nymphs, especially new first instar hatchlings. After that, spittle protects the nymphs and jumping protects the adults fairly effectively, though with enough mortality to allow, for example, for selection on color patterns and escape ability in adults. Relative immunity to predation no doubt contributes to the stunning success of some spittlebug species as pests, which is the subject of the next several chapters.

6

SPITTLEBUG PESTS IN NEW WORLD SUGARCANE

An Ongoing Drama

This chapter presents an overview of sugarcane spittlebugs, the most important pests of sugarcane over large parts of the tropical and subtropical Americas. It begins with the history of sugarcane spittlebugs in Trinidad. The Trinidad case is important because it was the first recognized occurrence of spittlebugs as economically important pests and the first instance in which the full force of science was brought to bear seeking effective countermeasures. By the 1930s, the ecology of the Trinidad pest spittlebug, Aeneolamia varia, *was better documented than any other tropical insect (Myers 1931), but it was still "the second worst insect pest in the world . . . responsible in certain years for the loss of a greater proportion of a single country's production of a particular commodity than any other insect with the exception of the Cotton Boll Weevil" (Pickles 1937). The Trinidad infestation also served as an incubator for every twentieth-century trend in the search for relief from damaging insect pests, ranging from pioneering experiments in biological control to on-again off-again engagement with modern chemical pesticides after the Second World War.*

The Trinidad saga, and it certainly qualifies as such, is followed by an overview of present-day spittlebug sugarcane pests from Mexico to Brazil. In aggregate, they pose the greatest challenge to sugarcane cultivation in the Americas. In Brazil in particular, they are now, after a dramatic increase over recent decades, the most prominent pests in the world's largest producer of cane.

The First Recognition of Spittlebugs as Sugarcane Pests: Belize, 1882

In 1882, Daniel Morris, a British colonial administrator in Jamaica, visited British Honduras, now Belize, to collect botanical specimens. While touring a large sugarcane plantation in the Seven Hills District south of Port Honduras, he observed that

> A 'fly,' or rather a moth, appeared to be troublesome to canes at certain seasons of the year. . . . About the end of June or July, a white froth, similar to what is known in England as 'cuckoo spittle,' or 'goat spittle,' appears at the roots of the canes, both under and above ground. On examination, a small greenish wingless insect is found embedded in the froth, which remains in this state till about August, when it matures into the 'fly.' While affected . . . the canes become stunted in growth, but even afterwards, with the mature insect, they are not free from injury, as the 'fly' attacks the leaves and causes them to be spotted and eventually to die off. The lower leaves appear to wither first, but the upper leaves are soon attacked, and sometimes so severely as to cause the whole to fall off, leaving nothing but the bare cane standing. Even among canes not severely attacked, the joints are short and poor. About the end of September and October the 'fly' disappears. The canes after this time, relieved from the attacks of the 'fly,' make very fair growth; the joints become long and full, and they continue to develop, up to cutting time. (Morris 1883, 35–36)

This is the first recorded observation of a spittlebug infestation in sugarcane. Although Morris's higher order taxonomy is off the mark, he got the culprit right and provides a remarkably accurate account of the nature and seasonal trajectory of the negative effects on cane. Morris was evidently talented. Knighted in 1903, he eventually became assistant director of the Royal Botanic Garden at Kew and later served as commissioner of the Imperial Department of Agriculture in the West Indies.

The Emergence of Spittlebugs as Serious Sugarcane Pests: Trinidad, 1906

Spittlebugs are still pests of sugarcane in Belize (Cawich 1981), but the drama of spittlebugs on sugarcane unfolded in full force in Trinidad in the early twentieth century. While cultivation of sugarcane was central to Caribbean early colonial development, largescale production came to Trinidad relatively late, in the mid-1700s. By 1900, production had consolidated into fifty large estates and by 1975 into only four (Wilson and Parmasad 2014). Through most of the 200 years to the end of the twentieth century, sugar vied with cacao as Trinidad's largest export. In 1960 it still accounted for one-fifth of the island's private sector employment (Auty 2017). Prior to the development of oil and gas reserves in the late twentieth century, sugar and sugarcane were mainstays of the economy and supported a large sector of rural society.

It is not surprising then, that the emergence of spittlebugs as a serious pest received focused attention in Trinidad. Williams (1921) gives a detailed account of the crisis that unfolded. The realization that a serious problem was afoot dawned slowly, beginning with a minor report of a "*Tomaspis*" damaging cane in 1889 by attacking roots in the Chaguanas area near the coast, south of Port of Spain. *Tomaspis* is a spittlebug genus of the family Cercopidae, used at the time as a taxonomic holding pen for a diverse collection of Central American and Mexican species (Fowler 1897), which were later parceled out among multiple genera (Distant 1909; Fennah 1948). It is not certain when the Trinidad spittlebugs first became a pest. Outbreaks of blight that might have been due to spittlebugs occurred as early as 1862, 1869, 1878, and 1886. One old planter remembered seeing spittle masses on cane roots in 1878. What is certain is that starting in 1906 there were major, destructive outbreaks of spittlebugs every few years. At the worst point in 1906, flying adults "got entangled in the horses' manes by dozens" (Williams 1921) at the Brechin Castle estate. Things were bad enough that Barbados (1910) and St. Lucia (1912) passed quarantine laws expressly aimed at keeping the sugarcane froghopper off their islands, and American entomologist George Wolcott (1913), working out of Puerto Rico, said that the Trinidad froghopper had "achieved as much notoriety as . . . the boll weevil in in the United States."

Early Efforts to Understand the Bug Behind the Blight

These outbreaks triggered substantial efforts at research and control. In 1909 the recently formed Trinidad Board of Agriculture appointed an entomologist, Frederick Urich, and a mycologist, James Rorer, to investigate the causes of the blight. The teaming of an entomologist and a specialist in plant pathogenic fungi reflected the fact that it was unclear at the outset whether the spittlebugs, associated fungi, or some combination of the two were to blame. Urich (1913) eventually published the first thorough account of froghopper biology and agronomics, including the first economic loss estimates and the first observations on reflex bleeding (Chapter 5). By 1918 his work was cited by the US Department of Agriculture in a manual on dangerous insects likely to be introduced to the United States (Pierce 1918). He also made early observations on spittlebug pests in Mexico. In 1910, the Board appointed an assistant entomologist, Plantagenet Guppy, eldest son of Robert Guppy, source of the name for aquarium guppies, which originated in Trinidad.

A particularly destructive outbreak of blight in 1912 led to the additional hiring of John Kershaw for the year 1913. Kershaw was a self-trained British entomologist. He had already made a reputation assisting the Hawaiian Sugar Planters Association in their successful hunt for a parasitic fly that saved the Hawaiian sugarcane industry from an introduced weevil. During his short engagement in Trinidad, Kershaw made important contributions to spittlebug biology. He published the detailed anatomy of *A. varia* and, in parallel with P. Émile Licent (1912), one of two early descriptions of the spittlebug alimentary system (Kershaw 1914) with its unique water shunting filter chamber (Chapter 3). He also made the critical observation that the watery excretions of the *Aeneolamia varia* digestive tract had no sugar despite the fact that "sap extracted directly from the leaves had plenty" (Williams 1921).

These early collaborators worked out the basic biology of the first important spittlebug pest of sugarcane and, in the process, foreshadowed much work to come later in the twentieth century. One of the first tasks was identifying the insect. In a couple of initial missteps, it was erroneously identified (in modern nomenclature) as *Aeneolamia contigua* (Plate 14f), a mainland Central American and Mexican species, or as *Prosapia bicincta* (Plate 10e), a species confined to the United States, Cuba, and Jamaica. Then Distant (1909) described it as a new species, *Tomaspis saccharina*. However, in 1911 Urich sent specimens to American spittlebug taxonomist Elmer Ball. He identified them as *Tomaspis varia* (Plate 15b), a species described from French Guiana by Fabricius in 1787. Ball's determination has stood the test of time. In 1948 Fennah included the species in the new genus *Aeneolamia*. The Trinidad sugarcane pest spittlebug is now known as *Aeneolamia varia saccharina*, a subspecies of *Aeneolamia varia*, which occurs in a variety of subspecific forms from French Guiana to Colombia. Fewkes (1969a) gives the most recent comprehensive overview of the biology of the species.

Nature of Spittlebug Attacks on Cane

Aeneolamia varia attacks sugarcane in two stages. The nymphs hatch from eggs deposited in the soil or on trash (dead debris from the cane) and live on the roots of the cane at or below the surface of the soil, forming conspicuous foamy spittles scattered over the soil surface among the plants (Carvalho and Webb 2005, plate 14a). They suck xylem sap from the roots, diverting plant nutrients and sometimes causing wilting or stunting of growth. Occasionally, in later instars, they move up the plant a bit to adventitious roots growing out of the stem, but they never feed high on the plant or on the leaves. The last instar nymphs make a special hollow spittle mass on the soil surface or a bit up the plant and transform to the adult stage within it. The adults, which emerge to feed on leaves, do the really destructive work, about five times more damage per feeding individual than nymphs (Ramirez-Gi et al. 2023).

Like the nymphs, adults feed by piercing xylem vessels, diverting water and nutrients (Chapter 3). In the process they inject an as yet unidentified toxin or toxic substances. Starting around the points of feeding, leaves develop semi-transparent pale areas around the punctures. These patches elongate along the leaf axis and form long, narrow, pale streaks, typically about 1.25 mm wide. Over the course of a few days the streaks get wider, turn reddish, and then an opaque, silvery, dead brown, reaching up to 12.5 cm in length. A single puncture can eventually kill three square centimeters or more of leaf tissue. If the original punctures are dense enough, the dead streaks merge to form large dead areas or kill the entire leaf. Leaves often have 50 to 100 punctures and, in bad attacks, 50 to 100% of the leaf area may be destroyed. Since the outside leaves suffer the greatest damage, the whole plant from a distance looks dead brown, "as if damaged by fire" (Williams 1921). In severe attacks whole fields turn brown. No wonder the blight caused consternation and set off alarm bells.

In Trinidad, *A. varia* exhibits an annual cycle initiated by the onset of the rainy season. When the rains come, a first generation hatches from *diapausing* (dormant) eggs, followed by three or four later generations. The early sugarcane workers confirmed that visible attacks came in waves, corresponding to adult emergence. The first generation of adults is generally small in numbers, causing little visible damage. The second generation causes the destruction of leaves and smaller canes. As the second-generation adults die off after two or three weeks, the plants get a respite and start to recover and put out new leaves. However, a third generation of adults then emerges, generally more destructive than the second. A fourth generation of adults follows, leading to another wave of destruction, until the whole cycle is ended by the onset of the dry season. The cane plants, which are perennial and typically harvested for four or five cuttings over multiple years, usually do not die, but cane yield and sugar production are severely curtailed in years of bad outbreaks.

Early Searches for Biocontrol Agents

Initially, there was suspicion that the blight was caused by a pathogenic fungus with the froghoppers as vectors. Although some fungal pathogens do attack cane, particularly the roots, fungal attack, as it turned out, was not the cause of the blight. But the investigators did discover one froghopper-associated fungus of lasting importance, the Green Muscardine fungus, *Metarhizium anisopliae* (abbreviated hereafter as *Metarhizium*). This fungus was noted to infect froghoppers in the first written report of the blight in 1889. *Metarhizium* occurs naturally in soil and is spread by spores. When spores germinate on spittlebugs, the hyphae penetrate living nymphs and adults (Mei et al. 2021), ramify throughout their bodies, and eventually kill them before sporulating on the exterior to spread to other victims. The mycologist Rorer (1913), following up on initial work by Gough (1910), conducted large-scale experiments with mass artificial culture of *Metarhizium* spores for field application as a biological control agent. In the short run, these efforts to augment natural *Metarhizium* levels proved unsuccessful as a control method, but in the long run, *Metarhizium* has become a major instrument of control in Trinidad, Mexico, Costa Rica, Venezuela, and, most importantly, Brazil (discussed later in this Chapter).

In addition to studying the entomopathogenic fungus, a tremendous amount of effort went into attempts to identify natural insect enemies. This led to the early discovery of a syrphid fly, *Salpinogaster nigra* (Fig. 4.5), whose larvae are specialized predators of spittlebug nymphs. It was first observed by Gough, bred to the adult stage by Urich, and studied in detail by Guppy (1913), who estimated that a single maggot destroys 30 to 40 nymphs over the course of its development. The maggots enter the spittle, pierce the nymphs, and suck out the body contents. Despite their efficiency as predators, natural *S. nigra* populations do not effectively control spittlebug populations. At the beginning of the rainy season, when *A. varia* first emerges, *S. nigra* populations are very low. By the time they build up in relation to spittlebug populations, the damage to cane is already done. To get around this, Kershaw suggested that Guppy attempt mass rearing in advance, during the dry season, an initiative that failed for want of nymphs to feed the dry season maggots. So, although *S. nigra* is widespread and abundant from Mexico to Brazil and attacks a wide range of spittlebug hosts, it has never proven useful for controlling pest spittlebugs (see more on *S. nigra* as a predator in Chapter 4).

A good deal more effort went into a search for egg parasitoids, tiny wasp larvae that devour embryos in eggs before they hatch. Guppy discovered the most promising candidate, a bright vermillion chalcid wasp, *Oligosita giraulti*. However, it proved to be locally common, parasitizing several Trinidad spittlebug species, as well as widespread on the Central and South American mainland. This made it an unlikely control agent. Had *A. varia* been susceptible to control by local parasites, it would not have become a pest in the first place. Instead, on Kershaw's recommendation, it was decided to search for promising parasites further afield. For this work, the British colonial government tapped Carrington B. Williams.

> C. B. was in training at the London School of Tropical Medicine, to assist the R.A.M.C. in identifying pathogenic bacteria and protozoa; largely engaged in 'that unheroic branch of warfare', the examination of the stools of dysentery patients. Then one day he was called from his microscope to see a representative of the Colonial Office who asked him to go out immediately to the West Indies to investigate a serious insect pest of sugar cane. In those days there was no English sugar beet crop and cane sugar was urgently needed . . . Five weeks later he landed in Trinidad and was to live in the West Indies for nearly six years. (Wigglesworth 1982)

Williams arrived in 1916, fresh from his work with stools of dysentery patients, to take up his new work with stools of sugarcane, stools being the ground-level clumps from which the canes grow. He was instructed "to visit such neighboring colonies and foreign countries as might be likely to contain fresh parasites of the froghopper." World War I had already begun to restrict travel, so he chose destinations close at hand. After a one-day stop in Surinam, where he looked at a mildly injurious local spittlebug pest, *Mahanarva tristis* (Plate 15e), he moved on to British Guiana (now Guyana), where he spent four months in the coastal Demerara District studying the local spittlebug sugarcane pest, *Aeneolamia flavilatera* (Williams 1917, 1918). He also visited Trinidad's neighboring island of Tobago, which harbors a different subspecies of *A. varia*, and Grenada, which had a small infestation of *A. varia saccharina*, probably introduced from Trinidad. In early 1917 he went to the Canal Zone in Panama with a day stop in Venezuela. And then he went from the Canal Zone to the Bocas-del-Toro region on the Panamanian border with Costa Rica, where he worked for four and a half months. After wartime disruptions to shipping derailed plans to go on to British Honduras, he returned to Trinidad via Martinique, St. Lucia, and Cartagena, Colombia.

With the exception of Martinique and St. Lucia, where he found no spittlebugs, he collected and observed spittlebugs and spittlebug parasites all along the way. In the end, though he made numerous natural history observations, the work was disappointing. Mostly, he found parasites that were already known in Trinidad. He decided it would be better to search farther afield and eventually recommended searching in Africa, but the war precluded this, and Williams settled down to extended studies of *A. varia saccharina* and blight in Trinidad (Williams 1931), producing towards the end of his appointment a 170-page magnum opus, *Froghopper-Blight of Sugar-Cane in Trinidad* (Fig. 6.1) (Williams 1921), which is still one of the two most comprehensive monographs on a spittlebug pest. It is rivaled only by the work of Weaver and King (1954) on *Philaenus spumarius* (Fig. 8.2). The search in Africa did not happen until the 1960s and proved unsuccessful (Fewkes and Buxo 1964; Greathead 1970).

As a final irony, in his last year in Trinidad Williams discovered by chance that Mosaic virus disease of sugarcane had recently been introduced to the island, leading to its suppression before it could spread. "So," he said, "the problem that was my official work was largely unsolved, but I saved the island many thousands of pounds by an accidental discovery outside my proper sphere. What bad organization!" (Wigglesworth 1982).

Williams went on to a career as a researcher and colonial administrator, next serving in Egypt and then in Tanganyika (now Tanzania), where he studied migratory locusts, at one point lying full length across the cowcatcher of a moving locomotive to see close up how crushed locusts greased the wheels and brought trains to a standstill for lack of traction. He later joined the Rothamsted Experimental Station in England, where he collaborated with the famous statistician Ronald Fisher and served as department head of Entomology for 23 years.

In the absence of effective biological controls, the sugarcane estates in Trinidad resorted to other methods. Light trapping, conducted using kerosene lanterns fitted out with water pans to trap incoming insects, captured as many as 250,000 individuals per night but suffered the immense disadvantage that more than 98% of the captured individuals were males, leaving the females to feed and lay eggs. Gangs of children were set to collecting nymphs by hand at the rate of 400 to 600 per day per child at five cents per hundred ($1.42 in 2022 dollars). At the Brechin Castle estate, 700,000 were collected in one season at a cost of $312 per acre (2022 dollars) (Kershaw 1913). Pity the overseers who had to count those slimy nymphs one by one. Even with child labor, this method was not cost-effective. There were also experiments at capturing adults by netting. One instance in British Guiana gives a sense of pest densities: 3,000 adult *Aeneolamia flavilatera* spittlebugs were captured in two minutes by two men wielding a two-foot by six-foot net (Williams 1921). Like the planters in Trinidad, the British Guinea planters exploited child labor, with gangs of boys collecting 533,540 adults and 540,000 nymphs in 22 days (Williams 1918). These methods also proved impractical for control. Other attempted control methods included soil improvement, selection of resistant varieties of cane, reduction in the

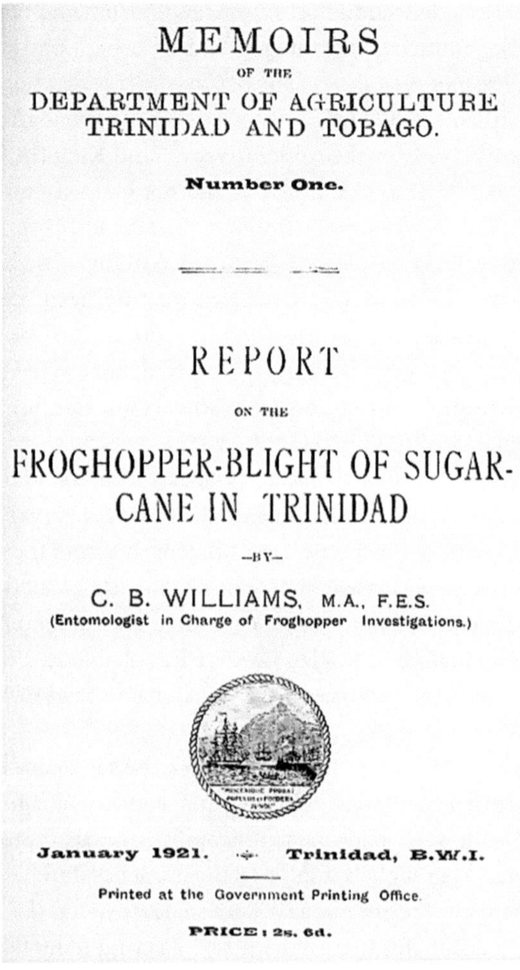

MEMOIRS
OF THE
DEPARTMENT OF AGRICULTURE
TRINIDAD AND TOBAGO.

Number One.

REPORT
ON THE
FROGHOPPER-BLIGHT OF SUGAR-CANE IN TRINIDAD
—BY—
C. B. WILLIAMS, M.A., F.E.S.
(Entomologist in Charge of Froghopper Investigations.)

January 1921. Trinidad, B.W.I.

Printed at the Government Printing Office.

PRICE: 2s. 6d.

FIGURE 6.1. Cover of the 1921 monograph by C. B. Williams, *Report on the Froghopper-blight of Sugar-cane in Trinidad*, printed by the colonial government. This, along with Weaver and King's monograph on the meadow spittlebug (Fig. 8.2), is one of the two most comprehensive studies ever published on single spittlebug species. The cover identifies Williams as "Entomologist in Charge of Froghopper Investigations" and carries the colonial badge of Trinidad and Tobago with the motto "Miscerique probat populos et foedera jungi," a quotation from Virgil's Aeneid, translated as "He approved the mingling of peoples and their being joined together by treaties." The report sold for 2 schillings 6 pence, equivalent to about $10 USD today.

number of seasons a single planting of cane was cropped, crop rotation with legumes, removal of dead leaves from plants, weeding out alternative grass hosts, burning the dead leaves (trash) on the fields, and spraying with kerosene, copper sulfate, and potassium cyanide (Williams 1921).

J. G. Myers: The Indiana Jones of Spittlebug Biocontrol Exploration

In the end, these methods proved ineffective or prohibitively costly. This led in 1925 to the formation of the Froghopper Investigation Committee, funded for five years to the tune of £2,500 per year (about $190,000 in 2022 dollars). Half of the funding was from the Colonial Government of Trinidad and half from the private sugar industry. It was no small commitment for the time and place (by 1929 they were spending the equivalent of $300,000 per year). The Committee engaged entomologists Alan Pickles and John G. Myers to continue the hunt for biological control agents. Myers, British by birth, New Zealander by upbringing, Harvard trained, brought a strong background in Hemiptera (he was a cicada specialist) and biological control.

At first, Myers, often with Pickles in tow, picked up where Williams left off, looking for spittlebug parasites and predators on several Caribbean islands and portions of the Caribbean rim (Myers 1931). His most notable find was *Carabunia myersi*, an encyrtid wasp parasite of *Clastoptera undulata* nymphs in Cuba (Chapter 4). It proved not to attack *A. varia* or other cercopid pests but later was used in an attempt to control an outbreak of *C. undulata* on *Casuarina* in Bermuda (Bennett and Hughes 1963; Cock 1985). For the most part, however, Myers found the same parasites as his predecessors, none with promise. Reasoning that he needed to go further afield, he undertook three expeditions that qualify him as the Indiana Jones of spittlebug field investigations. He recorded his adventures in understated manner in the Minutes and Proceedings of the Trinidad and Tobago Froghopper Investigation Committee, reproduced in delightful detail by Cock and Bennett (2011).

On his first expedition he travelled extensively in the interior of Venezuela, finishing on the Venezuelan coast close to Trinidad. His wife was "invalided back to England towards the end of 1931, with severe malaria resulting from the conditions of work in the unhealthy Delta of the Orinoco" (Myers 1935). On the second, he travelled to the mouth of the Amazon in Brazil, collected extensively near Belem and on Marajo Island, travelled upstream via Santarem and Manaus, and crossed by land to the Rupununi savannah, grasslands north of the rain forest, where he suspected he would have the best opportunities to find an effective parasite of grass-feeding spittlebugs. This required overland travel of 1300 km, on foot, with indigenous Native American carriers. He wrote that "of the three main difficulties in South American travel, namely rapids, disease and shortage of food, the writer at least has had his share" (Myers 1935). He did run out of food on occasion, and, walking from one indigenous village to another late one evening, he collapsed from malaria. He sent his party for help, but rescue was delayed by nightfall. Still too weak to walk, he crawled into the village the next morning on hands and knees (Cock and Bennett 2011). All his expeditionary work depended on non-European employees, "Latin, negro, East Indian, aboriginal Indian and mixtures of several," who he praised as "trustworthy, hardworking and intelligent" (Myers 1935).

On the third expedition, he motored upriver from Georgetown, British Guiana in an eight-meter launch, portaged it 42 km overland to a Rio Negro tributary, and sailed down to the Amazon, first to Manaus, then Santarem. En route, he lost most of his scientific equipment in a boat wreck and had to resupply at the Fordlandia hospital on the Rio Tapajoz in Brazil. After a few months collecting on the lower Amazon, he motored back upriver, went back to the Rupununi savannahs to collect again, and finally returned to Trinidad via Georgetown. Near Santarem, he collected a tachinid fly, *Metagonistylum minense*, that was successfully introduced in multiple localities to control the highly destructive sugar-cane borer *Diatraea saccharalis.*

In a 1933 essay titled "Men versus insects," no less a personage than the British intellectual Bertrand Russell, alluding to the work of entomologists who study parasites, paid indirect homage to Myers, saying that

> Official reports of their activities are fascinating; they are full of such sentences as: 'He proceeded to Brazil, at the request of the planters of Trinidad, to search for the natural enemies of the sugar-cane Froghopper.' One would say that the sugar-cane Froghopper would have little chance in this contest. (Russel 1935, 231–234)

Alas, despite Myers' genuinely heroic efforts and Bertrand Russell's optimistic forecast, the sugarcane froghopper had more than a chance. Myers made extensive observations on spittlebug parasites, host plants, and natural history, but never found a useful biological control agent for *A. varia.* As Cock and Bennett note, keeping track of his whereabouts and expenditures through it all must have been an administrator's nightmare. He finished his South American journeys in 1934, joined the Imperial College of Tropical Agriculture (now the University of the West Indies in St. Augustine, Trinidad), and published a summary of his far-flung investigations of South American spittlebugs (Myers 1935). In 1937 he shipped out to Anglo-Egyptian Sudan, where he was appointed to survey the agricultural prospects of Equatoria Province, now the country of

South Sudan. He died there in 1942 at age 44 in a motor accident, by one account while sweep-netting for insects from a moving vehicle. An obituary in *Nature* deemed him "the outstanding economic entomologist of his generation" (China 1942). So ended the short life of a prolific scientist and the most intrepid field biologist in the annals of spittlebug studies.

The Short-lived Postwar Utopia of Chemical Insecticides

Myers's sometime field companion, Pickles, carried on the froghopper control effort in Trinidad and was at the center of a seismic shift in that work with the advent of modern chemical pesticides right after the Second World War. In an early tremor of what was to come, he tested out the plant-derived insecticide pyrethrum on *A. varia* at the Woodford Lodge estate in 1940. It worked there but not when tests were extended to other estates. Nevertheless, the Woodford Lodge success stimulated 1945 experiments using the new synthetic insecticide DDT and a second, plant-derived insecticide, Sabadilla. Pickles (1946) concluded that both insecticides were "able to reduce froghopper populations . . . to a level at which economic injury to sugarcane does not take place," adding that "the probable residual effect of DDT is likely to make it a most valuable ingredient of insecticidal mixtures for this purpose." This set off a wave of interest in the new postwar chemical insecticides and in 1949 Potter and Carrington reported complete, prolonged control of nymphs with BHC (benzene hexachloride), leading them to declare "that a practical, economic and effective means of controlling the sugar cane froghopper in Trinidad has been determined." The "days of unpredictable and uncontrollable froghopper blight" seemed to be over and there was a resurgence in the Trinidad sugar industry (Blackburn 1954).

The euphoria was short-lived. In 1954 reports appeared that nymphs had developed resistance to BHC, in some cases approaching 100%. Malathion gave promise as a substitute, but the period in which chemical insecticides seemed to be an unmitigated panacea lasted a short five years. Fewkes (1969b) gives a good overview of the shifting variety of chemical treatments used during this period. By the late 1960s increasing resistance had led to an alarming increase in aerial spraying (Evans and Buxo 1972). There was also collateral damage. Excessive use of insecticides against *A. varia* was later blamed for triggering increased damage by the cane borer moth *Diatraea saccharalis* (Khan and Isaac 2020), apparently as a result of suppressing that caterpillar's natural enemies while it sheltered from insecticides inside the canes (Box 1953; Evans and Buxo 1972).

Through the 1990s, Trinidad agronomists made do with a shifting set of chemical insecticides, changing treatments to stay ahead of developing resistance

(Blades et al. 1997). During this period, they also revived interest in using *Metarhizium anisopliae* as a biological control agent, stimulated in part by its successful use in Brazil (Allard 1987; Allard et al. 1990). By the mid-1990s they had evolved a balanced system, using chemical insecticides on bad breakouts and *Metarhizium* treatments on mild or early season infestations. In 1997 this balance collided head-on with bad weather patterns and national politics.

Drastic Changes in the Trinidad Sugar Industry and an Ill-Fated Decision

When Urich, Guppy, and Rorer started their froghopper investigations, there were about 50 independent sugarcane estates in Trinidad. In 1997, for all practical purposes, there was one. The industry had consolidated in the 1930s in the hands of the British firm Tate and Lyle under the name Caroni (1937) Limited. In 1970, following the Black Power Movement domestic uprising, pressure to decolonize the industry led the Trinidad government to purchase a controlling interest in Caroni. In 1975 it purchased the rest (Wilson and Parmasad 2014). The new entity, Caroni (1975) Limited, was the largest employer in Trinidad. It lost money from the beginning, about $28 million USD in 1996 (Shah 2016).

In January 1997, at the direction of the Minister of Agriculture, the Caroni Board of Directors decided that no chemical insecticides were to be used on cane for the 1997 season, only *Metarhizium*. The Caroni staff objected. Based on their previous experience, *Metarhizium* by itself was not up to the task. In July the National Farmers and Workers Union observed that the early season *Metarhizium* treatments were not working. The Board persisted. The result was a fiasco, documented in detail by an independent committee convened by the Minister to survey the damage (Blades et al. 1997). Quoting from August reports by the Caroni technical staff, "Despite the relatively high frequency of treatments of *Metarhizium* applied . . . adult froghopper population levels have continued to rise dramatically . . . An optimistic yield loss due to froghopper of 70–80 percent can reasonably be anticipated with some fields suffering as much as a 100 percent loss . . . An immediate management decision in the judicious use of chemical insecticides is required to prevent any further damage to the 1998 sugarcane crop" (Blades et al. 1997). Finally, in September, the Board approved malathion treatment, but the supply ran out after the first spray.

Losses amounted to about 8500 tons of sugar in fields directly owned by Caroni and 5600 tons in fields owned by subsidiary independent, small-scale farmers. Also, as predicted, the 1998 crop was badly affected by damage to the plants in 1997. There were extenuating circumstances. Due to unusually wet weather, the 1996 froghopper infestation extended to a fourth and even a very unusual fifth brood, raising the number of eggs available to hatch at the beginning of the 1997 rainy season. Then the 1997 rainy season was unusually dry, presenting suboptimal conditions for the germination of sprayed *Metarhizium* spores. In the midst of this, Caroni annual losses grew to $39 million USD in 1997 and $48 million USD in 1998 (Shah 2016), and the froghopper control debacle spread to newspaper articles and exchanges in parliament, where in December 1997 Senator Daniel Teelucksingh, addressing a bill about a longstanding division in the independent cane farmers' union, declared that the factional parties were "all seeking the soul of the cane farmer, not caring whether his body is devoured by the froghopper, or whether he is cremated in the January cane fires" (Teelucksingh 1997).

Following the debacle, the United Nations Food and Agriculture Organization (FAO) made a two-year grant of $247,000 USD in 1999 to support the development of an effective, integrated pest management program for *A. varia* in Trinidad. Caroni still maintained a research station, a last vestige of Trinidad's rich history of support for spittlebug research, and, in 1999, even provided a consultancy on *Metarhizium* in Nicaragua and was mass producing *Metarhizium* for export (Boodram 1999). None of this was enough to save the industry. By 2002 the Trinidadian government was spending $92 million USD annually just to keep the company in business (Shah 2016). On July 31, 2003 it liquidated Caroni (1975) Limited. The main mill closed. Nine thousand direct employees were terminated. Including suppliers and the 5000 self-employed but dependent farmers, 20,000 people lost their livelihoods. The large-scale Trinidad sugar industry came to an end. So too did a way of life for tens of thousands of rural residents, most of them descended from East Indian indentured laborers brought in to work the cane fields after the abolition of slavery (Wilson and Parmasad 2015). More was lost than just wages. Caroni had been deeply entwined with daily life. For example, it had sponsored 67 playgrounds and 29 cemeteries, now orphaned, and maintained 47 public roads (Fernandéz T. and Benjamin 2010).

At the turn of the twenty-first century, Trinidad's sugar-related acreage accounted for 20% of the total national land area (Fernandéz T. and Benjamin 2010). By 2017, sugarcane did not even make the list of Trinidad and Tobago's principal agricultural products, which were, in order, cocoa, dasheen, pumpkins, cassava, tomatoes, cucumbers, eggplants, hot peppers, pommecythere, coconut water, and poultry (Khan and Isaac 2020). By 2019 Trinidad had sunk to fifty-seventh in world sugarcane production (Table 6.1). There are still farmers who cultivate cane but on a limited scale (Ayub Khan, personal comm. 2022). During the 1960s Trinidad produced about 2.4 million tons of cane per annum. By 2019, the total had dropped to 810,000 tons (Table 6.1), and the government had spent $1.5 billion USD winding down Caroni's affairs (Sorias 2019). Trinidad now imports sugar for domestic needs. Bertrand Russell notwithstanding, the froghoppers are still there, living on grasses and what little cane remains.

TABLE 6.1. World sugarcane production by country (top 60 producers in 2019)

RANK	COUNTRY	PRODUCTION (TONS)	PER CAPITA (KG)	AREA PLANTED (HECTARES)	YIELD (KG/HECTARE)
1	**Brazil**	752,895,389	3,593	10,081,170	74,683
2	India	405,416,180	303	5,061,090	80,105
3	Thailand	131,002,173	1,894	1,835,085	71,388
4	China	109,963,447	79	1,422,655	77,295
5	Pakistan	66,880,011	331	1,039,777	64,322
6	**Mexico**	59,334,059	476	795,984	74,542
7	**Colombia**	32,662,952	654	457,783	71,350
8	Australia	32,415,352	1,294	433,154	74,836
9	Indonesia	29,100,000	110	443,569	65,604
10	**Guatemala**	29,087,086	1,681	270,144	107,673
11	USA	28,972,760	88	369,560	78,398
12	Philippines	20,719,291	195	379,253	54,632
13	South Africa	19,482,246	337	298,611	65,243
14	Argentina	17,652,814	397	476,176	37,072
15	Egypt	16,316,134	167	140,969	115,743
16	Vietnam	15,269,716	161	233,387	65,427
17	Myanmar	11,846,176	220	182,158	65,032
18	**Peru**	10,929,341	350	87,095	125,488
19	Bolivia	9,558,472	845	174,630	54,736
20	Iran	9,284,942	114	113,150	82,059
21	**Ecuador**	9,257,700	542	121,812	76,000
22	Cuba	8,725,032	778	209,419	41,663
23	**Nicaragua**	7,371,811	1,173	77,134	95,572
24	**El Salvador**	7,177,803	1,080	80,972	88,646
25	Paraguay	5,819,500	825	103,000	56,500
26	Swaziland	5,665,898	4,888	58,762	96,421
27	Sudan	5,449,000	134	71,100	76,639
28	**Honduras**	5,409,000	600	62,286	86,841
29	Zambia	4,994,302	296	48,336	103,325
30	Dominican Rep.	4,895,891	477	103,523	47,293
31	Uganda	4,892,047	126	70,964	68,937
32	Kenya	4,606,100	90	71,900	64,063
33	**Costa Rica**	4,421,210	884	62,630	70,593
34	**Venezuela**	4,380,304	138	73,339	59,727
35	Mozambique	4,106,873	142	59,059	69,539
36	Tanzania	3,589,459	66	51,344	69,910
37	Zimbabwe	3,562,000	240	46,000	77,435
38	Nepal	3,557,934	122	71,625	49,675
39	Mauritius	3,405,250	2,692	45,054	75,582
40	Madagascar	3,174,161	121	99,799	31,806
41	Bangladesh	3,141,923	19	81,062	38,760
42	Malawi	3,136,918	175	29,164	107,561

TABLE 6.1. *(Continued)*

RANK	COUNTRY	PRODUCTION (TONS)	PER CAPITA (KG)	AREA PLANTED (HECTARES)	YIELD (KG/ HECTARE)
43	France	2,860,410	43	37,775	75,722
44	**Panama**	2,670,909	642	37,548	71,133
45	Congo-Kinshasa	2,319,282	29	47,260	49,075
46	Laos	1,970,566	283	31,826	61,917
47	Côte d'Ivoire	1,968,353	79	23,953	82,176
48	**Belize**	1,794,029	4,532	40,486	44,312
49	Haiti	1,689,520	152	25,535	66,165
50	Fiji	1,514,918	1,712	35,120	43,136
51	Ethiopia	1,499,134	14	32,069	46,747
52	Nigeria	1,456,039	7	87,330	16,673
53	Cameroon	1,285,769	54	135,814	9,467
54	Japan	1,164,593	9	21,289	54,704
55	Senegal	1,093,701	70	9,657	113,255
56	**Guyana**	1,042,088	1,332	17,896	58,230
57	**Trinidad/Tobago**	810,000	597	15,000	54,000
58	Jamaica	774,200	284	15,700	49,312
59	Congo-Brazzaville	707,136	131	19,843	35,637
60	Cambodia	660,919	41	29,703	22,251

Source: https://www.kaggle.com/datasets/rajkumarpandey02/list-of-countries-by-sugarcane-production
Note: Highlighted countries are subject to consequential spittlebug attack.

My Visit to Trinidad in December 2003

I visited Trinidad for a week in early December 2003 to look at sugarcane spittlebugs where they had been studied longest and had done the most proportional damage. Availability of a sabbatical determined the timing of the trip, but it happened to bring me to Trinidad in the immediate aftermath of the summer 2003 announcement of the dissolution of Caroni (1975) Ltd. As a result, I was able to interview several participants when things were still unsettled. To a person they were pessimistic, rightly so in retrospect, about the future of the industry. The Ministry of Agriculture, Land, and Marine Resources had set up a "Working Committee for the Establishment of the Sugar Industry Authority" to chart a path for the industry, but in the absence of a clear sense of direction and strategy it labored in vain. I interviewed one of its members, Balmattee Sukha, a chemist and head of the technical arm. She reported that the new Authority was supposed to have been created by October 5, 2003, but that in the absence of parliamentary action it had not happened. The target date had been pushed to January 2004, but she saw no action toward that end. In January, she wrote to say that the Working Committee had been disbanded as of December 31, 2003. No authority was ever created.

I interviewed Cynthra Persad, then the Deputy Director of Research at the Ministry of Agriculture. She had been chosen by the Minister of Agriculture to head a biocontrol program against the Hibiscus mealybug. The success of this campaign influenced the Minister's fateful decision to ban chemical pesticides against froghoppers. She thought he had been right to move against froghopper chemical pesticides, which accounted for half the pesticide use in Trinidad and had been used indiscriminately in aerial spraying, endangering human and environmental health. There were also continuing problems with chemical pesticide resistance. In her view, the Caroni Board's narrow interpretation of the Minister's mandate had created the chaos. She also emphasized that 1997 was a unique year, very dry, and that everything converged to add to the froghopper damage. In 2004 she was promoted to Ministry Director of Research, but by then biocontrol research in direct support of the sugarcane industry had ended.

The Minister in question was Reeza Mohammed. In 1999 he became Environment Minister, following a cabinet reshuffle. By chance, my entomologist host, Graham White, and I ran into him at a restaurant on my second day in the country. Graham thought it better not to approach him and I deferred, but it might have been an interesting conversation. He died in 2022, with obituaries full of praise from all political sides and not a word about the 1997 fiasco. Three days after that chance encounter, we had lunch in San Fernando with Shaheed Baksh, a cane farmer who had shown us his fields earlier in the day. A few months later I heard from Balmattee that Shaheed's nephew had recently been kidnapped and murdered, perhaps a symptom of the disintegration of rural civil society in the wake of the sugarcane upheavals. The wounds from the Caroni closing still run deep. In 2022, almost 20 years after the shutdown, organized former Caroni workers were still agitating for land promised in compensation, a point of national political contention.

Perhaps my most poignant interview in Trinidad was with Bibi Ali, who became, inadvertently, the last in the long and distinguished line of biologists devoted to froghopper control in Trinidad. She had hired on at Caroni in 1996 to oversee their *Metarhizium* program. They had been spraying spores by airplane and the results had been disappointing, with initial infection levels of only 15 to 20%. She pioneered a method using mycelially infected rice grains instead of spores and application at ground level where the adults would be emerging from spittles. This achieved infection levels of 70%, killed infected females in three days, and cut egg production by 50%. These were very promising results that appear never to have been fully tested at scale due to the collapse of the industry, though her approach appears to have become the primary method of application in Brazil (Mascarin et al. 2019). At the time of collapse, her unit had been geared up to sell *Metarhizium* expertise to other countries.

But the 2003 Caroni closure put an end to Trinidad's long standing as a center of spittlebug sugarcane pest research. None of the scientists and technical workers I met in 2003 continued in that tradition.

Spittlebugs Are Not Sugarcane Pests Elsewhere in the Caribbean or in the United States

Excepting the small, stray *A. varia* infestation in Grenada, apparently accidentally introduced from Trinidad, spittlebugs are not sugarcane pests in the Caribbean outside the islands of Trinidad and Tobago. This is perhaps because *Aeneolamia*, the primary sugarcane pest genus in the northern Neotropical region, has never spread to the other islands. Both Trinidad and Tobago are continental islands, connected during recent geological history to mainland South America. As a result, they share adjacent South America's fauna and flora, including its spittlebug fauna. However, they have never been linked by land directly to one another. Apparently as a consequence of their independent island origins, they harbor distinct subspecies of *Aeneolamia varia*, *A. varia saccharina* in Trinidad and *A. varia carmodyi* in Tobago.

One species in the spittlebug pasture grass pest complex, *Prosapia bicincta* (Plate 10e), did make it to Cuba and Jamaica in the forms, respectively, of subspecies *P. bicincta fraterna* and *P. bicincta basalis*. However, although *P. bicincta* is a pasture grass pest in the Southeastern United States and Cuba (Chapter 7), it does not damage cane in Cuba, Jamaica, or the mainland United States. There are many other spittlebug species in the Caribbean and the southern United States, but, excepting *A. varia* and *P. bicincta*, none of them are pests of cane or grasses. There is an isolated report of *Prosapia simulans* (Plate 14c) on potted greenhouse cane in southern Texas (Legaspi and Irvine 1996). Although *P. simulans* occurs on cane further south in Mexico and Central America and is now clearly resident in Texas in the lower Rio Grande Valley, there have been no further reports of *P. simulans* on cane in the United States.

Spittlebugs are Major Sugarcane Pests in Mexico and Central America

Spittlebug damage to sugarcane is widespread and often serious in Mexico and Central America. Belize, where spittlebugs were first recognized as pests, still has economically important infestations (Cawich 1981). It is a small producer

TABLE 6.2. Major spittlebug pests of sugarcane by country

COUNTRY	PRIMARY PEST SPECIES
Belize	*Aeneolamia contigua*
Brazil	*Mahanarva fimbriolata*
	Mahanarva liturata
	Mahanarva posticata
Colombia	*Mahanarva andigena*
	Mahanarva bipars
	Mahanarva trifissa
Costa Rica	*Aeneolamia albofasciata*
	Aeneolamia contigua
	Prosapia plagiata
Ecuador	*Mahanarva andigena*
El Salvador	*Aeneolamia contigua*
Guatemala	*Aeneolamia contigua*
Guyana	*Aeneolamia flavilatera*
Honduras	*Aeneolamia contigua*
Mexico	*Aeneolamia contigua*
Nicaragua	*Aeneolamia contigua*
Panama	*Aeneolamia lepidior*
Suriname	*Aeneolamia flavilatera*
Trinidad & Tobago	*Aeneolamia varia*
Venezuela	*Aeneolamia varia*

but ranks second in the world in per capita sugarcane production (Table 6.1). At 29 million tons per year, Guatemala is a major sugarcane producer, tenth in the world. Belize, El Salvador, Honduras, and Nicaragua together produce about 22 million tons. Mexico produces 59 million tons per year and ranks sixth in world production, with 800,000 hectares under cultivation. Spittlebugs occur at pest levels in Mexico, Guatemala, El Salvador, Nicaragua, and Honduras. In each country, the primary and probably only economically significant spittlebug pest of sugarcane is *Aeneolamia contigua* (Plate 14f) (Table 6.2), also found in the literature under the synonym *Aeneolamia postica*.

Mexican production is spread over fifteen states, including the Pacific coastal plain, but is concentrated on the Gulf of Mexico in Tabasco and Veracruz States. Veracruz State alone accounts for about half of national production (García-González et al. 2017). In the drier Pacific Region, where the annual rain cycle supports only a single spittlebug generation per year, they are not important sugarcane pests (Castro 1981). In the Gulf area, *A. contigua*, with three generations per year, is the most damaging insect pest of cane, reducing yield up to 60 percent, and considerable effort is devoted to its control (Bautista-Gálvez et al. 2012). The species *Aeneolamia albofasciata* (Plate 14d) and *Prosapia simulans* also occur there, but usually account for two percent or less of the individuals sampled and are not of economic significance. Spittlebugs have been recognized as pests in Mexican cane since 1903 (Urich 1913) and are known locally as *mosca pinta* (spotted fly), but it appears that *A. contigua* became a serious problem only in the 1940s and 1950s (Hernandez and Flores 1956). As in Trinidad, Mexican producers experienced increasing problems with chemical insecticide resistance and now make widespread use of *Metarhizium* (Bautista-Gavlvez et al. 2012). This has stimulated research on improving the efficiency of *Metarhizium* treatment, concentrated on identification of effective, locally adapted strains (Bautista-Gavlvez et al. 2012).

To the south, Costa Rica has been the center of sugarcane spittlebug biocontrol efforts in Central America, an effort spearheaded by the Dirección de Investigación y Extensión de la Caña de Azúcar (DIECA), created in 1982 by a private consortium of growers and processors to undertake research, technology transfer, and technical assistance for cane farmers. Although Costa Rica is a relatively small producer, thirty-second among countries in tons produced (Table 6.1), it punches above its weight in sugarcane integrated pest management. In 1994, I spent a week with Carlos Sáenz, Coordinator of DIECA's Entomology Department, and other DIECA staff, collecting spittlebugs at multiple sugarcane plantations and observing aerial spraying of *Metarhizium*. We also visited the DIECA field station and biocontrol production facilities in Grecia, Alajuela Province.

In contrast to countries to the north, where a single spittlebug species, *A. contigua*, is by far the most important pest of cane, Costa Rica has multiple spittlebug sugarcane pests. In Guanacaste Province, at the lowland Finca Taboga and Finca CATSA estates, I collected *Aeneolamia albofasciata* in abundance in and around cane. It was clearly the primary pest species. *Prosapia simulans* was present but much less abundant and *A. contigua* was absent. In the highland sugarcane area around Juan Viñas, Cartago Province, I collected *A. contigua*, *Zulia vilior* (Plate 10f), and *Prosapia plagiata* (Plate 10c, 10d) as well as a few *P. simulans* in and around cane, but *P. plagiata* was said to be the important pest species there (Thompson and León 2005).

In both areas spittlebugs were treated with *Metarhizium*, either aerially or by ground blowers, building on expertise gained from Brazil. Carlos Sáenz had done his graduate studies in Brazil and had worked there after graduation in an agency dedicated to spittlebug control (Chaves 2004). Bibi Ali of Caroni visited

Costa Rica around 1997 to share some of her *Metarhizium* expertise. DIECA produces and markets *Metarhizium* spores on a large scale. In 1999 it produced 1.6 metric tons of pure fungus (Rodríguez and Sáenz 1999). A decade later, in 2019, it produced 20.9 tons, so the use of *Metarhizium* has continued to increase, presumably because it has proved economically effective.

The Costa Rican agricultural cycle is governed by alternation of a May-through-November rainy season with a December-through-April dry season. DIECA's integrated pest management system, which they actively promote to member farmers in simple publications, including comic book format, is geared to these seasons. As in Trinidad, the spittlebugs ride out the dry season in the soil as diapausing (dormant) eggs. January through March farmers monitor the density of these eggs as an early warning indicator of infestation levels. Around the onset of the rainy season, they work on weed control to eliminate alternative host grasses. In the early rainy season, they put out sticky traps to monitor adult infestation levels. This is followed, if necessary, by application of *Metarhizium* and mass sticky trapping of adults in the mid-to-late rainy season, followed, again if necessary, by chemical insecticide treatments in the late rainy season. The use of yellow sticky traps (*trampas ahesivas amarillas*) for mass capture of flying adults appears to be unique to Costa Rica. DIECA reported capture of 81,464,555 individuals in 262,900 sticky traps from 1991 to 1998, calculated to have averted the deposition of about three billion eggs (Sáenz et al. 1999). Others have used sticky traps to monitor spittlebug populations (Dongiovanni et al. 2023), but to my knowledge only Costa Rica has made mass use of traps to reduce levels of infestation.

Aeneolamia reducta (Plate 14e) and *A. lepidior* (Plate 15h) are present on cane in Panama (Pérez Milián et al. 2018; Atencio et al. 2019; Guerra et al. 2023), but, judging by the absence of an extensive literature on damage and control, they do not seem to do significant harm. This parallels the situation in the Southern Zone of the Pacific coast of Costa Rica, adjacent to Panama, where spittlebugs also do not seriously damage cane despite their harmful effects in the north of the country.

Spittlebugs Are Also Important Sugarcane Pests in Parts of Northern South America

In Venezuela, the closest part of mainland South America to Trinidad, *A. varia* is the second most important pest of cane (Box 1950; Ferrer 2021). Severe pest status appears to be localized in the central-west part of the country, where it is the most important insect limiting production (Box 1950; Vreugdenhil 1981; Figueredo et al. 2021). As in Trinidad, there are usually four generations per year, with the major difference that, unlike Trinidad, virtually all the cane is irrigated and grown year-round, with no defined planting or harvesting season (Pickles 1944). This apparently leads to more weakly delineated generational cohorts (Vreugdenhil 1981), complicating control. Spittlebug infestations, and, by extension, the bugs themselves, are known as *candelilla* (little fires), referring to the burned appearance of affected fields (Box 1950; Guagliumi 1954). Recorded infestations date to the 1940s and became serious by the early 1950s (Box 1950; Guagliumi 1954), although Pickles (1944) was told that damage dated back to the nineteen-teens. Like Trinidad and other cane growing countries, Venezuela has endeavored to find the right balance of chemical insecticides and *Metarhizium* in the face of insecticide resistance, combining the use of both (Ferrer and Torres 1984). Perhaps uniquely, they have also promoted the widespread use of a lab-reared entomopathogenic nematode, *Heterorhabditis bacteriophora*, first imported from Cuba, as a control measure against nymphs (Ferrer 2021).

In the 1950s, Italian entomologist Pietro Guagliumi made detailed studies of the taxonomy and distribution of Venezuelan sugarcane spittlebugs, as well as their alternative host plants. Guagliumi, born in Modena and trained in Bologna, worked for the Venezuelan Ministry of Agriculture from 1947 to 1963 (Comitê Organizador 1976). He produced an 850-page, two-volume opus on "*Las Plagas de la Caña de Azúcar en Venezuela*" (Guagliumi 1962). He also wrote substantial works on the biology of *Aeneolamia flavilatera*, *Aeneolamia reducta*, and *Aeneolamia lepidior* (Guagliumi 1956a, 1956b), none of which achieve pest levels on cane in Venezuela, and a longer paper on the major Venezuelan cane pest, *A. varia* (Guagliumi 1955a). Faced with variability in color pattern and minute details of their male genitalia, he recognized no fewer than twenty Venezuelan *A. varia* subspecies, several of his own creation.

To the east of Venezuela, *A. flavilatera* has been an on-again, off-again pest of sugarcane in Guyana (Williams 1918) and Surinam (Van Dinther 1960), dating back to the early twentieth century. In some years it is inconsequential. In others it does significant damage (Wiedijk 1982). This seems to depend on annual variation in rainfall. Occasional periods of drought are associated with subsequent outbreaks, perhaps as a result of suppression of natural enemies by the unusually dry periods. Unlike *A. varia* and *A. contigua*, *A. flavilatera* does not produce dry-season diapause eggs.

Surinam was also the site of the first reported attack on cane by the spittlebug *Mahanarva tristis* (Plate 15e) (Williams 1917), with a later attack said to cause losses of at least 20% in several fields (Van Dinther 1960). This species was later responsible for an isolated but severe attack on cane at one estate in Trinidad in 1939 (Pickles 1939) and has been reported to attack cane in Venezuela (Guagliumi 1955c). Although *M. tristis* has not been a regularly recurring pest of cane, these attacks were harbingers of the present role of *Mahanarva* species as the most important South American pests of sugarcane. The genus name, conferred by Distant (1909), apparently derives from Mahanarva, an early nineteenth century indigenous leader, who pressed for the rights of the Carib Makushi people in what is now Guyana. Given the tenacity with which some *Mahanarva* species attack cane and grasses, the generic name was prophetic.

Despite its role as a major sugarcane producer (third in the Americas, seventh in the world, Table 6.1) and close proximity to countries with big cane spittlebug problems, Colombia made it through the twentieth century without a recognized spittlebug sugarcane pest (Peck 2001; Castro et al. 2009). This outlier status suggests that Colombia's cane crop is vulnerable to spittlebug outbreaks given the right combination of species and circumstances. Three species, *Mahanarva bipars* (Peck et al. 2004), *Mahanarva trifissa*, and *Mahanarva andigena* have, in fact, recently achieved pest status in panela cane, the source of panela, a brown, unrefined cane sugar popular in Latin America. Colombia is the world's leading panela producer. Both *M. trifissa* and *M. andigena* appear to have been recently introduced to Colombia (Rodriguez and Peck 2007; Castro et al. 2009). In neighboring Ecuador, *M. andigena* has emerged as an important pest of sugarcane (Mendoza et al. 2004; Valle R. et al. 2015; Valle-Ramírez et al. 2022).

Perhaps a greater long-term threat comes from the recent discovery of *A. varia* (Plate 15b) in the Cauca Valley, where 80% of Colombian cane is grown. *Aeneolamia varia* is now an established local sugarcane pest (Castro et al. 2009; Ramirez-Gi et al. 2023). It may have been introduced from the Colombian Llanos on the other side of the Andes, where it is a pasture pest (Chapter 7). *Prosapia simulans* (Plate 14c), a Mexican-Central American pasture grass pest that may play a small role as a cane pest in Costa Rica, has also been collected at multiple sites in the Cauca Valley (Peck et al. 2001). I had an indirect hand in its discovery. Visiting the International Center for Tropical Agriculture (CIAT) in Cali, Colombia in 1997, I recognized two Venezuelan *P. simulans* specimens in their collection, an alert that *P. simulans* had reached South America. Shortly thereafter, CIAT colleagues discovered the Colombian pasture populations. So far, this introduction, while threatening to pasture grasses (see Chapter 7), does not seem to have led to outbreaks on cane. Nevertheless, with multiple newly established pests, Colombia's outlier status as the only major sugarcane producer in the Americas free of spittlebug pests has apparently come to an end.

Spittlebug Pests in Brazil: The World's Sugarcane Superpower

Destructive as spittlebugs are to sugarcane production in Mexico, Central America, and parts of northern South America, the harm done barely holds a candle to the damage wrought by *Mahanarva* species endemic to Brazil, the world's sugarcane superpower (Table 6.1), accounting as of 2024 for 51% of world sugar exports. *Mahanarva liturata* (Plate 15c) was the first reported major Brazilian spittlebug sugarcane pest and received a lot of attention in the early literature (Moreira 1925; Filho 1944). Since the 1960s it has been eclipsed in importance by *M. posticata*, *M. fimbriolata* (Plate 14g), and possibly *Mahanarva indentata* (Chaves et al. 2014), but is still present in cane fields (Alves and Carvalho 2014). *Mahanarva posticata* (found in the earlier literature, a bit confusingly, under the synonym *Mahanarva indicata*), had been a minor pest of cane in the Rio de Janeiro region since at least 1918 (Guagliumi 1968). Around 1960 it was accidentally introduced into Northeastern Brazil's sugarcane region (Silva et al. 2009), source of about a third of the country's immense production, where it proliferated and achieved serious pest status. In 1964 a spectacular outbreak in the State of Pernambuco damaged thousands of hectares and threatened to destroy the local sugarcane industry. This stimulated a private industry group, the Sugar and Alcohol Institute (IAA), to look for outside help. In 1966, in collaboration with the Brazilian Foreign Ministry and the United Nations, they engaged FAO entomologist Pietro Guagliumi, the Italian spittlebug expert we met in Venezuela.

Guagliumi drew on the cumulative experience of the long struggle with *A. varia* in Trinidad, filtered through the lens of his extensive personal work with the same species in Venezuela. In his initial appraisal of the *M. posticata* outbreak in Pernambuco and the neighboring state of Alagoas, he advocated a full-scale *campanha de luta biológica* (campaign of biological struggle), based on a detailed understanding of the biology of the spittlebug, its ecology, natural enemies, and host plants (Guagliumi 1968). In this, he consciously set out to build on the foundational work of Urich, Williams, Myers, and Pickles. He also recognized from the beginning that despite the limitations of chemical control, success would likely require complementary applications of the two approaches. His initial recommendation included three prongs of attack: 1) Dissemination of Brazilian parasites and predators, perhaps imported from the original area of *M. posticata* distribution in the Campos area of Rio De Janeiro State; 2) utilization of entomopathogenic fungi like *Metarhizium*; and 3) searches for foreign natural enemies, perhaps from the Old World (Guagliumi 1968).

Begun in 1969, the *campanha de luta biológica* against *M. posticata* initially concentrated on egg parasitoids and predatory flies. Due to poor field results in other regions of the world, fungi were not included (Li et al. 2010). However, that first year, the Brazilian agronomists Arthur Mendonça and Edmilson Marques, who had joined Guagliumi's team, encountered an epizootic of *Metarhizium anisopliae* on *M. posticata* in Sergipe, Brazil's smallest state, just south of Alagoas and Pernambuco (Silva et al. 2009). This observation, reinforced by the simultaneous occurrence of a *Metarhizium* outbreak in Alagoas State, and building on a background of Brazilian *Metarhizium* studies dating back to the 1920s, led to successful laboratory and field trials and the eventual incorporation of *Metarhizium* treatments into an effective, integrated control program (Li et al. 2010).

The use of *Metarhizium* expanded rapidly. Some states in the Northeast sponsored mass production of spores. In the larger São Paulo state growing area, farmers founded cooperatives in the 1970s to produce fungus for field application. In 1972, 500 hectares in the Northeast were treated with fungus. A decade later 22,000 hectares were treated. Carlos Sáenz, later a key developer of the Costa Rican *Metarhizium* program, studied at the Federal University of Pernambuco (1975–1981) and then worked for the local spittlebug control commission (1981–1982) just as these developments took full hold (Chaves 2004). In the state of São Paulo, the area treated increased from 2,000 hectares in 1977 to 85,000 hectares in 1982. By 2008, the cane area treated nationally for spittlebug control reached an estimated 750,000 hectares, astonishing growth over four decades (Li et al. 2010). *Metarhizium anisopliae* had finally moved from being the most important natural control of *A. varia* (Pickles 1933) to being an effective instrument of spittlebug pest control in sugarcane.

As it happened, the development of successful *Metarhizium* treatment had come just in time. São Paulo state has the largest concentration of sugarcane in the world and accounts for over half of Brazil's national production. In the 1990s, concerns over air pollution led growers to begin phasing out the annual burning of sugarcane fields before harvest, a traditional practice necessary for hand harvesting with machetes. This change in practice, now codified in law, has had two large side effects. First, it has required a massive shift from hand cutting to mechanical harvesting, with major implications for rural labor markets. Second, it has led to large, chronic, and widespread infestations of *Mahanarva fimbriolata*. This spittlebug had been a minor pest in the region for decades, but its numbers had been kept under control by the annual fires, which destroyed diapausing eggs laid in or near the surface soil layer under the cane. With the cessation of burning, egg survival increased dramatically, and the nymphs that hatched had a happier, moister home in a cane root layer now permanently mulched by dead leaves and other cane detritus (Dinardo-Miranda and Fracasso 2013).

The result was a spectacular increase in *M. fimbriolata* numbers. Beginning in the mid-1990s, it became the most important pest in the largest sugarcane growing area in the world. Ironically, Brazil had imported a strain of *Metarhizium* from Trinidad in 1925 to test against *M. fimbriolata*, an effort that failed (Li et al. 2010) perhaps due to a mismatch between the exotic Trinidad isolate and the local pest. Now, with their successful experience in the Northeast against *M. posticata*, Brazilian agricultural scientists were well-positioned to unleash a campaign against the new *M. fimbriolata* eruption, and this led to the extraordinary expansion of *Metarhizium* application described previously.

Recently, *Mahanarva spectabilis* (Plate 10a), a serious pest of pasture grasses (Chapter 7), has appeared in cane fields in the states of Goias, Piauí, and Maranhão (Alves and Carvalho 2014; Silva et al. 2020) and may pose yet another spittlebug threat to the Brazilian sugar industry. This threat is compounded and complicated by the recent discovery of a new pest species, physically resembling and closely related to *M, fimbriolata* and *M. spectabilis*. Its geographical range overlaps the ranges of these species, raising the prospect that two or three of these species may co-occur in cane fields, particularly in São Paulo State and adjacent areas (Andressa Paladini, personal comm. 2025).

Brazil produces 42% of the world's sugarcane while using only 25% of the nitrogen fertilizer applied to cane, apparently due to Brazilian varieties' strong association with endophytic nitrogen-fixing bacteria (Robinson et al. 2011). The acute susceptibility of Brazilian sugarcane varieties to spittlebugs may be due, in part, to this association (Thompson 2004). It is a local manifestation of the more general attraction of spittlebugs to associative nitrogen-fixing C4 grasses (see more on sugarcane and nitrogen fixation in Chapters 3 and 10). Fortuitously, the use of *Metarhizium* for biological control may itself reinforce nitrogen supplies. Some entomopathogenic fungi, including *Metarhizium anisopliae*, live in both soil and plant tissue and convey nitrogen from dead insects to plants (Behie and Bidochka 2014). This raises the intriguing possibility that *M. anisopliae*, which associates with sugarcane roots, both protects sugarcane against spittlebugs and feeds the sugarcane nitrogen from some of the bugs it kills (Thompson 2023b).

Why Did Spittlebugs Only Emerge as Sugarcane Pests in the Twentieth Century?

We don't know. Columbus brought sugarcane to Hispaniola in 1493 and it spread throughout the New World Tropics hundreds of years before suffering major attacks by spittlebugs. The earliest reports of spittlebug outbreaks on cane did not appear until 1913 in Mexico, 1918 in Guyana, 1925 in Brazil, and 1929 in Argentina and Peru (Myers 1931), as well, as we have already seen, in 1883 in Belize and 1889 in Trinidad. If it is correct that sugarcane was not introduced to Belize until 1848 (Latha and Bautista 2020), then it took less than forty years for the local pest, *Aeneolamia contigua*, to reach pest levels. This is in contrast to other countries where spittlebugs became pests only in the twentieth century though sugarcane had been introduced many centuries earlier (for example, to Mexico in 1520, Brazil in 1532, and Trinidad in 1632). Why did it take so long? It was not because it took time for the spittlebugs to find the cane. All the spittlebug cane pests are localized and indigenous, following the general pattern of sugarcane insect pests, which are overwhelmingly local and of limited geographical distribution (Strong et al. 1977). However, it appears anomalous that it took the local spittlebugs so long to reach pest status when most introduced crops achieve saturation with local pests within a few decades of introduction (Strong et al. 1977). The pattern in sugarcane is in sharp contrast to the pattern in pasture grasses, where spittlebugs became pests shortly after the introduction of new African species that proved to be acutely vulnerable to spittlebug attack (Chapter 7).

Economic Damage Estimates

Since the early days of the antifroghopper campaign in Trinidad, there have been attempts to quantify losses due to spittlebug attacks on cane. Urich (1913) projected average annual losses of 10% of the sugar yield, or £54,613 per annum, $8.8 million USD adjusted for inflation. In the bad year of 1917 Williams (1921) estimated a loss of about 18%, at an economic value of $30 million USD, similarly adjusted. Since then, losses have been estimated in multiple ways, including loss of sugarcane tonnage, decline in cane quality, and loss of sugar and ethanol production. Trinidad chemical control era yield loss estimates range from 38% (Fewkes et al. 1965) to 50% (Norton and Evans 1974). In Belize yield losses to *A. contigua* are estimated at 10% (Latha and Bautista 2020). In Venezuela, yield losses for severe *A. varia* infestations are 30% or more in second-generation outbreaks (Ferrer and Torres 1984), while in Colombia losses have been measured at 17% (Ramirez-Gi et al. 2023). In Ecuador losses due to *Mahanarva andigena* have been as great as 40 to 60% (Valle-Ramírez et al. 2022).

In Brazil about half of all sugarcane production is fermented to ethanol as a gasoline fuel supplement. *Mahanarva fimbriolata* reduces sugarcane productivity by 20 to 40 tons per hectare (Kassab et al. 2014), which, given an average yield of 74.7 tons per hectare (Table 6.1), represents a yield loss of 26.7 to 53.5%. But, in addition to lowering yield, *M. fimbriolata* lowers sucrose accumulation and raises the level of fiber and phenolic compounds, which hinder fermentation. As a result, the loss of sugar and ethanol production is greater than the loss due to simple reduction in yield (Ravaneli et al. 2006, 2011; Madaleno et al. 2008). Correspondingly, in Guatemala, *A. contigua* infestation reduced sugar extracted per ton of cane by 14.5% or more, even though cane yield was not affected (Salguero-Navas et al. 2000). Twenty years ago, I summarized local loss estimates and calculated a rough global loss estimate of $350 million per annum (Thompson 2004, table 1). Given inflation, a large increase in Brazilian acreage over time, and the concomitant increase in *M. fimbriolata* infestation, the magnitude of this loss has undoubtedly increased considerably.

Old World Sugarcane Spittlebug Pests Are Few and Far Less Destructive

India is Brazil's only serious rival in sugarcane production (Table 6.1). However, in contrast to Brazil and the tropical Americas overall, spittlebugs are minor pests of cane in India. Only two species, *Aphrophora sigillifer* (*Peuceptyelus sigillifer*) and *Callitettix versicolor* (Plate 12a), achieve sporadic pest status (David and Nandagopal 1986), and a third species, *Ptyelus nebulosus*, has been reported at pest levels once (Tripathi 1986). *Callitettix versicolor* is also reported to be a pest in China, the fourth largest sugarcane producer (Liang 2020), and has been listed as a sugarcane pest in Myanmar and Thailand but apparently not of any severity. *Abidama producta* (Fig. 8.3a) is listed as a minor pest of sugarcane in Thailand and parts of Myanmar, but as a significant local pest in Sagaing Region, Myanmar (Asia and Pacific Plant Protection Commission 1987; Morris and Waterhouse 2001). In Bangladesh, a *Clovia* species is a local sugarcane pest (Abdullah and Mannan 2011; Ahmed et al. 2019), and in Malaysia a *Clovia* species (perhaps *C. sarawakana*) is listed as a "major pest" but said to cause only "slight damage" (Lim and Pan 1980). There is an old report of *Poophilus costalis* (Fig. 8.3f, Plate 13g) damaging sugarcane in Taiwan (Schumacher 1915).

In Australia, *Euryaulax carnifex* is a "minor, restricted, irregular" pest of sugarcane (Bailey 2007). Africa appears to be free of reported spittlebug sugarcane pests. With the exception of *C. versicolor*, Old World spittlebugs are not important, widespread pests of sugarcane.

Controlling Spittlebug Infestations in Sugarcane

The large economic losses due to spittlebug infestations have evoked considerable effort to hold their populations to economically sustainable levels. We reviewed some of those efforts in Trinidad toward the beginning of this chapter. They include adjusting cultivation practices, application of chemical insecticides, selecting and breeding resistant varieties, treatment with entomopathogenic nematodes, and, most prominently, treatment with *Metarhizium* (Ferrer and Torres 1984).

Controlling weedy grasses in and adjacent to sugarcane fields is an important first step toward reducing spittlebug infestations. All spittlebug sugarcane pests are native to the areas where they attack cane or to areas not too far away. They all have alternative grass hosts, either the native grasses on which they evolved or, in the Americas, introduced Old World pasture grasses (Chapter 7). Consequently, minimizing weedy grasses in the sugarcane field ecosystem is a simple but effective first line of attack. In addition to weed control, removal of some or all of the trash in mechanically harvested fields can markedly reduce *M. fimbriolata* infestation, though with tradeoffs for loss of positive effects of trash mulching (Castro et al. 2019).

A second line of attack is the use of chemical insecticides, but, as we saw in the antifroghopper campaign in Trinidad, chemical control agents have serious limitations. They are expensive to apply, require careful monitoring and timing to synchronize with the cycle of multiple generations, and are subject to the development of genetic resistance by local spittlebug populations. Moreover, they may cause collateral damage to agricultural field workers and local residents, and, by killing natural enemies of spittlebugs, they undermine the natural limitations on spittlebug populations. This is especially important in sugarcane, a semiperennial crop, which, once planted, goes through several annual cuttings, with "ratoon" cane growing back from the base of cut cane. Such an environment favors a multiyear ecosystem of ants, spiders, wasps, predatory flies, and heteropteran bugs that help keep pest populations in check. Indiscriminate chemical treatments that kill them all can result in "flashback," or resurgence, in which spittlebugs, freed from these checks, return in even greater virulence.

From the beginning, the weaknesses of chemical control led to interest in biological control alternatives. Much of the pre-Second World War effort in Trinidad was devoted to seeking natural insect predators and parasites that would hold the populations in check. Neither these efforts nor the later efforts by Guagliumi and his associates in Brazil toward the same end succeeded. They encountered two big problems. First, it was difficult to raise control agents like predatory flies out of season. Larvae of the spittlebug-specific syrphid fly *Salpingogaster nigra* are voracious predators of spittlebug nymphs, but this makes them difficult or impossible to raise during the dry season when nymphs are rare or absent (Ferrer and Torres 1984). Second, despite diligent searches by Williams, Myers, Pickles, and others, including extension of the search to Africa (Greathead 1970), they never found introducible parasites with the desired host specificity.

Metarhizium anisopliae filled the gap. It has achieved widespread adoption throughout the Neotropics as an effective, economical, ecologically sustainable, spittlebug biocontrol agent. In Mexico, three doses of *Metarhizium* reduced *Aeneolamia* infestation by 81 to 99% and permitted a reduction of 75% in insecticide use (Bautista-Gálvez and González-Cortes 2005). In Ecuador, two doses reduced infestation of *Mahanarva andigena* by 91% (Cruz and Cajilima 2012). In Brazil, the large-scale use of *Metarhizium* against spittlebugs has become the largest pest control program in the world based on entomopathogenic fungi (Li et al. 2010; Mascarin et al. 2019). In most cases, *Metarhizium* has been used to best effect in conjunction with modest, narrowly targeted chemical applications. For example, a detailed study combining *Metarhizium* and chemical treatments of *M. fimbriolata* showed that although chemical treatments often provide greater control, the two most economically effective combinations were *Metarhizium* alone and *Metarhizium* plus low-dose chemical treatment (Kassab et al. 2014). Nymph-destroying nematodes have been tested against spittlebug pests (Chapter 4), but so far, with the exception of Venezuela (Ferrer 2021), they do not seem to have been applied widely in the field.

There has also been substantial effort to identify sugarcane cultivars (genetically uniform, vegetatively reproduced varieties) that are resistant or tolerant to spittlebug attack. Some cultivars decrease spittlebug reproduction and development (making them more resistant). Others better maintain yield in the face of attack (making them more tolerant). For example, Dinardo-Miranda et al. (2018) tested twenty cultivars for resistance and tolerance to *M. fimbriolata*. They found substantial variation in the degree to which varieties attract spittlebugs, stimulate egg deposition, and promote survival of nymphs to maturity. Cultivars under spittlebug attack also differed in their ability to add aboveground biomass, roughly equivalent to cane yield. Nonetheless, they found no strains that could be considered truly resistant, since all supported *M. fimbriolata* proliferation. Nor were any truly tolerant, since all experienced substantial yield reduction. Their results are characteristic of screening tests for resistance and tolerance to spittlebug sugarcane pests. Cultivars do vary in spittlebug susceptibility, suggesting utility in planting less susceptible strains, but so far none offer an economical alternative to chemical and biological control. Recent work suggests that *Saccharum robustum*, the closest wild relative of the primary ancestor of cultivated sugarcane (Zhang et al. 2019), has resistance traits that drastically reduce *M. fimbriolata* population growth (Orozco-Restrepo et al. 2017). If so, this species offers a promising genetic resource for breeding truly resistant varieties.

Many Sugarcane Spittlebugs Are Also Serious Pests of Pasture Grasses

This chapter has reviewed the important role spittlebugs play as widespread, economically destructive pests of sugarcane in the Neotropics. From Mexico to Brazil, many sugarcane spittlebugs double as important pests of pasture grasses. The dual sugarcane-pasture grass pest group includes *A. varia*, *A. contigua*, *P. simulans*, *P. plagiata*, and *M. spectabilis*, as well as other species in the genera *Aeneolamia* and *Mahanarva*. The complex of species attacking pasture grasses is even more diverse than those attacking cane, and damage is more widespread. These spittlebug pasture pests are the subject of the next chapter.

7

THE WORST PASTURE PESTS IN LATIN AMERICA

Grazed pastures planted with introduced African grasses constitute the single largest use of land in the Neotropics. Spittlebugs, including some of the same species that attack sugarcane, are by far the most important pests of these pastures. This chapter begins with the history of spittlebugs and pasture grasses in Monteverde, one small but well-studied locality in Costa Rica. It continues with a historical overview of the relatively recent "Africanization" of Neotropical grasslands, taking Brazil as an example, and putting the evolution of Latin American spittlebug grass pests in the context of larger social and economic developments. This is followed by a review of the major New World spittlebug pasture pests by genus and major geographical area, from the Southeastern United States to Northern Argentina.

Monteverde, Costa Rica: Where Spittlebugs Almost Sank the Dairy Industry

Monteverde, Costa Rica and surrounding highland areas were settled relatively late, largely in the early 1950s, by migrating Costa Rican farmers seeking land and expatriate American Quakers seeking refuge from the draft and military service. Costa Rica had abolished its military following a civil war in 1948. Monteverde is far up in the mountains at about 1,400 m. Roads were rudimentary. As in much of Latin America, cattle were an answer to the problem of farming economically marginal habitats distant from markets. They do not require large cash investments, just clearing land, planting grass, and buying initial stock. In the absence of workable roads, the product can walk all or part of the way to market. On the other hand, the added economic value of beef on the hoof is modest. The early Monteverde settlers, many of whom were raised on dairy farms in the United States, realized that economics favored an easily transported, value-added product that complemented the other virtues of cows. They built a cheese factory that became the nucleus of a thriving dairy industry (Griffith et al. 2000).

Dairy cows need abundant grass. Monteverde farmers planted Kikuyu, *Cenchrus clandestinus* (formerly *Pennisetum clandestinum*), an African highland grass that had recently been introduced and popularized in several high-elevation areas in the American tropics (Parsons 1972). It thrived in the local environment, but by the late 1950s spittlebug blight appeared in scattered pockets. By the late 1960s whole pastures were damaged, and by the early 1970s milk production was in serious decline (Griffith et al. 2000; Stuckey and Stuckey 2021). Spittlebug damage to grass is similar to their damage to sugarcane (which itself is a large grass). The nymphs typically feed on the roots at the plant/soil interface. At high density they divert nutrients and stunt plant growth. But, as in cane, the adults do the bulk of the damage, creating dying streaks from the puncture points where they feed on leaves. In heavy attacks they may kill the entire above-ground portion of the plants, turning pastures yellow, or *quemado*, which is Spanish for "burned" (see Byers and Wells 1966 for a detailed description of the effects caused by *Prosapia bicincta* adults feeding on coastal Bermuda grass).

By the early 1970s, Monteverde dairy farmers responded to spittlebug attacks on Kikuyu by switching to Estrella, *Cynodon nlemfuensis*, which is more tolerant of spittlebugs but less productive. It is also less friendly to cows and visiting entomologists, tangling feet in its creeping stolons. This innovation was introduced and promoted by Kenny Wallace, an Alabama Quaker who moved to Monteverde to farm in 1961 and eventually became manager of the cheese factory. Along with the simultaneously introduced practices of rotational grazing of cows among paddocks and application of chemical fertilizers, the substitution of Estrella for Kikuyu "spread like wildfire throughout the area" (Stuckey and Stuckey 2021) because it solved an immediate problem and was easy to implement. All you had to do was plant seeds. This, as I discovered when I first visited Monteverde in 1987, saved the local dairy industry (see Preface).

We owe most of our knowledge of the Monteverde spittlebug pest complex to Daniel Peck (see sidebar: Daniel Peck's Contributions to the Biology Underlying Spittlebug Control), who carried out his PhD dissertation fieldwork in Monteverde from 1991 to 1994 (Peck 1998a, 1999). The most abundant, conspicuous, and destructive Monteverde spittlebug is *Prosapia tilaranensis* (Plate 14h), a species that was not formally named and described until many years after Peck's work (Castro-Valderrama et al. 2020). It shares the Monteverde pastures with three other spittlebugs: *Prosapia simulans*, *Prosapia plagiata*, and *Zulia vilior* (Plates 10c, 10d, 10f, 14c). Despite the replacement of Kikuyu, spittlebugs are still abundant in Monteverde pastures but adequately controlled by the planting of Estrella, the paddock system, and systematic fertilization.

Daniel Peck's Contributions to the Biology Underlying Spittlebug Control

Daniel Peck (Plate 16a, 16b) began studying spittlebug pests as a Cornell graduate student. I visited him in Monteverde in 1993 while he was carrying out fieldwork there. New PhD in hand, he went on to work from 1996 to 2002 at the Centro International de Agricultura Tropical (CIAT) in Cali, Colombia, where he and his collaborators produced a small avalanche of detailed studies on the biology of multiple spittlebug pests, especially those of pastures, but also emerging sugarcane pests. Peck's work benefited greatly from tying into CIAT's ongoing program to breed grasses resistant to spittlebugs and from a steady stream of local undergraduate thesis students ready to carry out detailed field studies. He also had the support of Cesar Cardona (1941–2013), commemorated in the species name *Ocoaxo cardonai* (Plate 15d), foresighted leader of the CIAT breeding program. Peck's work established baseline studies for spittlebug pest biology that remain a model. He is still active in the field, presently contributing to combat against a spittlebug pest recently introduced to Hawaii (Wilson et al. 2023, 2024).

The Monteverde dairy industry had its own twist on changing patterns of processing plant ownership, which was more or less opposite that of the Trinidad sugar industry described in Chapter 6. In 2013 a Mexican subsidiary of an international company bought the cheese factory—which had been locally owned—cut the labor force in half, and expanded production. So, in contrast to Trinidad, where an iconic company integral to the life of the community was bought by the government from foreigners, run at a loss, and precipitously closed, the Monteverde cheese factory was bought by private sector foreigners from local producers. It is still in operation, tied to the regional agricultural community as an employer and purchaser of milk.

The Bigger Picture: Spittlebugs and the "Africanization" of Neotropical Pastures

Monteverde is a small place. It now relies far more heavily on ecotourism than dairying, but the local outbreak of spittlebugs on Kikuyu was not an isolated development. In the same mid-twentieth century period, spittlebugs badly damaged Kikuyu plantings in Brazil (Lepage and Monte 1942), in Peru (Bullón 1956), and closer to Monteverde on the slopes of Costa Rica's Poás Volcano (Washbon 1968; Vargas Picado 1970). These outbreaks were a harbinger of the second of two dramas featuring Neotropical spittlebug pests of introduced grasses. The first drama, starring sugarcane, commenced a few decades earlier and played out in Trinidad before moving to other areas, including Mexico, Central America, Northern South America, and Brazil (Chapter 6). The second drama, with a revolving cast of introduced African grasses, played out on an even broader stage, stretching from the Southeastern United States to Argentina. The spittlebug plague of Kikuyu grass in Monteverde has been reenacted many times over as native spittlebugs of the genera *Aeneolamia*, *Deois*, *Mahanarva*, *Notozulia*, *Prosapia*, and *Zulia* have wreaked destruction on pasture yield and carrying capacity. The Monteverde outbreak offered in microcosm the most important characteristics of spittlebug pests in pasture forage grasses in the tropical and subtropical Americas. One, the spittlebugs are native to the localities where they occur. Two, they attack introduced grasses, most originating in Africa. Three, the adults do the heavy damage. Four, multiple spittlebug species usually co-occur in the same pastures. And five, damaging attacks have driven recurring replacement cycles in the grass species planted.

To understand this second drama, we need to understand the singular nature of Latin American cattle pastures. Nothing like them exists on a comparable scale anywhere else in the world. First, the planted area is enormous. In 2020 in Brazil alone, planted pasture covered 1,132,243 km^2 (IBGE 2022), equaling 113 million hectares or 280 million acres. Second, the vast majority of these pastures are planted in introduced grasses. In sharp contrast, planted pastures play almost no role in cattle production in the African continent from which these grasses originate (Ndikumana and Leeuw 1996). Third, spittlebugs have evolved as the major constraint on the productivity of Neotropical pastures (Peck 1999; Valério et al. 2001; Jank et al. 2014; Tolotti et al. 2018; Espitia B. et al. 2022), making spittlebugs "the most damaging pasture pests in tropical America" (Valério et al. 1996). Last, there have been successive efforts to breed or introduce grass cultivars that resist spittlebug damage, leading to substantial turnover in the grasses planted over time. The net result has been both increased pasture productivity and an enduring search for more resistant cultivars (multiple articles in Miles et al. 1996; Jank et al. 2014; Pereira et al. 2018; Tolotti et al. 2018).

Brazil as a Case in Point

Brazil is the world's largest beef exporter and, after the United States, the second largest cattle producer. About 200 million cattle occupy Brazilian pastures. They are divided between the 113 million hectares of planted pasture and roughly 60 to 70 million hectares of unimproved native pasture (Jank et al. 2014; Pereira et al. 2018). Introduced African grasses have been important pasturage since the colonial period. The first introduction, Guinea grass (*Megathyrsus maximus*, but known widely in the literature as *Panicum maximum*), was accidental, apparently in the 1700s, as possibly part of straw bedding in ships transporting enslaved people from Africa (Parsons 1972; Jank et al. 2014). It proved to be more palatable to livestock than native grasses, more productive, and better adapted to grazing. It spread quickly, both as an aggressive weed and by human agency. Other African grasses followed in the colonial period, including Pará grass, *Urochloa mutica* (formerly *Brachiaria mutica/Panicum purpurascens*), Molasses grass (*Melinus minutiflora*), and Jaraguá (*Hyparrhenia rufa*). They too proved superior to native grasses and were more aggressive, likely the result of long evolution under the hooves of the African megafauna (Parsons 1972; Jank et al. 2011). All of the imported African grasses have C4 photosynthesis and most, if not all, give evidence of root-associated nitrogen fixation, probably accounting for part of their superiority to native grasses and their attractiveness to spittlebugs (Thompson 2004, and see Chapter 3 on the import of C4 grasses and nitrogen fixation). Over time, their advantages led to the "Africanization" of the grass cover in the Neotropics (Parsons 1972). Combined with large-scale deforestation, this created the most familiar sight in rural Latin America: grazing cows in green pastures, catalyzed by the advent of productive African grasses. By one estimate, such grassland covered nearly 40% of the tropical Americas, excluding Amazonia, by the early 1970s (Parsons 1972).

Table 7.1 summarizes the major planted forage grasses involved in this drama. Taxonomic revision by botanists over the past few decades has resulted in important name changes, most notably the shift of *Brachiaria* species into the genus *Urochloa*. I give the earlier scientific names for reference, since many are still more familiar to agronomists and, at least in the case of *Brachiaria*, have crept into common parlance among farmers and ranchers. The older names are also most common in the literature. I would fault the botanists for a

TABLE 7.1. Important introduced pasture grasses of the New World tropics and subtropics

SPECIES	FREQUENT NAME IN EARLIER LITERATURE	COMMON CULTIVAR	COMMON NAMES: ENGLISH	SPANISH	PORTUGUESE
Andropogon gayanus			Bluestem	Andropogon	Andropógon
Bothriochloa pertusa			Hurricane grass	Colosuana	
Cenchrus ciliaris			Buffel grass	Buffel	Búfalo, Bufel
Cenchrus clandestinus	*Pennisetum clandestinum*		Kikuyu grass	Kikuyu	Quicuio
Cenchrus purpureus	*Pennisetum purpureum*		Elephant grass	Elefante	Elefante
Cynodon dactylon			Bermuda grass	Bermuda	Bermuda
Cynodon nlemfuensis			African star grass	Estrella	Estrela africana
Cynodon plectostachyus			Giant star grass	Estrella	Estrela africana
Dicanthium aristatum			Angleton grass	Dicantio	
Digitaria didactyla			Blue couch	Suazi	Suázi
Digitaria eriantha	*Digitaria decumbens*		Pangola grass	Pangola	Pangola
Hemarthria altisima			Limpo grass	Canita	Gamalote
Hyparrhenia rufa			Jaragua grass	Jaraguá	Jaraguá
Megathyrsus maximus	*Panicum maximum*	Colonião	Guinea grass	Guinea	Colonião
Megathyrsus maximus	*Panicum maximum*	Mombaça	Mombaça guinea grass		Mombaça
Megathyrsus maximus	*Panicum maximum*	Tanzânia	Purple guinea grass	Guinea	Tanzânia
Megathyrsus maximus	*Panicum maximum*	Zuri	Zuri grass		Zuri
Megathyrsus maximus hybrid		Quênia			Quênia
Megathyrsus maximus hybrid		Tamani			Tamani
Melinis minutiflora			Molasses grass	Gordura	Melado, Gordura
Urochloa arrecta	*Brachiaria arrecta*		Tanner grass	Tanner	Tanner
Urochloa brizantha	*Brachiaria brizantha*	Marandu	Palisade grass	Marandu	Marandú
Urochloa brizantha	*Brachiaria brizantha*	Paiaguás	Palisade grass		Paiaguás
Urochloa brizantha	*Brachiaria brizantha*	Piatã	Palisade grass		Piatã
Urochloa brizantha	*Brachiaria brizantha*	Xaraés	Palisade grass		Xaraés
Urochloa decumbens	*Brachiaria decumbens*	Basilisk	Signal grass	Braquiaria	Braquiária Decumbens
Urochloa dictyoneura	*Brachiaria dictyoneura*	Llanero	Coronivia grass		
Urochloa humidicola	*Brachiaria humidicola*	Tully		Humidicola	Humidícola Quicuio, Agulha
Urochloa mutica	*Brachiaria mutica* *Panicum purpurascens*		Para grass	Pará	Pará, Angola
Urochloa ruziziensis	*Brachiaria ruziziensis*	Kennedy	Congo grass Ruzi grass	Congo	Congo Ruziziensis
U. arrecta × *U. mutica*			Tangola grass	Tangola	Tangola
U. ruziziensis × *U. brizantha*		Ipyporã			Ipiporã

Note: I am indebted to José Raul Valério for helpful input on newer Brazilian cultivars and Portuguese common names.

proliferation of confusing name changes, but nomenclatural revisions are a normal part of taxonomy, and spittlebug taxonomists have caused their own confusion with name changes in the past. Most of the spittlebug species attacking sugarcane and pasture grasses were originally included in the catch-all genus *Tomaspis*, only to be parceled out to newly created genera over time (Fennah 1948; Carvalho and Webb 2005).

The developing African grass economy fits a rural Brazil dominated by poorly capitalized small holders. For them, cattle are an ideal investment (Smith et al. 1995). Once land is cleared, pastures are easy to sow, usually after a first year of annual crops. While crops can spoil in the field, cows are a low-risk, liquid investment, cashable when a crisis or opportunity arises. When the market is down, they can be held indefinitely. When good, they can be herded to point of sale despite bad roads. Compared to agricultural fieldwork, they require minimal labor, and ranching is culturally more prestigious than farming. Furthermore, a few milking cows are good for the family and provide a bit of cash income. No wonder then, that cattle and cattle pastures are an entrenched part of Brazilian life.

The relative stability of pasturage based on the early introductions was upended mid-twentieth century by the advent of new and even more productive African grasses. Pangola grass (*Digitaria eriantha*, formerly *Digitaria decumbens*) was introduced to Brazil from Africa via Florida in the 1950s (Parsons 1972). This was followed by the introduction of *Urochloa decumbens* (formerly *Brachiaria decumbens*) from Uganda via Australia in the late 1960s (Keller-Grein et al. 1996; Jank et al. 2014). In many environments these grasses outproduced the earlier African imports, but, like Kikuyu in Monteverde, they proved to be exquisitely susceptible to spittlebug attack. More new grasses were to come. *Urochloa humidicola* (formerly *Brachiaria humidicola*), originally from South Africa, was released for commercial use in 1980. It is spittlebug tolerant, meaning that it can sustain a modest spittlebug infestation without serious damage, but it allows spittlebug populations to grow. The same is true of *Urochloa dictyoneura* (formerly *Brachiaria dictyoneura*), released in 1987 and originally from Zambia. *Urochloa brizantha* cv. Marandu (formerly *Brachiaria brizantha* cv. Marandu) stood out from the rest. Originally from Zimbabwe and released in 1984, it proved to be spittlebug resistant. By killing or retarding the development of nymphs, Marandu greatly reduced spittlebug populations (Keller-Grein et al. 1996). The abbreviation cv. stands for "cultivar," a true breeding, genetically uniform line.

The cycle of introductions drove a cycle of replacement of dominant planted grasses, the details determined by local ecology. *Urochloa decumbens* cv. Basilisk was the first grass planted in the Savannah-like Cerrados, where its adaptability to acid, low phosphorus, high aluminum soils and persistence in a green state in the dry season gave it advantage (Valério 1988). However, its susceptibility to

TABLE 7.2. Thirty-year turnover in planted pasture grasses in one area of Brazilian Amazonia (1965–1995)

FIRST GENERATION 1965–1975	SECOND GENERATION 1975–1982	THIRD GENERATION 1983–1995
Megathyrsus maximus	*Urochloa humidicola*	*Urochloa brizantha*
Urochloa decumbens		*Andropogon gayanus*
Hyparrhinia rufa		

Source: Smith et al. 1995.

spittlebug attack led to its subsequent displacement by the spittlebug resistant Marandu. In wetter environments there was often an intermediate stage in which the spittlebug tolerant grass *U. humidicola* displaced *U. decumbens*, only to be displaced itself by Marandu (Valério et al. 1996, 2001). Smith et al. (1995) give a timeline for pasture species turnover during a 30-year period in one area of Brazilian Amazonia (Table 7.2). Fazolin and Costa (2009) present a more recent example in Acre State, where earlier *Urochloa* species plantings were replaced by a combination of grasses, including Tanner (*Urochloa arrecta*) and Tangola (a Tanner-Pará hybrid).

The situation on the ground is complicated by the fact that each grass species has different strengths and weaknesses. *Urochloa decumbens* is very good at coping with poor soils. Marandu is resistant to spittlebugs but requires better drainage and higher soil fertility, limiting it to more favorable areas. *Urochloa humidicola* tolerates poor soils but has lower forage quality than *U. decumbens* or *U. brizantha*, and its spittlebug tolerance has limits (Keller-Grein et al. 1996; Valério et al. 1996, 2001). As a result, multiple species persist in use, and regional plantings are a mosaic of grasses, including species other than *Urochloa*. However, *Urochloa* species accounted for 85% of Brazil's planted pastures in 2014, with Marandu alone accounting for 50 million hectares. It was said to be "the world's largest monoculture in terms of area" (Jank et al. 2014).

Despite the spittlebug challenges, the spread of *Urochloa* cultivars had direct and tangible effect on ranching productivity. In 1940 it took, on average, about 2.5 hectares to support a cow. At the beginning point of wide introduction of *U. decumbens* in 1970, that number had dropped to about 2.2 hectares. Following that introduction and the mass planting of multiple *Urochloa* species, it dropped to 1.1 hectares in 2006 (Jank et al. 2014), an astonishing increase in productivity over a vast area (Fig. 7.1). Such an extraordinary development must have big ramifications for the rural areas where pasture dominates, but the ultimate economic and ecological consequences are unclear (see sidebar: Does Spittlebug Control Mitigate or Exacerbate the Deforestation of Amazonia and the Cerrados?). From a biological point of view, the existence of monocultures over vast

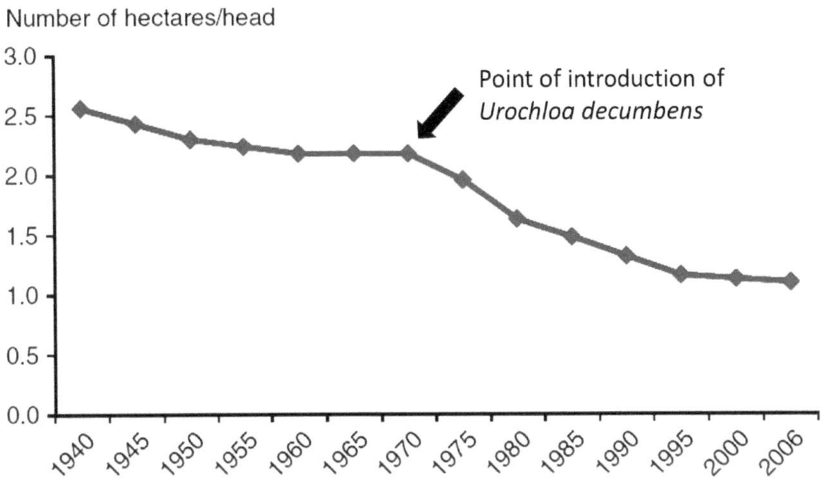

FIGURE 7.1. Number of hectares of pasture per head of cattle in Brazil 1940–2006, a measure of pasture productivity. There is a distinct inflection point following the widespread introduction of *Urochloa decumbens* in the late 1960s, which led to a major increase in the carrying capacity of Brazil's pastures. These data illustrate the power of pasture grass introductions to alter pasture productivity over a vast area in a relatively short period of time. Reproduced from Jank et al. (2014) with permission of CSIRO Publishing. Arrow and text indicating inflection point added.

areas raises warning signals (Valério et al. 2001). Almost all of the widely planted grass cultivars are *apomictic polyploids*, meaning that they reproduce clonally by asexual seed production, making them genetically uniform (Jank et al. 2011). This makes the plantings potentially vulnerable to plant pathogens and to evolving adaptation by spittlebugs.

The result has been an ongoing search for new forage species and cultivars, either through the screening of new accessions from wild populations in Africa or the breeding of new apomictic strains in the lab. It has also led to new interest in the genetic resources of grasses native to the Neotropics, particularly in the genus *Paspalum* (Jank et al. 2011), which are less susceptible to the native spittlebugs, probably because they coevolved with them (Thompson 2004). Almost all of this work has been carried out at a handful of research centers, including CIAT (Centro International de Agricultura Tropical) in Cali, Colombia, and multiple Empresa Brasileira de Pesquisa Agropecuária (Embrapa) facilities in Brazil, including Embrapa Beef Cattle in Campo Grande, Embrapa Dairy Cattle in Juiz de Fora, and Embrapa Cerrados in Brasilia (Jank et al. 2011, 2014). CIAT is an independent nonprofit organization. Embrapa is a government owned corporation affiliated with the Brazilian Ministry of Agriculture. Valério et al. (2001) provide a snapshot of these efforts at the turn of the century.

Does Spittlebug Control Mitigate or Exacerbate the Deforestation of Amazonia and the Cerrados?

In the short term, most newly cleared forest and Cerrado land is converted to planted pasture. To the extent that spittlebug-resistant grass cultivars increase the value of newly cleared land, spittlebug control would seem to drive the economics toward the loss of natural habitat. On the other hand, more productive land requires less acreage per cow (Fig. 7.1), potentially allowing for a more efficient cattle industry on less land, weakening the long-term pressures for conversion. These contradictory influences are playing out at the local level over vast reaches of rural Brazil. The latest detailed government land use analysis shows conflicting patterns (IBGE 2022). There is actually an ongoing net *loss* of presently planted pasture to crop and agroforestry uses (Jank et al. 2014), but it is countered by conversion of forest patchworks to planted pasture, leading to an increase in planted pasture from 1.125 million km^2 in 2018 to 1.132 million km^2 in 2020, a 0.6% gain. Crop cultivation is expanding at the expense of ranching, but deforestation goes on, adding to the pasture ecosystem at a slightly faster rate than the loss to crop farming. The largest part of the loss of pasture to crop use is taking place in the Cerrados, with a substantial additional amount in the border areas of Amazonia (IBGE 2022, map 4). These phenomena are also characteristic of other areas in Latin America (Graesser et al. 2015). In the best of outcomes, effective spittlebug control will lead to more intensive and productive ranching on less land, relieving pressure on remaining natural habitats (Smith et al. 1995; Jank et al. 2014; Pereira et al. 2018). In the worst, it will reinforce a "grass rush," incentivizing continuing relentless destruction of remaining natural habitats (Sternberg 1973).

Major New World Spittlebug Pasture Grass Pests

Now that we have the big picture, we can review the major New World spittlebug pasture pests. There are many, and the literature on their occurrence and control is vast, most of it in Spanish or Portuguese. As noted above, all of the significant pests fall into six genera. I have arranged these genera one by one from north to south in order of geographical occurrence. Species within genera are arranged likewise. This is a somewhat arbitrary approach to sorting out an enormously complicated reality on the ground, but it serves as a useful framework, keeping in mind that Neotropical pastures often harbor multiple genera and species of spittlebug pests, sometimes as many as six species in appreciable numbers (Sá 1981). Espitia B. et al. (2022) provide a hemispheric overview of Neotropical spittlebug pasture and sugarcane pests. For country-by-country overviews of pasture pests, see Guagliumi 1954, 1955a, 1956a, 1956b, (Venezuela); Peck 2001 (Colombia and Ecuador); Thompson and León 2005 (Costa Rica); Valério 2009 (Brazil); Foieri 2017 (Argentina); and Tolotti et al. 2018 (Brazil). Table 7.3 presents a summary list of important pasture pest spittlebugs by country. Readers without an interest in species-by-species detail may skip to the end sections of the chapter without fault.

The Genus *Prosapia*

Except for recent introductions in South America and Hawaii, *Prosapia* is confined to the North American continent and the Caribbean Islands of Cuba and Jamaica. It has the northernmost distribution among the New World Cercopidae, including two species in the United States north of the US–Mexico border area (Thompson and Carvalho 2016). Its center of diversity lies in Mexico (Armendáriz-Toledano et al. 2022). It includes four significant pasture pest species. Major taxonomic references include Hamilton 1977, Carvalho and Web 2005, and Armendáriz-Toledano et al. 2022.

Prosapia bicincta: *A Pest of the American Southeast, Cuba, and Hawaii*

Prosapia bicincta (Plate 10e) is a pasture pest, sometimes serious, in Florida and Georgia, where it attacks Coastal Bermuda grass (*Cynodon dactylon*), Limpograss (*Hemarthria altissima*), and Pangola grass (Byers and Wells 1966; Fagan and Kuitert 1969; Newman et al. 2003). It is also a pest of turf grasses, including St. Augustine grass (*Stenotaphrum secundatum*) and Centipede grass (*Eremochloa ophiuroides*) (Braman and Pendley 1993). Mid-1990s data collected in Georgia comparing chemically treated and untreated plots suggests annual forage yield losses attributable to *P. bicincta* of 5 to10% and damage of $1.9 million to lawns and turf. Control has focused on identifying resistant cultivars (Stimmann and Taliaferro 1969) and, in the case of turfgrasses, resistant alternative species (Shortman et al. 2002).

TABLE 7.3. Confirmed spittlebug pasture grass pests by country

COUNTRY	PRIMARY PEST SPECIES
Brazil	*Aeneolamia flavilatera*
	Aeneolamia selecta
	Deois flavopicta
	Deois incompleta
	Deois schach
	Notozulia entreriana
	Mahanarva spectabilis
	Mahanarva tristis
Colombia	*Aeneolamia reducta*
	Aeneolamia varia
	Zulia carbonaria
Costa Rica	*Aeneolamia albofasciata*
	Aeneolamia contigua
	Aeneolamia reducta
	Prosapia simulans
	Prosapia plagiata
	Prosapia tilaranensis
	Zulia vilior
Cuba	*Prosapia bicincta*
French Guiana	*Aeneolamia flavilatera*
Honduras	*Aeneolamia contigua*
Mexico	*Aeneolamia albofasciata*
	Aeneolamia contigua
	Prosapia simulans
Nicaragua	*Aeneolamia contigua*
Peru	*Notozulia entreriana*
United States	*Prosapia bicincta*
Venezuela	*Aeneolamia varia*
	Aeneolamia lepidior
	Aeneolamia reducta

Note: This includes only species that have been explicitly identified in the literature as damaging pasture within the country or, in the case of Nicaragua and Honduras, observed by the author at pest levels. It is almost certain that spittlebugs occur in many other Central and South American countries at pest levels and that many of the pests listed here occur more widely than this list would suggest.

Prosapia bicincta occurs in Cuba in the form *P. bicincta fraterna*, which has distinctive broken, wavy dorsal wing bands. As in Georgia and Florida, it is a pest of *Cynodon dactylon*, with yield losses up to 56% (Barrientos et al. 1991). In addition, it attacks Kikuyu (Barrientos 1983), with old reports as a pest of Pará (Myers 1931) and *Echinochloa colonum* (Williams 1921). *Prosapia biciincta fraterna* occurs on sugarcane, but not at pest levels (Williams 1921; Fernández and Ramos 1986). A second Caribbean form, *P. bicincta basalis*, occurs in Jamaica but appears to have no economic importance.

Recently, *Prosapia bicincta* has been introduced to the Kona area of Hawaii Island, where it has spread rapidly, destroying pasture grasses and devastating the beef cattle industry (Wilson et al. 2023, 2024). Pastures that had verdant Kikuyu and Pangola grass plantings have been turned into scruffy mixtures of invasive weeds and low-quality forage. First detected in 2016, *P. bicincta* occupied 72,000 hectares by 2021 and threatens damage to more than 800,000 hectares of range and agricultural lands across the state. Hawaii has no native spittlebugs, but *P. spumarius* and *Clastoptera xanthocephala* were introduced during the twentieth century (Table 10.1) (Thompson et al. 2023). Neither of these species has caused discernible harm.

Prosapia bicincta adults exhibit an unusual host preference. While its nymphs are strictly grass feeders and feeding adults do the serious damage to grass hosts, adult males also regularly feed on hollies (genus *Ilex*) and cause commercial harm to nursery stock (Braman and Ruter 1997; Nachappa, Guillebeau, and Braman 2006). Peck (1998b) observed a similar phenomenon in Monteverde, where immense numbers of *Prosapia tilaranensis* and *Prosapia simulans*, apparently all male, congregated on native *Ilex haberi* trees left for cattle shade in Estrella pastures. Something unknown, innate to hollies, attracts feeding male *Prosapia*.

Prosapia simulans: *Northern Mexico Through Panama and Recently Introduced Into Colombia and Venezuela*

Prosapia simulans (Plate 14c) occurs in pastures from the Lower Rio Grande Valley through Mexico and Central America to Panama. It reaches its northern limit in the southernmost tip of Texas but is not associated with economic damage in the United States. It generally co-occurs with one or more other spittlebug species in mixed species assemblages. In Mexico through Nicaragua, it coexists with *Aeneolamia albofasciata* (Oomen 1975; Clark et al. 1976) and/or *Aeneolamia contigua* (Cruz-Zapata et al. 2016) to form the *mosca pinta* spittlebug pest complex (a specifically Mexican term, Coronado 1978). In Costa Rica it co-occurs with *A. albofasciata*, *A. contigua*, *Aeneolamia lepidior*, *Aeneolamia reducta*, and *Zulia vilior* in pastures at low to medium elevations in combinations of two to four species (Thompson and León 2005).

Recently, *Prosapia simulans* was introduced into South America, a fact that came to light when I stumbled across specimens from Venezuela in the CIAT insect collection in Cali in 1997. Shortly thereafter, it was found in numbers in Colombia, including at potentially pest levels in one *U. decumbens* pasture where it co-occurred with *Zulia carbonaria* and *Zulia pubescens* (Peck 2001; Rodriguez et al. 2003). This is the only known instance in which a tropical spittlebug has been introduced into a new continent (see Chapter 10).

Prosapia tilaranensis: *The Primary Monteverde Pest*

This species (Plate 14h) is covered above in the section on Monteverde. Though it occurs in other high-elevation areas in central Costa Rica, its agronomic importance seems to be confined to higher elevations in the Tilarán mountain range surrounding Monteverde (Thompson and León 2005).

Prosapia plagiata: *A Pest of High-Elevation Costa Rican Pastures*

In 1949 the Costa Rican Ministry of Agriculture received reports of Kikuyu pastures on the slopes of Poás Volcano turning yellow and then brown during the rainy season. The blight was diagnosed to be due to *Prosapia plagiata* (Plate 10c, 10d) a spittlebug that occurs at high elevations and is limited to Costa Rica and perhaps adjacent Panama (Washbon 1968; Vargas P. 1970; Fagan and Vargas Picado 1971; Thompson and León 2005). Less studied than the other *Prosapia* pest species, it is one of the secondary spittlebugs in pastures in Monteverde but occurs in abundance as the dominant spittlebug in high elevation pastures in the middle of the country. A closely related form of uncertain taxonomic identity is abundant in high-elevation pastures in Coto Brus in southern Costa Rica and in adjacent Panama (Thompson and León 2005). In contrast to *P. plagiata*, which is typically brown (Plate 10c) with 1 to 10% of the individuals of both sexes bright red (Plate 10d), some of the Coto Brus populations are 100% red.

The Genus *Aeneolamia*

Aeneolamia includes seven major pasture pests, spanning North and South America from southern Arizona to Brazil. Several species are destructive of both pasture grasses and sugarcane (Chapter 6). Major taxonomic references include Carvalho and Webb 2005 and Armendáriz-Toledano et al. 2022, 2023.

Aeneolamia albofasciata: *Scourge of Mexican Buffel Pastures*

Aeneolamia albofasciata (Plate 14d) occurs from southern Arizona to Costa Rica and reaches further north than any other *Aeneolamia* species. It is a serious pest of Buffel (*Cenchrus ciliaris*), an African/Asian grass introduced to Mexico and adjacent parts of the United States in the 1950s, supplanting vast areas of native range with planted pasture (Enkerlin and Morales 1980). While Buffel is an order of magnitude more productive than native range, it is highly susceptible to spittlebugs (Clark et al. 1976). In Sonora and Northeast Mexico, *A. albofasciata* is the primary and sometimes sole pest of Buffel (Ooman 1975), especially in Sonora where it was first reported in 1981 (Martin et al. 1995). Substantial effort has gone into screening for resistant strains, but control by summer burning seems to offer the most practical control (Martin-R. et al. 1999).

At the opposite end of its range in Guanacaste Province, Costa Rica, *A. albofasciata* is a pest of sugarcane but apparently not an important pasture pest (Thompson and León 2005). Why it attains serious pest status in just two regions, on different hosts, at dead opposite ends of its large distribution, is a mystery. The two environments, Pacific lowland Costa Rica and Sonoran Desert Mexico, are united only in having prolonged and acute dry seasons. Across the rest of Central America and Mexico, *A. albofasciata* is just one of three species in the "*mosca pinta*" complex, and rarely, if ever, a consequential pest on its own.

The Texas Range Border Invasion That Did Not Happen

In May 1974, *The Cattleman*, a trade publication of the Texas and Southwestern Cattle Raisers Association, published a screaming headline: "Spittle Bug Plague Threatens South Texas Range" and predicted that "an insect reported less than 50 miles south of the Texas-Mexican border could turn out to be the most economically devastating plague to hit South Texas in this century . . . threatening to leave large expanses of South Texas range looking like it had been ravaged by a prairie fire" (Brown 1974). This reflected the fact that, like Northern Mexico, large tracks of South Texas range had been planted in Buffel and could be vulnerable to the same sort of attacks that had been ravaging Northern Mexico. It never happened. While iNaturalist records leave no doubt that *A. albofasciata* has jumped the border into Arizona and been recorded as far north as Tucson, there is no record of the species in Texas, even while *P. simulans* has made it, harmlessly, so far, to the US side of the lower Rio Grande Valley. So, at least for now, the Texas range spittlebug nightmare has been averted.

Aeneolamia contigua: *Widespread Mexican and Central American Pest of Pastures and Sugarcane*

A bit further south in Mexico, *P. simulans* and *A. albofasciata* overlap in distribution with *Aeneolamia contigua* (Plate 14f), the primary pest of sugarcane from Mexico through Nicaragua (Chapter 6). *Aeneolamia contigua* (*A. postica* in much of the older literature) has been recognized as a grass pest in Mexico since at least 1876 when local "growers reported that when the pest was at its height they could not dine in peace, because the insects, attracted to light, frequently landed in their saucers" (Coronado 1978, my translation). It was first studied as a pest in the early 1900s and came under serious scrutiny as a result of pasture outbreaks in the 1950s (Coronado 1978). It is now the most important constituent of the *mosca pinta* pasture spittlebug complex through most of eastern and southern Mexico. The 1876 record makes *A. contigua* the first documented Neotropical spittlebug pest, beating Belizean cane spittlebugs by six years (Chapter 6).

Although cited as the most injurious insect pest of Mexican Gulf Coast area pastures (Flores and Velasco 1974; Velasco 1975), *A contigua* has received relatively little separate attention as a forage pest, perhaps in part because its similar banded pattern makes it difficult for a nonexpert to separate from *A. albofasciata*. This has led to many studies performed on *Aeneolamia* spp., or just *mosca pinta*. Recently, *A. contigua* has been studied on weedy grasses abutting sugarcane (García-González et al. 2017), where it generally coexists with *P. simulans* (Cruz-Zapata et al. 2016). Color polymorphism also muddies the waters. In Mexico (Armendáriz-Toledano et al. 2022, fig. 13a, 13b) and Central America north of Costa Rica (my observations), *A. contigua* occurs in multiple color forms, sometimes with the typical yellow-orange bands, sometimes with red bands, and sometimes polymorphic for the two. In Costa Rica, *A. contigua* is common in pastures over much of the country excluding the Pacific Coast (Thompson and León 2005) and, as already noted for Monteverde, is invariably mixed with one or more other species, including *A. albofasciata*, *P. simulans*, *P. plagiata*, and *Z. vilior*.

Aeneolamia reducta *and* Aeneolamia lepidior: *A Commonly Co-occurring Species Pair in Costa Rica, Panama, Colombia, and Venezuela*

Aeneolamia reducta (Plate 14e) and *A. lepidior* (Plate 15h) both range from southern Costa Rica (Thompson and León 2005) through Panama to

Colombia and Venezuela and often co-occur in pastures. The pair has been studied by Guagliumi (1956b) in Venezuela and Peck and his colleagues in Colombia (four articles in a special number of *Pasturas Tropicales*, Vol. 24, No. 1, 2002). Guagliumi notes that both species co-occur in Venezuela with *A. varia* and/or *A. flavilatera* and that both have been observed to cause occasional localized but severe damage to planted pastures. In Venezuela, Khan and Mark (1983) report that *A. reducta* attacked a wide range of introduced grasses planted in preparation for establishment of a pasture-seed industry, doing severe damage to Buffel, while Linares and Pérez (1985) report that it damaged *Digitaria eriantha*, *U. decumbens*, and *Cynodon plectostachyus* at economic levels.

In Colombia, *A. reducta* and *A. lepidior* are the only pasture spittlebugs in the Caribbean Coast region, where *A. reducta* causes serious damage to Hurricane grass (*Bothriochloa pertusa*) and also feeds on Angleton grass (*Dichanthium aristatum*) as a preferred host (Peck et al. 2002). Interestingly, these two grasses are native to India and adjacent parts of Asia, making exceptions to the general rule that grasses badly damaged by Neotropical spittlebug pests are African in origin. *Aeneolamia lepidior*, in contrast, favors Pará. Of the two, *A. reducta* is clearly the more widespread and important pest and, as such, has been used in multiple studies of grass resistance to spittlebug damage (Aguirre et al. 2013). Holmann and Peck (2002) estimate that it causes economic losses of up to $176 million per year in the Caribbean Coast region.

Recently, in a major extension of its known range, *A. reducta* has been recorded for the first time in the Cauca Valley region of southeastern Colombia (Hernández et al. 2021) and in the north Brazilian state of Roraima (Fidelis et al. 2021). Though widely separated, these areas are both well south of the previously recognized distribution. In Roraima, *A. reducta* severely attacked *U. humidicola* and *U. brizantha* cv. Xaraés.

Aeneolamia varia: *Not Only a Pest of Sugarcane but a Serious Pasture Pest in Colombia and Venezuela*

As we saw in Chapter 6, *Aeneolamia varia* (Plate 15b) had a long and storied history as a sugarcane pest in Trinidad and, to less drama, in Venezuela. In Trinidad, it has done notable damage to pasture grasses only twice: once in 1918 to an unidentified grass on one estate (Williams 1921) and in 1960 to a large Pangola pasture (Wilson et al. 1962). Any lingering threat to Trinidad's pastures is inconsequential. Only 1.4% of its land area is devoted to permanent pasture. Not so on the adjacent mainland, where, by the 1950s, *Aeneolamia* pasture pests had become more important economically than *A. varia* in Trinidad sugarcane (Guagliumi 1954).

Guagliumi (1955a) devoted a 59-page paper to the distribution and characteristics of 20 named *A. varia* subspecies in Venezuela, where the species often co-occurs with *A. reducta* and *A. flavilatera*. *Aeneolamia varia* is well-named. In modern terms, these subspecies are best interpreted as an exuberance of local variants in an abundant, widespread, and highly variable species. Guagliumi (1955a) reports that half of his named subspecies caused significant damage to pastures, without specifying the grasses damaged. Linares and Pérez (1985) report that *A. varia* causes economic damage to *U. decumbens*, *Cynodon plectostachyus*, and an *Axonopus* species in central western Venezuela. *Axonopus* and *Paspalum* are the only native genera commonly planted in Neotropical pastures. The pest spittlebugs rarely do serious damage to these native grasses with which they coevolved (Thompson 2004), although they may damage *Axonopus catharinensis* in Santa Catarina state in southern Brazil (Ribeiro et al. 2022).

The oldest record of *A. varia* damage to pasture grasses dates to Colombia in 1913. Pastures near Bogota were reported to have suffered "for many years" from the ravages of "*el mion*" (bedwetter), the local word for spittlebugs, specifically *A. varia bogotensis* (Williams 1921). In the 1970s and 1980s, when large swathes of the Colombian Llanos were planted with *Urochloa* species, livestock productivity initially increased more than tenfold, but spittlebug attack soon diminished these spectacular returns (Rudel et al. 2015), and *A. varia* emerged as a serious regional pasture pest. Spittlebug damage made "large areas planted with *U. decumbens* totally unusable for grazing, and in some cases they [were] completely lost" (Jiménez 1978). Along with *Zulia pubescens*, *Aeneolamia varia* is also one of the two primary spittlebugs damaging grasses in the Colombian Amazonia Piedmont, contributing to estimated annual losses of up to $37 million (Holmann and Peck 2002).

The alarming susceptibility of *Urochloa* species and other grasses to spittlebug attack stimulated three decades of intensive work at CIAT, near Cali, Colombia, to screen and breed for forage grasses resistant to spittlebugs, particularly to *A. varia*, but also *A. reducta* and *Zulia carbonaria*. This work, led for many years by Cesar Cardona, John Miles, and Stephen Lapointe, resulted in the successful development of several *Urochloa* cultivars with resistance based on antibiosis to nymphs of multiple pest spittlebugs. They drastically reduced nymphal survival. However, none of the cultivars reduced adult survival, and the quest for cultivars of commercially important grasses that can fully resist the Colombian suite of pasture pest spittlebugs at all life stages is open ended (Cardona et al. 2004; Aguirre et al. 2013; and references therein).

Collecting in the Shadow of the *Narcotraficantes*, Cali 1997

I visited the CIAT spittlebug and forage grass breeding operation outside Cali in July 1997, an occasion memorialized in a photo taken with Daniel Peck (Carvalho and Webb 2005, plate 16c). Colombia was going through a difficult period. Competing drug groups vied for power and influence. The two most notorious were the Cali and Medellín cartels. The Cali Cartel started out as a kidnapping racket that reinvested profits into cocaine trafficking. At one time it controlled a large majority of cocaine sales in the United States and Europe (Washington 2001). Its local presence necessitated security precautions unusual for entomologists in the field. One day, we made a chancy excursion to Popayán, a beautifully preserved colonial city south of Cali, and collected in pasture at an agricultural field station along the way. It is the only time I have netted spittlebugs under the immediate protection of armed guards carrying automatic weapons. On another day we set out to collect in the vicinity of Cali but had to weigh the risks of falling prey to freelance kidnappers. For security, we elected to collect along a road high above Cali, home to the mansions of leading members of the Cartel. Freelancers, we conjectured, would think twice about nabbing victims who might, by virtue of location, be connected to the Cartel. The collecting went without incident. I still have the specimens from both adventures.

Aeneolamia flavilatera: *A Widespread Species in Northern South America but Rarely a Pasture Pest*

Aeneolamia flavilatera occurs from eastern Venezuela to French Guiana. Like *A. varia*, it is morphologically variable. Guagliumi (1956a) recognized six subspecies. In Venezuela it is usually mixed with but less numerous than *A. varia* and occasionally co-occurs with *A. reducta* or *Mahanarva tristis* but never seems to be responsible on its own for significant damage to grasses. Although it is a recognized pest of sugarcane in Guyana and Suriname (Wiedijk 1982), it is not a significant pasture pest in those countries, and any damage done in Venezuela appears to be in an auxiliary role to *A. varia*. However, it has recently surfaced in Roraima State in Brazil, where it caused severe damage to a large planting of *U. brizantha* cv. Xaraés (Fidelis et al. 2021). This expansion to a new area combined with the destruction of a cultivar bred for resistance to spittlebugs merits close attention. In French Guiana, *A. flavilatera* has been reported to attack the African grass *Digitaria didactyla* (Rémillet et al. 1982).

Aeneolamia selecta *and* Aeneolamia colon: *Outliers in Brazil*

Aeneolamia selecta and *Aeneolamia colon* (Plate 15f) are outliers in the genus *Aeneolamia*—the only species that reach Northeast Brazil, the only ones confined to a single country, and the least studied of putative *Aeneolamia* pasture pests. *Aeneolamia selecta* coexists in numbers with *Deois schach* in Pangola and *Urochloa* pastures (Térano et al. 1989) and has severely damaged Pangola, with spillover damage to adjacent sugarcane (Guagliumi 1970). It is also an important pest in parts of Amazonia (Fontes et al. 1995). *Aeneolamia colon* has received less study. It has been reported in numbers on *U. decumbens* in Bahia and on *U. decumbens* and *M. maximus* in Pará State (Pereira 2013; Sousa et al. 2021), but has not been directly associated with damage in either case.

The Genus *Zulia*

The genus *Zulia* is distributed from southern Mexico to northern South America. The name derives from the Venezuelan state of Zulia (Fennah 1948). The only recent taxonomic treatment is in Carvalho and Webb (2005).

Zulia vilior: *The Most Widespread Costa Rican Pasture Species*

Zulia vilior (Plate 10f) is common in pastures all over Costa Rica except at high elevations and in the coastal area of Guanacaste Province. It co-occurs with other Costa Rican pasture spittlebugs and probably contributes widely to pasture damage but has received little targeted study (Thompson and León 2005). It likely also occurs in parts of Nicaragua and Panama adjacent to Costa Rica. In a test of *Metarhizium anisopliae* for spittlebug control, it was the dominant spittlebug present in a southern Costa Rican *Urochloa* ruziziensis pasture, accompanied by *A. lepidior* and small numbers of *A. reducta* (Gomez 2005).

Zulia pubescens: *A Widespread Species of Northern South America*

Zulia pubescens (Plate 14a) is found across much of northern South America, with the possible exception of Venezuela, and seems to be the most widespread spittlebug in Colombia, occurring over a range of habitats (Peck 2001). It has been studied in the lab for life cycle (Rodriguez et al. 2002) and susceptibility to antibiosis by pasture grass cultivars (Cardona et al. 2004), but it has not received concentrated attention in the field. Like *Z. vilior*, it is an ecologically versatile but understudied pest.

Zulia carbonaria: The Dominant Pasture Pest in the Cauca Valley, Colombia

In contrast to *Z. vilior* and *Z. pubescens*, *Zulia carbonaria* (Plate 14b) has been subject to considerable study, both of its biology (Arango and Calderón 1981; Rodríguez et al. 2002; Castro et al. 2005; Rodríguez and Peck 2006) and as a test subject for pasture grass resistance to spittlebugs (Cardona et al. 2004; Aguirre et al. 2013). In the Cauca Valley, where it is the dominant spittlebug species, it has damaged *Urochloa* pastures since at least 1978 and achieves strikingly high densities, up to 984 nymphs/m^2, one of the highest densities ever recorded for any spittlebug (Thompson 2021). Its main occurrence in Colombia is in the inter-Andean valleys, but it occupies a localized area in Caquetá and is also found in Bolivia, suggesting that it might occur in Peru and Ecuador.

The Genus *Mahanarva*

Mahanarva is a sizable genus with thirty-eight species, ranging from Southern Mexico to Argentina (Schöbel and Carvalho 2021). Its species are bigger than other Neotropical pasture spittlebugs, and historically, *Mahanarva* species have been much more important as sugarcane pests than pasture pests (Chapter 6). In recent years, *Mahanarva* gained prominence as pests in Brazilian pastures, making particularly serious attacks on Marandu, the most widely planted forage grass (Valério 2009; Pedreira et al. 2014). Apparently, *Mahanarva* species are undeterred by Marandu's toxic effects on the nymphs of other spittlebugs. In a greenhouse experiment, Congio et al. (2020) found that a mixture of *Mahanarva fimbriolata* and *Mahanarva liturata* adults reduced Marandu yield by 66%. In a parallel Marandu field trial, they found that an unidentified *Mahanarva* species, mixed roughly 9:1 with unidentified spittlebugs of other genera, produced a spectacular 74% reduction in beef productivity. There is speculation that by suppressing competing spittlebug species, Marandu actually gives *Mahanarva* species an ecological opening to proliferate (Valério et al. 2001; Valério 2009). Only two *Mahanarva* species have been recorded specifically as pasture pests, and only one, *Mahanarva spectabilis*, has received substantial research attention. As noted in Chapter 6, the genus is named for a South American indigenous leader. The only recent general taxonomic treatment is Carvalho and Webb (2005).

Mahanarva tristis: Widespread but Rarely a Pest

Mahanarva tristis (Plate 15e) is distributed across lowland South America from Trinidad to Uruguay (Schöbel and Carvalho 2021), which is probably the widest distribution of any New World Cercopidae species. It seems to have achieved pasture pest status only once, in the state of Acre, Brazil, on intercropped Tangola and Tanzania grass, in combination with *Notozulia entreriana*, after Marandu declined locally due to a fungus-like root infection (Fazolin and Costa 2009).

Mahanarva spectabilis: A Newly Emerging Major Pest in Brazil

Mahanarva spectabilis (Plate 10a) has emerged, seemingly out of nowhere, to become, over the last two or three decades, a major pasture pest in Brazil. Previously uncommon, it has exploded in numbers and done serious damage to the otherwise spittlebug resistant grass cultivar Marandu (Pedreira et al. 2014). It is now the most important limit on productivity in dairy and beef farms that rely on Elephant grass, *Cenchrus purpure* (formerly *Pennisetum purpureum*), for cut fodder and rotational grazing systems (Auad et al. 2007). This has stimulated a large amount of work on its biology and control (refs. in Alvarenga et al. 2019). In one example, *M. spectabilis* has been reported to occur at damaging levels in Acre on two cultivars of *U. brizantha*: cv. Xaraés, now the most widely planted local grass due to disease-driven declines in Marandu (Santos et al. 2021) and cv. Piatã (Santos and Silva 2023). Both these cultivars are resistant to the common pasture spittlebugs *Z. entreriana* and *D. flavopicta*, confirming that resistance measured against particular spittlebug species does not automatically imply resistance against all species.

In response to the growing vulnerability of Marandu, Embrapa is promoting Ipyporã, a *U. ruziziensis X U. brizantha* hybrid cultivar that resists attack by *Mahanarva* as well as *N. entreriana* and *D. flavopicta* (Valle et al. 2017). Likewise, one of the latest Guinea grass cultivars, cv. Zuri, is highly resistant to all tested spittlebugs, including *M. spectabilis* (Jank et al. 2022). Work with *Mahanarva* infestations has been complicated by the difficulties of identifying *Mahanarva* species. *Mahanarva spectabilis*, *M. fimbriolata*, and *M. liturata* share some very similar color forms (compare Plates 10a, 14g, 15c), and they are about the same size (Tolloti et al. 2018), making it hard to separate them without dissection of male genitalia. This may have led to misidentification of *M. spectabilis* as *M. fimbriolata* in some studies. In a further complication, a newly discovered species closely related to *M. fimbriolata* and *M. spectabilis* overlaps their ranges and probably co-occurs with them locally in pastures (see Chapter 6).

Spittlebugs that have been identified as *Mahanarva spectabilis* are found throughout most of Brazil (Tolloti et al. 2018; Schöbel & Carvalho 2021).

Mahanarva fimbriolata: *A Sugarcane Pest That Sometimes Spills over to Pastures*

Mahanarva fimbriolata (Plate 14g) is the major pest of sugarcane in Brazil but also occurs in grasses. It has received little attention as a potential pasture pest, but Bernardo et al. (2003) report that it co-occurs with *Deois flavopicta* in *U. brizantha* pastures in middle-north Mato Grosso State, representing 13 to 37% of the spittlebugs in those populations. However, it appears this might have been a mistaken identification and that these are actually *M. spectabilis* (Jank et al. 2022), or perhaps the newly discovered look-alike species for both (see Chapter 6). *Mahanarva fimbriolata* has been collected in small numbers in an ungrazed *U. decumbens* pasture in Bahia, but these were interpreted as spillover from an immediately adjacent sugarcane planting (Pereira 2013). Grisoto et al. (2018) have tested *M. fimbriolata* for resistance to a variety of forage grasses in anticipation that the species might become a significant pasture pest in the future.

The Genus *Deois*

In concert with *Notozulia entreriana*, with which they regularly co-occur, *Deois* spittlebugs are by far the most important pasture pests in Brazil, where they infest millions of hectares of African grasses, primarily species of *Urochloa*, planted in the Cerrados and deforested areas of Amazonia. They are the major pests of these grasses and the single major limiting factor in regional pasture productivity. They are also important pests in other parts of Brazil. The genus is large, with 24 species (Carvalho and Texeira 2011), all confined to South America, including three important pest species and others that may have pest potential. Valério 2009 and Tolotti et al. 2018 illustrate *Deois* species and their geographical distributions for Brazil, Foieri 2017 for Argentina. The only recent general taxonomic treatment is Carvalho and Webb 2005.

Deois incompleta: *A Major Pasture Pest in Brazil, Particularly in Amazonia*

Myers (1935) recorded *Deois incompleta* (Plate 15g) on weedy grasses near houses in the famous Fordlandia settlement in the lower Amazon. With the introduction of *Urochloa* grasses it became the major pasture pest in the Amazon basin (Penny and Arias 1982; Valério 1988) and in the humid forest area of

Pernambuco (Ramos 1988). Despite its ongoing importance, it has been the subject of remarkably little study as a separate species, those studies limited to the 1980s (Magalhães 1982; Magalhães et al. 1984; Ramos 1988). In that respect it has been far eclipsed by *D. flavopicta* and *N. entreriana.*

Deois flavopicta*: One of the Two Most Important Pasture Pests in Brazil*

Almost invariably sharing pastures with *Notozulia entreriana, Deois flavopicta* (Plate 15a) is the most important pasture pest of the Cerrados of Central Brazil (Valério 1988; Fontes et al. 1995; Sujii et al. 2002). It is one of the two most widespread, damaging pasture spittlebugs in South America, contending for that title with *N. entreriana.* As a consequence, it is the subject of an enormous body of literature, often in conjunction with *N. entreriana* (Valério and Koller 1982; Gusmão et al. 2016). Pires et al. (2000) give an excellent overview of its biology and its evolution from a very low-density, harmless spittlebug of natural Cerrado grasses to a high-density pest of planted *Urochloa* pastures.

Deois schach*: A Major Pasture Pest in Northeast Brazil*

Deois schach is an important pasture pest in parts of Amazonia and Northeast Brazil (Netto et al. 1989; Fontes et al. 1995). It is also abundant in Santa Caterina state, where it shares pastures with *D. flavopicta* and *N. entreriana* (Chiaradia et al. 2013; Ribeiro and Cazarotto 2019). Along with *D. flavopicta,* it has been used to test the resistance of native *Paspalum* grass species to spittlebug attack (Bernardes et al. 2022).

The Genus *Notozulia*

Notozulia entreriana*: The Second of the Two Most Important Pasture Pests in Brazil*

Notozulia entreriana (formerly *Zulia entreriana*) (Plate 10b) is the only member of its genus. Along with *D. flavopicta,* it is one of the two worst pasture pests in Brazil, as well as the most intensively studied. It occurs throughout central and southern Brazil and adjacent countries and is the subject of an enormous and scattered literature, much of it involving joint studies with *D. flavopicta.* In addition to presently attacking many species of *Urochloa,* it damaged Kikuyu in Peru near Machu Picchu in 1953 (Bullón 1956), anticipating by a handful of years the outbreak on Kikuyu in Monteverde, and in 1984 it was reported as a pest of Buffel in Minas Gerais state (Reis et al. 1984). It was the subject of an influential experiment showing that an infestation of 25 adults/m^2 on *U. decumbens* reduced per hectare carrying capacity from 6.65 to 4.41 cows, a walloping 34% reduction (Valério and Nakano 1988). This became a key statistic for economic damage calculations by Holmann and Peck (2002) and by Valério (2009), who estimated spittlebug damage to the Cerrados at $100 to $800 million per year. Because of its key pest status, *N. entreriana* has also been used in molecular studies to seek *U. decumbens* genes governing resistance to spittlebugs (Ferreira et al. 2019), as well in as cultivar screenings to find native *Paspalum* grasses with spittlebug resistance (Valério et al. 2015).

Pasture Spittlebugs as Rice and Maize Pests

In Brazil, pasture spittlebugs have been reported as significant pests of rice and, to a lesser extent, maize. The primary rice pest is *D. flavopicta* in the adult stage, moving onto young dryland rice seedlings from nearby *Urochloa* plantings, sometimes in mass movements. In the 1980s, in conjunction with *N. entreriana*, it caused the loss of tens of thousands of hectares of rice in the states of Goiás and Mato Grosso do Sul and continues to significant damage to rice plantings (Nilakhe 1985; Nilakhe and Buainain 1988; Ferreira et al. 2003; Paladini, Domahovski et al. 2018). *Deois schach* and *D. flexuosa* have also been observed to damage rice (Rossetto et al. 1978), and some spittlebug nymphs can grow on rice seedlings in standing water (Hickel 2023, fig. 1h, 1i). Nymphs and adults of an unidentified species of *Mahanarva* are reported to attack rice in Peruvian Amazonia (Couturier et al. 1996).

Migrating adult spittlebugs, particularly *D. flavopicta*, also damage maize plantings, though less often and with less intensity than rice. As in rice, their impact on maize is a function of immediate proximity to plantings of *Urochloa* (Cruz et al. 2010). Maize is a very poor host for spittlebug nymphs, at least for *N. entreriana*, with lab survival of less than 3% (Foieri et al. 2020).

Pasture Spittlebug Control

Because pasture is a relatively low-value and very extensive crop, it is seldom, if ever, economical to apply chemical control (Valério et al. 2001; Begnami et al. 2024). Persistent pesticides could also pose problems up the food chain, especially for dairy cows. As in sugarcane (Chapter 6), the entomopathogenic fungus

Metarhizium anisopliae might be useful as a biological control, but so far experiments in that direction have been disappointing and "the method has fallen into disrepute." (Valério et al. 2001). Also, *Metarhizium* poses substantial costs and is not well-suited for small producers because its effective use would require timely application during critical periods of the spittlebugs' annual life cycle, well before any visible damage appears in the pasture. However, *Metarhizium* has the ability to live as a persistent endophyte in some pasture grasses. This might provide a more practical mode of dissemination (Netto et al. 2024), an idea that has shown promise in one field test (Campagnani et al. 2024).

For now, this narrows practical control to a search for grass species or cultivars that are resistant to or tolerant of spittlebugs. Resistant grasses exhibit antibiosis toward spittlebugs, meaning that they reduce survival (in nymphs or adults) or lengthen the nymphal stage, lowering pest loads in pastures (Resende et al. 2024). So far, despite a lot of screening, no one has found a cultivar truly resistant to adult spittlebugs, but grasses do vary in tolerance, in the level of adult feeding they can sustain without serious loss of productivity, and in their ability to recover following spittlebug damage. The serial substitution of tolerant (e.g., *U. humidicola*) or resistant (e.g., Marandu) grasses for more easily damaged predecessors has driven much of the succession of planted pasture grasses in the Neotropics. The search for additional alternatives goes on, and there is increasing interest in native grasses, especially of the genera *Paspalum* and *Axonopus* (Begnami et al. 2024), which are intrinsically less susceptible to damage, probably because they have coevolved with the native spittlebugs (Thompson 2004).

There is unlikely to be a magic bullet solution. Sometimes, tolerant grasses permit the buildup of large spittlebug populations, undoing their original purpose. Something to this effect seems to have happened in some localities with *U. humidicola*. Also, as seen in the rise of *M. spectabilis*, cultivars resistant to one spittlebug species may not be resistant to all. Valério (2009) gives another example. A new *Urochloa* cultivar that exhibited strong antibiosis toward *A. varia* in Colombia exhibited none against *N. entreriana* in Brazil. Finally, the development of cultivars truly resistant to both nymphs and adults would put strong selection pressures on the pest populations to genetically adapt, with unpredictable evolutionary and agronomic consequences. Even when resistant cultivars are found or bred, they have to meet many other criteria to be useful and adopted. They must be highly productive, adapted to poor soils or wet or dry habitats, and resistant to other pests and diseases. Combined with the need to tailor local countermeasures to local spittlebugs and circumstances, this makes screening and breeding for resistance complex and time-consuming. Several sources offer thoughtful, comprehensive overviews of available approaches to control (Valério et al. 2001; Peck 2001; Valério 2009; Espitia B. et al. 2022; Wilson et al. 2023).

Finally, new spittlebug pests are probably in the offing. Throughout Latin America, many spittlebugs that are not now pests are waiting in the wings. Given the right hosts and circumstances, they could break out as new threats. The rise of *M. spectabilis* and its newly discovered close relative in Brazil is one cautionary tale, and the introduction of *P. simulans* into South America another. During the past few years, the central Mexican spittlebug *Prosapia ignifera* has spread widely and quickly in turf grasses in the San Diego–Los Angeles area of California, brought to the attention of researchers by a parallel outbreak of iNaturalist postings (see sidebar, Chapter 10: Citizen Science in Action: The Identification of an Invasive Spittlebug in Southern California). So far, this species is not a recognized pest, but its rapid spread hundreds of kilometers from its native distribution in central Mexico demonstrates the latent potential of grass-feeding spittlebugs to proliferate into new areas.

An Overview of Neotropical Grass Pests

This chapter reviewed the important role of spittlebugs as pests of New World planted pastures. They are destructive pests of the dominant introduced grasses, paralleling their negative effects on sugarcane (Chapter 6), another introduced Old World grass. Taken together, spittlebug damage to pastures and cane likely totals in the billions of dollars annually (Holmann and Peck 2002; Thompson 2004). This makes Neotropical grass-feeding spittlebugs one of the most important pest complexes in the world. No other spittlebug pests rival these Neotropical Cercopidae for scale and intensity of damage, but many others are economically important. These will be the subject of Chapter 8.

8

SPITTLEBUG PESTS OF RICE, PINES, ALFALFA, AND OTHER TREES AND CROPS

This chapter covers significant spittlebug pests other than those attacking sugarcane and pasture grasses, which are treated in Chapters 6 and 7, respectively. All of the pests presented are at least locally important. None has achieved the notoriety or economic severity of the Neotropical sugarcane and pasture pests. The examples covered exclude innumerable instances in which spittlebugs are listed as "pests" of particular crops or trees without any specificity as to damage or relative importance. Some of these minor occurrences may prove to be harbingers of economically significant infestations yet to be documented, but most involve laundry lists of species taken on or around the crop in question. In a dozen or so cases there are credible, isolated instances in which spittlebugs have achieved local pest status on a particular economic plant or tree. To economize space and preserve the reader from an overly tedious list of examples, I omit most of these instances. Pests of annual and perennial crops are treated first, pests of fruit and nut trees second, and pests of forest trees third.

Spittlebug Pests of Annual and Perennial Crops

Philaenus spumarius: A Historically Important Pest of Alfalfa in the North American Midwest

Following intensive work on the Trinidad sugarcane froghopper in the first half of the twentieth century (Chapter 6), the next big round of effort devoted to a single spittlebug species took on the meadow spittlebug, *Philaenus*

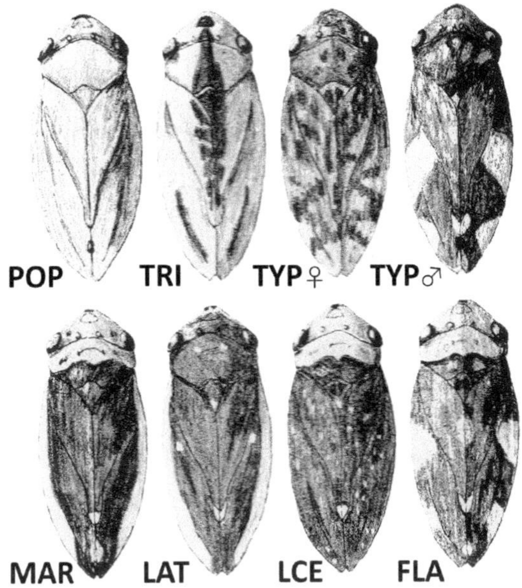

FIGURE 8.1. Illustrations of *Philaenus spumarius* color forms, from Weaver and King (1954). Color form terminology is updated. The three-letter abbreviations stand for populi (POP), trilineatus (TRI), typicus (TYP), marginellus (MAR), lateralis (LAT), leucocephalus (LEU), and flavicollis (FLA). The expression of TYP is characteristically different in females and males. Markings and ground color in the forms POP, TRI, and TYP are in brownish shades ranging from cream to dark brown. Markings in MAR, LAT, LCE, and FLA are black and white. In females the forms POP and TYP blend in a continuum, requiring arbitrary separation. See Fig. 5.1 for schematic versions of these color forms and Stewart and Lees (1996) for an excellent discussion of the subtleties of color form variation. For color photos of TYP, MAR, and TRI, see Plate 11a–c.

spumarius (Figs. 1.7, 5.1, 8.1, Plate 11a–c), a damaging pest of leguminous forage crops in the American Midwest. This species was introduced in North America from Europe (Hamilton 1982; Seabra et al. 2010, 2021), the timing of introduction unknown. Asa Fitch, the founder of American economic entomology, did not report its presence in a mid-nineteenth-century compendium of insects of New York state (Fitch 1851), but by the early twentieth century *P. spumarius* was widespread and abundant in the eastern United States (Osborn 1916). It apparently reached pest abundance in hay meadows in the mid-1930s, and by the late 1940s it achieved recognized status as a serious pest (Weaver and King 1954; Everly 1958). This precipitated a series of studies, including several PhD dissertations, that culminated in "Meadow Spittlebug" (Fig. 8.2), a 99-page

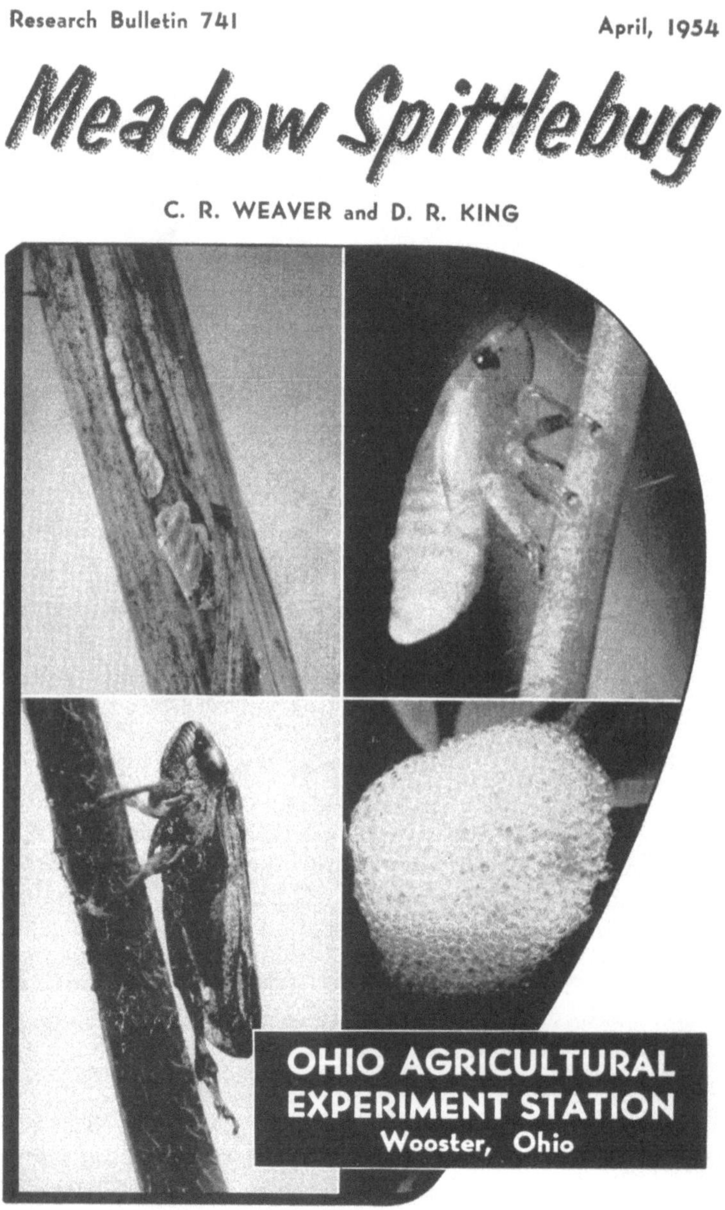

FIGURE 8.2. Cover page for Weaver and King's classic 1954 monograph on *Philaenus spumarius* as a forage legume pest. The images, clockwise from upper left, show *P. spumarius* eggs embedded in a dead plant stem, a nymph, a spittle mass, and an adult.

comprehensive overview of *P. spumarius* biology and agricultural ecology by C. Richard Weaver and Donald R. King (1954). This publication ranks with C. B. Williams's monograph on the Trinidad sugarcane froghopper as one of the two most thorough works ever published on a single spittlebug species.

The spittlebug pasture and sugarcane pests we examined in the last two chapters do their major damage through destruction of leaf tissue by adults. In contrast, *P. spumarius* damages alfalfa and other leguminous forage crops simply by diverting xylem stream nutrients in both the adult and, especially, nymphal stages. Weigert (1964a) estimates that for every calorie of energy ingested, the host plant loses about five calories in photosynthetic capacity, compounding the damage. Harm to forage crops is substantial and economically significant. Weaver and King (1954) report experimental first cutting yield losses of 27 to 31% in alfalfa, 23% in bird's-foot trefoil (*Lotus corniculatus*), and 22 to 45% in red clover (*Trifolium pratense*). Full-season percentage damage is substantially less because effects on the second cutting, subject only to adult attack, are much smaller. Other workers also report substantial, though smaller, effects on yield (Everly 1958; Wilson et al. 1979; Parman and Wilson 1982). *Philaenus spumarius* has big negative effects on wild hosts as well as forages. Tall goldenrod (*Solidago altissima*) exposed to insect herbivory, primarily by *P. spumarius*, had 85 to 90% less end-of-season biomass than goldenrod protected by insecticide treatment (Carson and Root 1999).

Entomologists Often Live to a Ripe Old Age

Clyde Richard Weaver spent most of his career at the Ohio State University agricultural branch in Wooster, where he carried out extensive work on the meadow spittlebug as a forage crop pest in the early 1950s and coauthored the major treatise on the species (Weaver and King 1954). After the spittlebug work, he shifted course and spent the rest of his career in the creation and operation of a statistics laboratory that served all departments. He was an avid tennis player and continued playing into his late nineties with Lowell Nault, an Auchenorrhyncha vector specialist. He died in 2019 at age ninety-nine. One of my minor regrets in life is that I did not seek him out when I visited Wooster for an Auchenorrhyncha Congress in 1990 after he had formally retired.

The predilection of *P. spumarius* for legume forage crops is a clear manifestation of xylem feeding ecology driving spittlebug host plant preferences and abundance (Chapter 3). These hosts are all nitrogen-fixing plants, well-positioned to provide spittlebugs with ample xylem-borne amino acids as primary nutrients (Thompson 1994). This permits *P. spumarius* to reach nymphal densities as high as 1280/m^2 in alfalfa fields (Wiegert 1964a) and even higher levels (> 3500/m^2) in experimental cages seeded with unnaturally high densities of egg laying adults (Parman and Wilson 1982). It may achieve extraordinary population levels in part because, as an introduced species in North America, it has escaped control by Palearctic natural enemies. Presently, *Philaenus spumarius* is often locally abundant but seems no longer to be a major pest of forage crops in North America, apparently because of effective insecticides and the introduction of spittlebug-resistant alfalfa cultivars (Halkka and Halkka 1990). In Finland, like North America, it achieves its highest densities on a legume forage crop, red clover (Halkka et al. 1967), but it is not a serious pest of forages in Europe.

Philaenus spumarius Is Also a Strawberry Pest in North America and Europe

Philaenus spumarius has also been a serious pest of strawberries in both North America and Europe (Rosenstiel and Vaughan 1952; Weaver and King 1954; Zajac and Hall 1986). Nymphs have heavily infested commercial plantings in Finland, and greenhouse experiments there demonstrated a strong negative effect on fruit yields (Halkka et al. 1967). In Ohio *P. spumarius* caused reductions in yield of 12 to 36% (Zajac and Hall 1986). Old reports describe damage to strawberries from nymphs of *Aphrophora saratogensis* in Louisiana and *A. permutata* (Fig. 1.9l) in Oregon, no doubt in the immediate vicinity of pine adult hosts (Plakidas 1928; Rosenstiel and Vaughan 1952). There is a report of *P. spumarius* on strawberries in the remote Indian Ocean island of Reunion (Quilici et al. 1998), but this seems to have been an accidental introduction that did not take hold.

Poophilus costalis and Multiple *Locris* Species Attack Sorghum, Rice, Millet, and Maize in Africa

Several species of the genus *Locris* (Cercopidae) and the widespread Afro-Asian species *Poophilus costalis* (Aphrophoridae, Fig. 8.3f, Plate 13g) damage local grains in West, Central, and East Africa. In Nigeria, sorghum suffers serious attack from *Locris rubra* and *P. costalis*, with experiments suggesting yield reductions of up to 34% (Ajayi and Oboite 1999, 2001). Elsewhere, *Locris auripennis* is recorded as a pest of sorghum in Ethiopia (Abate 1988), and *P. costalis* has

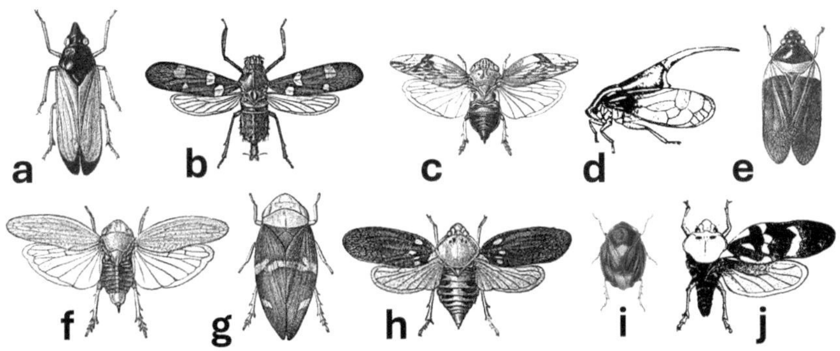

FIGURE 8.3. Some spittlebugs infesting crops, orchards, and plantations: a) *Abidama producta* (Cercopidae), b) *Callitettix versicolor* (Cercopidae, for color see Plate 12a), c) *Clovia conifera* (Aphrophoridae, for color see Plate 13d), d) *Machaerota ensifera* (Machaerotidae), e) *Phymatostetha deschampsi* (Cercopidae), f) *Poophilus costalis* (Aphrophoridae, for color see Plate 13g), g) *Ptyelinellus praefractus* (Aphrophoridae), h) *Surascarta tricolor* (Cercopidae, for color see Plate 12b), i) *Clastoptera globosa* (Clastopteridae), and j) *Cosmoscarta relata* (Cercopidae). Images a, b, c, d, e, f, g, and h from Distant (1908), image i from Fowler (1897), image j from Fletcher (1914). Note: not scaled to relative size.

been reported as a pest of sorghum in Burkina Faso, Uganda, Kenya, and Tanzania (Seshu Reddy and Omolo 1985). Spittlebugs are also reported to do serious damage to sorghum in Benin and Niger and to attack sorghum in Cameroon, Cote d'Ivoire, and Ghana. In Nigeria, *L. rubra*, *L. erythromela*, and *P. costalis* occur at extremely high infestations in pearl millet, causing yellow leaf blotch (Dike et al. 2002), and *Locris arithmetica* has attacked millet in Congo (Mayné and Ghesquière 1934). Outside Africa, *Ptyelus sexvittatus* (Fig. 1.9e) nymphs have been observed in numbers on sorghum in India (Verma et al. 1977).

In West Africa, *Locris rubra*, *L. maculata*, and *L. erythromela* attack rice, with preference for irrigated and lowland rice. They can cause leaf bronzing and wilting, but severe damage is not common. Though they are generally regarded as minor pests (Heinrichs and Barrion 2004), *L. rubra* and *L. erythromela* have been cited as major pests of lowland rice in Nigeria, with *P. costalis* and *Clovia minuta* included as minor pests (Emosairu and Shiyam 2000). *Locris rubra*, *L. maculata*, and *P. costalis* are all experimentally demonstrated vectors of rice yellow mottle virus, the main viral disease of rice in Africa (Koudamiloro et al. 2015), and *L. rubra* may play an important role in its transmission in the field (Nwilene et al. 1999). *Locris maculata* has also been cited as a pest of maize in Nigeria and a vector of Maize streak virus (Egwuatu and Ita 1982; Dike et al. 2002).

Callitettix versicolor: An Emerging Pest of Rice in Asia

While spittlebugs are significant rice pests in Africa and occasional spillover pests of rice after migrating from pastures in South America (Chapter 7), *Callitettix versicolor* (Fig. 8.3b, Plate 12a) (Cercopidae), has emerged in recent years as a major pest of rice in China. We met this colorful red, white, and black insect in passing in Chapter 6 as a minor Asian pest of sugarcane. In China, it has recently developed into a serious pest of rice and maize (Lei et al. 1992; Li et al. 2001), partly as a result of northward movement due to a warming climate and changes in farming practices (Chen and Liang 2012).

The nymphs develop on the roots of weeds and crops, such as sweet potatoes and soybeans, from eggs laid in soil at the margins of rice fields. The adults, which do the serious damage, spread into rice and maize fields and feed on the maturing crop. Damage to maize can be substantial, in severe cases killing whole plants and even leading to crop failure (Lei et al. 1992). In rice, yield losses in affected areas range from 5 to 40%, with a mean loss of 21% (Xu 1986; Li et al. 2001). In severe cases, the fields look scorched, as though they had been burned by fire (Xu 1986). The adults also attack sorghum. In contrast to the Neotropical grass and sugarcane spittlebugs, there is a single generation per year, but the adults hit the fields just as the plants are at a critical developmental stage for forming grain, exacerbating the damage (Li et al. 2001). *Callitettix versicolor* is now widespread in China (Yang et al. 2016), the world's largest producer of rice and second largest producer of maize, suggesting that it has the potential to become a world-class pest.

In addition to *C. versicolor*, a second cercopid pest spittlebug of rice and maize, *Abidama liuensis*, has emerged in China and may cause losses of up to 80% of local production (Wang, Bai et al. 2021). In India, two aphrophorid spittlebugs, *Clovia puncta* (synonym: *C. punctata*) and *Poophilus costalis*, are said to cause leaf yellowing in rice (Ghosh 2008). A brief report cites no fewer than five spittlebug species attacking rice in northern India: *C. versicolor*, *P. costalis*, *Abidama rufula*, *Abidama producta* (Fig. 8.2a), and *Cosmoscarta bispecularis* (Pawar and Bhalla 1977). In the absence of further substantial reports, it appears that none of these species are presently significant pests in India. There are old reports of *P. costalis* damaging rice in Taiwan (Schumacher 1915) and Japan (Clausen 1931), but again, in the absence of subsequent reports, it does not appear to be a persistent pest in either location.

Several Spittlebugs Attack Pigeon Pea in Africa and the Americas

Pigeon pea (*Cajanus cajan*), a widely grown annual or perennial nitrogen-fixing legume, seems to have a special attraction for spittlebugs. *Clastoptera variabilis* is an important pest of pigeon pea in Peru (Korytkowski and Torres 1966; Pflucker 1981), and pigeon peas are badly attacked by the legume-loving polyphage *Ptyelus grossus* (Plate 8a–c) in Cameroon (Rathore 1995). Pigeon pea also hosts unidentified *Clastoptera* species in Mexico (Smithsonian Museum pin label records) and Trinidad (Williams 1921), *Ptyelus goudoti* in Madagascar (Laurent 1964), and both *Poophilus adustus* (Fig. 1.9m) and *Ptyelus grossus* in Nigeria (Medler and Adenuga 1969). In each case, both nymphs and adults live on the host.

An Inordinate Fondness for Malvaceae, Once Including Cotton

While spittlebugs have a documented and physiologically explainable predilection for nitrogen-fixing and ectomycorrhizal plants (Chapter 3), they also exhibit an inordinate fondness for some groups of plants for which there is no obvious basis for special attraction. The Malvaceae fit this bill. There are several reports of spittlebug cotton pests in the first half of the twentieth century. These include *Machaerota spangbergii* (synonym: *M. planitiae*) (Fig. 1.9h, Plate 6a) as a serious pest in India (Fletcher 1920), *Machaerota ensifera* (Fig. 8.3d) as a minor pest in the Philippines (Otanes and. Butac 1939), *Huaina inca* as a pest in El Salvador (Bates 1942), and an unidentified *Clastoptera* species causing considerable damage in one area of Brazil (Costa Lima 1942). In addition, Mayné and Ghesquière (1934) report that *Hemitriecphora haglundi* (Cercopidae) (Fig. 1.9p) was common on cotton in the Congo. Probably not coincidentally, all of these reports predate the post-Second World War use of modern insecticides. Cotton is subject to heavy and repeated applications of insecticides. Their intensive use may have eliminated spittlebug cotton pests as a byproduct.

Other malvaceous crops still attract spittlebug pests. In India *Ptyelus nebulus* is reported to be a serious pest of two species of *Hibiscus* commercially grown for fiber, *H. cannabinus* (kenaf) and *H. sabdariffa* (roselle) (Pandit and Chakravorty 1986). *Locris maculata* has also been observed as a minor pest of roselle in Nigeria (Daramola 1984). Two spittlebugs have been listed as pests of the malvaceous vegetable okra: *P. nebulus* in Bangladesh (Ahmed et al. 2019) and *Machaerota humboldti* in Papua New Guinea (French 1956).

Spittlebugs are also abundant on many noncrop Malvaceae. *Hibiscus rosa-sinensis* is a prominent example. This Old-World ornamental shrub with long-blooming bright red or orange flowers is widely planted in the tropics and subtropics. In Costa Rica it hosts abundant populations of nymphs and adults of multiple *Clastoptera* species. It also hosts abundant *Clastoptera* in Brazil (my observations). Elsewhere in the Americas it has been reported in association with *Clastoptera* in Cuba and Trinidad (Myers 1930), as well as Mexico (Sánchez-Soto

1998) and Puerto Rico (Bennett 1965). In the Old World, closer to its origins, *H. rosa-sinensis* hosts *Machaerota rastrata* (Maa 1949) and *Ptyelus nebulus* (William and Ananthasubramanian 1991).

Other Malvaceae, including cacao (see the section on Spittlebug Pests of the Chocolate Tree in this chapter), host species of *Clastoptera*, *Machaerota*, *Tremapterus* (Plate 6f), and a few aphrophorids, leaving the strong impression that something makes Malvaceae especially attractive to spittlebugs. This apparent relationship merits further investigation.

Pests of Castor Oil Plant and Giant Milkweeds: Testimony to Spittlebugs' Ability to Overcome Formidable Chemical Defenses

It was not accidental that David Livingstone found *Ptyelus flavescens* "busily distilling on a branch of the *Ricinus communis*" (Euphorbiaceae), the castor oil plant (Livingstone 1857, see Chapter 3). *Ptyelus flavescens* (Plate 1c, 1d), *Ptyelus grossus* (Plate 8a–c), and *Ptyelus sexvittatus* (Fig. 1.9e) have all been identified as a castor oil plant pests or associates in Africa and India (Mayné and Ghesquière 1934; Darling 1946; Khan 1974). As in the case of Malvaceae, the basis of the castor oil plant's strong attraction for spittlebugs is not immediately apparent. The Euphorbiaceae in general and the castor oil plant in particular are fiercely chemically defended. The notorious poison ricin is a castor bean derivative. If nothing else, heavy feeding by these *Ptyelus* species is a strong demonstration of the ability of xylem-feeding spittlebugs to circumvent plant poisons.

The giant milkweeds *Calotropis gigantea* and *Calotropis procera* (Apocynaceae) also demonstrate spittlebugs' extraordinary ability to evade formidable chemical defenses. They are large plants, reaching three to four meters in height, and they are widely used in traditional medicine in South and East Asia. They exude latex sap heavily fortified with toxic compounds, so poisonous that they can cause temporary blindness. *Calotropis procera* even serves to make poison arrows in Africa (Philippe and Angenot 2005). Nevertheless, two aphrophorid spittlebugs (both misidentified as the cercopid *Abidama rufula*) attack and damage *C. gigantea* in Assam and Sri Lanka, respectively (Saikia et al. 2015; Wijeweera et al. 2022). In India, *Ptyelus nebulus* occurs in abundance on *C. procera* in West Bengal (Jana et al. 2012), as well as on *C. gigantea* (Dhileepan 2014). These host relationships are particularly interesting in light of the fact that *Philaenus spumarius*, the most promiscuously feeding insect herbivore, mostly avoids milkweeds of the related genus *Asclepius* (Thompson et al. 2023).

Spittlebug Pests of Nut and Fruit Trees

Not Quite Bye-Bye Miss American Pie: A Spittlebug Pest of Pecans

The pecan, *Carya illinoinensis*, is a nut-producing ectomycorrhizal tree native to and cultivated in the United States and Northern Mexico. Like other hickories (Genus *Carya*), it is attacked by the spittlebug *Clastoptera achatina* (Plate 5d), which can significantly damage production (Tedders 1995). In one experiment, only 29% of developing nuts exposed to *C. achatina* matured, compared to 55% of those protected by insecticide. However, overall damage to the American crop yield does not appear to be severe, and orchards can support high infestations with little apparent damage. A report from Georgia, the main producing state, attributes only $50,000 to spittlebug annual control costs (and none to crop losses), compared to total annual pecan insect damage and control costs of $15 million (Hudson and Dutcher 2008). For now, the iconically American pecan pie seems to be safe from the depredations of spittlebugs. South of the border in Mexico, *C. achatina* apparently causes more damage to pecans, leading to production losses as high as 43% if not controlled (Burrola-Morales et al. 2022). A machaerotid spittlebug of the genus *Taihorina* has been observed to damage another nut tree, the ectomycorrhizal Chinese chestnut, *Castanea mollissima* (Zhou and Zhong 1986), but this seems to be an isolated report.

Jak and Jujube: South Asian Fruit Trees Attacked by Spittlebugs

Jackfruit, borne by the Jak tree, *Artocarpus heterophyllus*, is the world's largest tree fruit, weighing up to 55 kg. *Cosmoscarta relata* (Cercopidae) (Fig. 8.3j) is a serious local pest of jak in India (Fletcher 1914; Kareem et al. 1971; Balaji Rajkumar et al. 2018), while *Clovia lineaticollis* (Aphrophoridae) attacks and withers its leaves (Ghosh 2008). Jak is also a secondary host of *Cosmoscarta thoracica* (Cercopidae), otherwise a pest of cardamom, which is, in turn, attacked by the cercopid spittlebug *Eoscarta nilgiriensis* (Ghosh 2008).

In India and Bangladesh, *Machaerota spangbergii* (synonym: *M. planitiae*) (Fig. 1.9h; Plate 6a) attacks jujubes, *Ziziphus mauritiana* and *Ziziphus jujuba* (Gravely 1915; Mann 1979; Prodhan et al. 2016; Pramanik and Liang 2019). The machaerotid *Hindoloides indicans* has also been observed on jujube, though not as a pest (Gravely 1915). A species resembling *M. spangbergii* attacks the Indian gooseberry, *Emblica officinalis* (Singh et al. 2016).

Spittlebug Pests of the Chocolate Tree

Cacao, *Theobroma cacao* (Malvaceae), has large seeds, the source of chocolate. While in Panama searching for parasites to control the Trinidad sugarcane spittlebug, C. B. Williams, whom we met in Chapter 6, stumbled across spittlebugs attacking cacao. Before Williams returned to England, he made one last contribution to the spittlebug literature, the first report of infestations damaging this important tropical crop (Williams 1922). He found abundant *Clastoptera* nymphs infecting and killing flower clusters and named the species *C. theobromae*, theobromine being a major chemical constituent of chocolate (and a close chemical relative of caffeine). This name later proved to be a synonym for *Clastoptera globosa* (Lara and Shenefelt 1961), a species that ranges from Panama to the Rio Grande Valley in Texas (Fig. 8.3i). *Clastoptera globosa* is a widespread pest of cacao in Costa Rica, causing considerable damage (Lara and Shenefelt 1961; Saunders 1981), and has been reported as a pest of cacao in the Peruvian Amazon (Delgado et al. 2023), though the species identification there is suspect due to the geographical location far outside the known range.

Cacao's strong attraction for *Clastoptera* is confirmed by the occurrence of *Clastoptera laenata* as a cacao pest in Mexico (Sánchez S. 1995; Sánchez S. and Sol S. 1998; Gerónimo L. et al. 2013) as well as *Clastoptera ochrospila* and an unidentified *Clastoptera* species as cacao pests in Brazil (Mendes and Garcia 1985; Ferraz 1985). In Africa the attraction apparently extends to other spittlebug groups. Boulard (1969), an auchenorrhynchist who worked from 1965 to 1970 at the French Institute of Coffee and Cocoa at Boukoko in the Central African Republic, reports that four local spittlebugs, none described as pests, occur as nymphs and adults on cacao: *Lehina apicalis* (Cercopidae) and *Poophilus camerunensis*, *P. costalis*, and *Clovia bigoti* (all Aphrophoridae).

Spittlebug Pests of Forest and Plantation Trees

Spittlebugs are frequent and sometimes serious pests of trees providing forest products. Most of the species attacked are ectomycorrhizal, including pines, *Eucalyptus*, and willow. Teak is a notable exception to this pattern.

Spittlebug Pests of Pines in the United States and Canada

There are two serious spittlebug pine pests in the United States and Canada: *Aphrophora cribrata* and *Aphrophora saratogensis*. *Aphrophora cribrata* lives on pines in both the nymphal and adult stages, the only spittlebug in Eastern

North America and one of only two in the United States and Canada to complete its whole life cycle on pines. I have also reared it from nymphs on balsam fir (*Abies balsamea*). It does the greatest damage to plantations of introduced Scotch Pine (*Pinus sylvestris*), where heavy infestations may kill 70 to 100% of the trees and wipe out all trees in pockets of an acre or more (Wilson 1991). Native pines suffer less damage.

Aphrophora saratogensis, also native to the eastern United States and Canada, is more destructive and on a wider scale. Its nymphs live on a wide variety of understory plants but favor the actinorhizal nitrogen-fixing plant *Comptonia peregrina*, so much so that the proportion of *C. peregrina* in ground cover is highly predictive of damage to overstory pines (Kennedy and Wilson 1971). Adults move to the pine canopy, where they feed on shoots among the needles and cause tissue injury, necrosis, and scarring. Eventually, heavily infested shoots die, with greater injury to younger trees. Severe infestation leads to tree death, and *A. saratogensis* routinely kills or ruins large portions of pine stands (Wilson 1987). Red pine (*Pinus resinosa*) and jack pine (*Pinus banksiana*) are the primary species damaged, with red pine being most vulnerable.

Ocoaxo Spittlebugs of Mexican Pines: An Emerging Pest Complex

In 2008 and 2015 reports emerged of a new and mysterious "pine decline" syndrome in Puebla and Oaxaca states in Mexico (Castro-Valderrama et al. 2017). Symptoms began with the rainy season drying of needles, the foliage becoming yellowish, orange, or brown, followed by needle fall. Affected trees recovered their foliage in the following rainy season, but after three years of recurring needle loss, many died (Castro-Valderrama et al. 2017). The results were serious. From 2014 through 2018 pine decline affected more than 7200 ha of Mexican pine forest. The cause was eventually traced to adult spittlebugs of the genus *Ocoaxo* (Cercopidae): *O. assimilis*, *O. cardonai* (Plate 15d), and *O. varians* (Cid-Muñoz et al. 2022). One or more species in this complex attack several species of pines in the states of Puebla, Veracruz, and Oaxaca, including *Pinus oaxacana*, *P. patula*, and *P. pseudostrobus*. Only *O. assimilis* has been studied in the nymphal stage. Its nymphs make spittles on the roots of a variety of understory plants, including a grass, four dicot herbs, a shrub, and, most commonly, two trees: the adult host *P. pseudostrobus* and an oak, *Quercus rugosa*. After the third instar, all nymphs migrate to pine roots to complete development (Cid-Muñoz et al. 2020). In July 2024, I visited a pine decline area near the town of Chignahuapan in Puebla state and observed firsthand swathes of dead and dying pines disfiguring forested hillsides.

TABLE 8.1. Spittlebug pests of crops[a]

COMMON NAME	SCIENTIFIC NAME	SPITTLEBUG PEST SPECIES	AREA
Alfalfa	*Medicago sativa*	*Philaenus spumarius*	North America
Strawberry	*Fragaria × ananassa*	*Philaenus spumarius*	North America
Sorghum	*Sorghum bicolor*	*Locris auripennis*	Ethiopia
		Locris rubra	West Africa
		Poophilus costalis	West, East Africa
Pearl millet	*Cenchrus americanus*, formerly *Pennisetum glaucum*	*Locris arithmetica*	Congo
		Locris erythromela	Nigeria
		Locris rubra	Nigeria
		Poophilus costalis	Nigeria
Rice	*Oryza* spp.	*Abidama liuensis*	China
		Callitettix versicolor	China
		Clovia minuta	Nigeria
		Locris erythromela	West Africa
		Locris maculata	West Africa
		Locris rubra	West Africa
		Poophilus costalis	Nigeria
Maize	*Zea mays*	*Abidama wuningiana*	China
		Locris maculata	Nigeria
Pigeon pea	*Cajanus cajan*	*Clastoptera variabilis*	Peru
		Ptyelus grossus	Cameroon
Cotton[b]	*Gossypium* spp.	*Clastoptera* sp.	Brazil
		Huaina inca	El Salvador
		Machaerota ensifera	Philippines
		Machaerota spangbergii	India
Kenaf	*Hibiscus cannabinus*	*Ptyelus nebulus*	India
Roselle	*Hibiscus sabdariffa*	*Ptyelus nebulus*	India
		Locris maculata	Nigeria
Okra	*Abelmoschus esculentus*	*Machaerota humboldti*	Papua New Guinea
		Ptyelus nebulus	Bangladesh
Castor oil plant	*Ricinus communis*	*Ptyelus flavescens*	Africa
		Ptyelus grossus	Africa
		Ptyelus sexvittatus	India
Cardamom	*Elettaria* or *Amomum*	*Cosmoscarta thoracica*	India

[a] Excludes sugarcane, which is covered in Table 6.2.

[b] All records of cotton pests precede the Second World War.

TABLE 8.2. Spittlebug pests of fruits, nuts, and forest trees

COMMON NAME	SPITTLEBUG PEST SPECIES	AREA
Pecan	*Clastoptera achatina*	North America
Jak tree	*Cosmoscarta relata*	India
Indian gooseberry	*Machaerota* nr. *spangbergii*	India
Jujube	*Machaerota spangbergii*	India, Bangladesh
Banana	*Phymatostetha deschampsi*	India
Cacao	*Clastoptera globosa*	Central America & Mexico
	Clastoptera laenata	Mexico
	Clastoptera ochrospila	Brazil
Pines	*Aphrophora cribrata*	Eastern North America
	Aphrophora flavipes	China, Japan
	Aphrophora saratogensis	Eastern North America
	Haematoloma dorsata	Europe
	Ocoaxo spp.	Mexico
Eucalyptus	*Cephisus siccifolius*	Brazil, Argentina
	Cosmoscarta spp.	China
	Leptataspis sp.	Thailand
	Ptyelus flavescens	Central African Republic
	Ptyelus sp.	Thailand
Teak	Cercopidae sp.	Ecuador
	Clovia conifer	India
	Leptataspis sp.	Thailand
	Machaerota elegans	Thailand
	Phymatostetha deschampsi	India
	Ptyelinellus praefractus	India, Thailand
Willows	*Aphrophora salicina*	Europe, China
Bamboo[a]	*Aphrophora notabilis*	China

[a] Bamboo is technically a grass but is treated here ecologically as a forest tree.

Haematoloma dorsata and *Aphrophora flavipes*: Old World Pine Pests

While the *Ocoaxo* spittlebugs of Mexico are the only Cercopidae pests on pines in the Americas, the striking red and black spittlebug *Haematoloma dorsata* (Plate 11e) is a long-established, important pest of pines in parts of Mediterranean Europe (Hernández A. et al. 1992; Roversi and Baccetti 1994). Only the adults do damage. The nymphs live on the roots of understory grasses. This species has expanded northward in the past few decades, achieving pest status in the Netherlands, apparently due to proliferation of the host grass *Deschampsia flexuosa* driven by increased atmospheric nitrogen deposition (Moraal 1996).

Aphrophora flavipes is an important pest of ectomycorrhizal conifers in parts of East Asia. It attacks *Pinus densiflora*, *P. tabuliformis*, *P. sylvestris* var. *mongolica*, and *Larix* spp. in China (Yan et al. 2019; Global Desertification Research Network 2022), and *P. densiflora* and *P. thunbergii* in Japan (Akiyama and Matsumoto 1986).

Nymph to Adult Host Shifts in Spittlebugs: A Common Phenomenon

Like *A. saratogensis*, *O. assimilis*, and *H. dorsata*, many spittlebugs have different nymphal and adult hosts. In many instances, the host change is dramatic, involving movement from herbs to woody plants and/or between distantly related plant taxa. *Haematoloma dorsata* (Plate 11e), for example, moves from feeding as nymphs on grass roots (herbaceous monocot angiosperms) to feeding as adults on pine foliage (woody gymnosperms). Other major host shifts are context-dependent, such as the common movement of *P. spumarius* from the herb layer to woody plants during the summer dry season in Mediterranean climates (Cornara et al. 2021). These movements are likely driven by a single factor, the need for an accessible, nutritionally sufficient xylem stream, prompting shifts to plants with deep root access to soil water and favoring ectomycorrhizal and nitrogen-fixing hosts that provide adequate, later-in-season, xylem-borne organic compounds.

Many Spittlebugs Attack *Eucalyptus* and Teak Plantations

The ectomycorrhizal genus *Eucalyptus* is host to several spittlebugs in its native range in Australia (tabulated in Thompson 2022), but it is subject to serious spittlebug damage only in areas where it has been introduced. These include Brazil, where some *E. urophylla* clones have been severely attacked by the spittlebug *Cephisus siccifolius* (Ribeiro et al. 2005), Argentina, where *Eucalyptus* is also subject to *C. siccifolius* attack (Golfari 1963), and China, where four species of the genus *Cosmoscarta* damage *E. grandis* plantations (Pang 2013). Hutacharern and Tubtim (1995) list a *Leptataspis* species (Cercopidae) and a *Ptyelus* species as pests of *E. camaldulensis* in Thailand. Lepesme (1947) reports that *Ptyelus flavescens*, the polyphagous spittlebug studied by Livingstone (Chapter 3), severely attacked *E. citriodora* in what is now the Central African Republic, weakening trees to the point of death. A species of the genus *Ptyelus* has also been reported to attack *E. smithii* in Cameroon (Brunck 1958).

Teak (*Tectona grandis*) is an important lumber tree native to India and Southeast Asia but raised in plantations throughout the seasonally dry tropics. Spittlebugs are locally important pests. In India teak is attacked by *Ptyelinellus praefractus* (Fig. 8.3g), *Phymatostetha deschampsi* (Fig. 8.3e), *Clovia conifer* (Fig. 8.3c), which causes severe damage, and an unidentified spittlebug that can even cause death (Balu et al. 1997; Roychoudhury et al. 2003; Basu et al. 2010; Kumar et al. 2014; Tripathy and Rout 2018). Hutacharern and Tubtim (1995) list four spittlebug pests of teak in Thailand: a *Leptataspis* species, *Ptyelinellus praefractus*, *Suracarta tricolor* (Fig. 8.3h, Plate 12b), and *Machaerota elegans*, the last of which can cause "serious damage" (Nair 2001). None of these species are specialists on teak. There is one New World report of a spittlebug pest of teak, an Ecuadoran species of Cercopidae incorrectly identified as *Aeneolamia postica* (synonym: *A. contigua*) (Flores 2005). In addition to attacking teak, *Phymatostetha deschampsi* is a minor pest of banana in Kerala (ICAR-National Research Centre for Banana 2024).

Aphrophora salicina is a Widespread Pest of Willows

Reports of *Aphrophora salicina* (Plate 2g) as a pest of willows span temperate Eurasia. This species, also referred to as *A. costalis*, *A. pectoralis*, and *A. salicis*, has been reported as a severe, localized pest of *Salix americana* in Spain (Ledesma 1979) and of *S. viminalis* and *S. rigida* in Romania (Pascovici 1975). In China they have been reported to attack *S. gracilior* and *S. saposhnikovii* (Huang et al. 2021) and to do serious damage to "Xinjiang willow" and "Xinjiang poplar" (Zhang et al. 1996). In Hungary, *A. salicina* (2g) and *Aphrophora alni* (Plate 11f) are the most dangerous pests of the basket willow, *S. americana* (Babos 1983). In 1923 spittlebugs caused damage so extreme to Hungarian willows that caterpillars of *Lymantria dispar* deserted the willows for other hosts (Webber 1927), a remarkable displacement of a moth species otherwise known as a relentless, defoliating herbivore.

A Spittlebug Pest of Bamboo in China

Aphrophora notabilis is an important pest of bamboos in parts of China, doing greatest damage to *Phyllostachys pubescens* (Lu and Xu 1992), the bamboo most commonly used in engineered flooring. Technically, bamboo is a grass (Poaceae), not a tree, but it is treated here as a forestry product, grown for its woody stems. Notably, *P. pubescens* has been implicated as a nontraditional nitrogen-fixing plant (Thompson 2004), and the genus *Phyllostachys* exhibits nighttime positive xylem pressure (Chapter 3), both factors that might make *P. pubescens* particularly attractive to a spittlebug pest. There has been some confusion over the identity of this spittlebug, originally identified as *A. horizontalis*, but corrected to *A. notabilis* (Liang, cited in Thompson 2013). Shih and Yang (2005) report that *A. horizontalis* lives on a bamboo species in Taiwan.

Overview and a Look Ahead

This completes our tour of diverse spittlebug pests on a wide range of hosts. Many but not all of the pest hosts are ectomycorrhizal or nitrogen-fixing, likely accounting for their attractiveness to spittlebugs and ability to sustain pest-size populations. All of these pests damage their hosts by feeding. In the next chapter, we take a close look at an acute and developing situation in which one spittlebug has caused catastrophic damage, not by feeding, but by transmission of a xylem-borne bacterial pathogen, *Xylella fastidiosa*.

9

XYLELLA FASTIDIOSA IN EUROPE

The *Sputacchina* and the Invisible Fire

The drama of the Trinidad sugarcane spittlebug (Chapter 6) *was protracted, playing out slowly over the course of a century, with ongoing reverberations in other countries. Likewise, the drama of Neotropical spittlebugs on African pasture grasses* (Chapter 7) *unfolded relatively slowly, starting in earnest in the 1950s and picking up in intensity over time. The drama of* Xylella fastidiosa *in Italy has played out fast and hard. It is an extraordinary story. As with sugarcane and pasture grasses, spittlebugs play a key role, this time in the form of a single species, the meadow spittlebug* Philaenus spumarius, *in its capacity as a vector. Like all xylem-feeding insects, it can transmit the xylem-borne plant pathogen,* Xylella fastidiosa. *In Europe, part of its native range,* P. spumarius *hosts include* Olea europaea, *the cultivated olive. This, combined with the introduction of a strain of* X. fastidiosa *ferociously destructive to the local olive cultivars, has led to an olive apocalypse in Salento, the boot heel region of Italy.*

A Disaster Unfolds in Salento

Sometime between 2008 and 2010 local farmers in the vicinity of Gallipoli in Salento, the boot heel of Italy, started to notice something odd. Branches of otherwise healthy olive trees had begun to dry out, leaves shriveling. In August 2013 one farmer brought a desiccated branch to the attention of his son-in-law, Donato Boscia, a scientist at the Institute for Sustainable Plant Protection of the National Research Council of Italy, in Bari, the capital of Apulia Region, of

FIGURE 9.1. *Xylella fastidiosa* infected olive orchard in Surano, Italy in 2019. The wholesale death of mature trees is typical of the epidemic and gives a sense of the apocalyptic landscape created by the spreading disease. Photograph by Sjor, through Wikimedia Commons CC BY-SA 4.0 https://creativecommons.org/licenses/by-sa/4.0/deed.en. Reproduced in black and white and cropped.

which Salento is a part. Boscia clipped it and showed it to his colleagues. One of them, plant virologist Giovanni Martelli, made a prescient guess. The symptoms, he thought, resembled those associated with the bacterium *Xylella fastidiosa*, a plant pathogen. He was familiar with *Xylella fastidiosa* from experience in the United States, where it causes serious diseases of grapes and other crops (Purcell 1997), and had helped diagnose its presence in almonds in Turkey in 2005 (Güldür et al. 2005). They tested. The test came back positive. It was indeed *X. fastidiosa*. Though the formal announcement appeared as a modest half-page note in an Italian journal, the last of twenty such notes in the issue (Saponari et al. 2013), drum rolls and French horns would have been appropriate. *Xylella fastidiosa* is one of two species in the genus *Xylella*. The other is *X. taiwanensis*, which infects pear plants in Taiwan, but about which very little is known. *Xylella fastidiosa*, shortened hereafter to *Xylella*, is native to the New World. It was feared in Europe, but with the exception of an isolated report in Kosovo (Berisha et al. 1998), it had not been found. Now it had arrived in force.

No one was sure of the implications, but they unfurled with frightening speed. Over the next few years, an epidemic of tree-killing blight expanded from a pocket around Gallipoli to the whole Salento peninsula. An olive apocalypse was underway, the social and agricultural consequences summarized by writer Daniele Rielli in a finely constructed and well-researched book, *Il Fuoco Invisibile: Storia Humana di un Disastro Naturale* (The Invisible Fire: Human

Story of a Natural Disaster) (Rielli 2023). Here is how Rielli, whose paternal line came from the region, opens the story:

> It was under the high vaults of the village house that my grandfather, shortly before dying, told of a strange dream. An olive tree in his countryside, a large tree that overlooked the local road, was completely burned. And it was burned without flame, as if consumed by an invisible fire. My grandfather knew it as one sometimes knows things in dreams, with an inexplicable but no less rigorous certainty: it wasn't a normal fire that killed it, but the result was the same . . . Today the branches of that tree are leafless, the wood is browned, it is a skeleton that sticks out like a black spire above the stone wall that divides it from the road. It wasn't a real fire that reduced it like this but an invisible fire.
>
> Standing in front of the dead trunk and looking all around, every other olive tree that comes into view appears burned in the same way. If a drone took off from that point, an enormous plant cemetery would stretch out underneath it, starting from Santa Maria di Leuca, the southernmost point of Puglia, almost up to Ostuni, in the province of Brindisi, a territory one hundred and fifty kilometers long and eighty wide, a single, almost uninterrupted expanse of desiccated olive trees, an unprecedented natural disaster. Twenty-one million trees, many of them centuries-old or a thousand years old, all burned by the same invisible fire. (Rielli 2023, 13, 16)

At this writing, Rielli's book is available only in Italian. I have leaned on it heavily to understand the framework of the epidemic and some of the key personalities involved in the response. Details about the unfolding of the epidemic not attributed to other sources are taken from Rielli's work. I cite only direct quotations.

Olives, as Rielli makes clear, are not just a commodity crop in Salento. They have been an integral and iconic part of the culture and a major component of the rural landscape. Many of the trees are hundreds of years old. By 2018 trees were dying en masse, and the affected area had grown from about 8,000 hectares in 2012 to about 715,000 hectares (Saponari et al. 2019), covering the better part of the boot heel. Looking out over desiccated trees as far as the eye could see, the French *Xylella* expert Joseph Marie Bové declared it to be "the worst phytosanitary emergency in the world" (Rielli 2023, 170). Here, I recount this disaster, focusing on the spittlebugs at the center of the action. In addition to Rielli's book, Saponari et al. (2019) give an excellent overview of developments in the early years of the epidemic, and Caimi and Piccinni (2022) provide a gripping photodocumentary of the people and land under siege. To help the reader keep track of the drama, which unfolded over the course of a decade, I have summarized the major developments in the form of a timetable (Table 9.1).

TABLE 9.1. Timeline for events around the *Xylella fastidiosa* outbreak in Salento olives

DATE	DEVELOPMENT
2008–2010	First signs of olive tree "desiccation" appear in Salento
October 2010	*X. fastidiosa (Xf)* strains imported to Bari institute for workshop
August 2013	D. Boscia, M. Saponari, and G. Martelli discover *Xf* in Gallipoli area olives
November 1, 2013	Saponari, Boscia, Nigro, and Martelli publish short note on *Xf* discovery
Fall 2013	D. Cornara confirms *Xf* positive *P. spumarius* in Gallipoli area olives
October 28, 2013	Boscia meets with angry farmers at Gallipoli area agricultural co-op
Fall 2013	Formal complaint against scientists filed with Lecce prosecutor's office
Fall 2013	Lecce prosecutor opens investigation of "culpable spread of plant disease"
April 2014	*Xf* attacking olives in Salento determined to be subspecies *pauca*
July 23, 2014	EU issues eradication order to prevent the spread of *Xf*
April 21, 2014	Saponari flies to Brazil for a month to learn *Xf* lab techniques
Late 2014	Salento *Xf* determined to be genetic match for Costa Rican strain
December 2014	*Xf* discovered on olives in Oria, 35 km north of epidemic frontier
February 2015	Italy declares state of emergency, appoints "Extraordinary Commissioner"
February 2015	Thousands of "Voice of the Olive Tree" farmers demonstrate in Lecce
March 29 2015	Thousands of "Olive Tree People" hold counter demonstration in Lecce
May 18, 2015	EU mandated eradiation measures officially replaced by containment
April 2015	Commissioner uproots handful of trees in Oria, protestors shut down effort
April 3, 2015	Judicial police visit labs in Bari, Boscia speaks with prosecutors in Lecce
April 7, 2015	Boscia offers prosecutors cooperation and guided tour of infected orchards
May 4, 2015	Police confiscate computers and documents from Bari labs
May 12, 2015	Italian agricultural scientists issue letter supporting Bari scientists
about May 18, 2015	Police seize documents from the Ministry of Agriculture in Rome
July 7, 2015	400 Commissioner troops reseize Oria orchard, but cut fewer than 50 trees
July 14, 2015	Lecce prosecutors interrogate Saponari in her lab for five hours
December 10, 2015	EU opens infringement procedure over Italy's failure to enforce containment
December 18, 2015	Saponari, Boscia, Commissioner notified they are under formal investigation
December 19, 2015	Extraordinary Commissioner resigns saying containment efforts blocked
Beginning of 2016	Olives experimentally infected with Salento *Xf* first exhibit desiccation
Spring 2016	All cutting and *Xf* sampling in orchards blocked by court order
March 2016	Paper puts end to speculation that *Xf* existed in Salento prior to epidemic
March 31, 2016	EU panel issues opinion that containment is proper scientific strategy
June 9, 2016	European Court of Justice determines EU-mandated control measures legal
July 2016	Under threat from EU, Lecce prosecutor lifts ban on cutting trees
August 2016	New containment plan finally implements compensation for farmers
late August 2016	Systematic monitoring of olives for *Xf* resumes, slowly
April 2017	The Oria area trees defended by protestors in 2015 are dying
May 31, 2017	EU issues scathing audit of Italian containment plans citing failures
May 2017	EU decrees that resistant olive varieties can be used to replant orchards
2018	About 715,000 ha infected with *Xf*, compared to about 8,000 ha in 2013
May 2019	Prosecutor drops all charges but perpetuates misinformation about epidemic
August 14, 2020	EU updates, broadens rules on *Xf* containment, recognizing wide presence

Source: Compiled from information in Rielli 2023 supplemented by information in European Commission 2014, 2015, 2020; Cariddi et al. 2014; Abbott 2015a, 2015b, 2017; EFSA 2016; Almeida 2016; Saponari et al. 2019.

Where Did *Xylella* Come From? How Did It Get to Italy, and How Does It Spread?

Xylella first came on the scene in connection with a crippling disease of grapes in Anaheim, California. Presently home to Disneyland, Anaheim in the late 1800s was host to German immigrant grape growers. But a mysterious disease was making it more and more difficult to grow grapes. Before it became impossible, the US Department of Agriculture sent Newton Pierce, a young Harvard researcher, out to investigate the problem. He suspected a bacterial cause but could not isolate a bacterium. Later, in the 1930s, the affliction was named Pierce's disease (PD). Absent a culturable bacterium, it was believed to be caused by a virus. In the mid-twentieth century, interest in PD revived with a wave of infestation elsewhere in California. Severin (1950) experimentally demonstrated that four spittlebug species, including *Philaenus spumarius*, could transmit PD from infected to healthy grapevines. In the 1970s, almost one hundred years after Pierce began his investigations, the cause of PD was finally recognized as a bacterium (Davis et al. 1978; Purcell 2013), eventually described as *Xylella fastidiosa* (Wells et al. 1987). The species name *fastidiosa* refers to the difficulty of culturing it in vitro. It is slow-growing and fastidious in its nutritional requirements.

In addition to PD in grapevines and an early recognized link to alfalfa dwarf disease, *Xylella fastidiosa* was eventually tied to several other diseases, including leaf scorches of almond, oleander, coffee, and pecan, as well as phony peach disease and citrus variegated chlorosis (Purcell 1997). Initially, all the infections were in the Americas, where *Xylella* first evolved and differentiated into three major genetic lineages, designated as subspecies: *X. f.* ssp. *fastidiosa*, *X. f.* ssp. *multiplex*, and *X. f.* ssp. *pauca*. Shortly after the discovery of *Xylella* in Salento, it was identified by DNA sequencing as *X. f.* ssp. *pauca* (Cariddi et al. 2014) and shown to match an isolate from coffee in Costa Rica, leading to the hypothesis that it had invaded Italy in asymptomatic imported ornamental coffee plants. *Xylella* had been on the EU quarantine list since 1980 (Stokstad 2015), but this importation evaded detection.

Xylella cannot live outside plant and insect hosts. From the moment they realized they had a local outbreak, the Bari scientists knew there had to be vectors. They tapped Daniele Cornara, a local doctoral student who had been working with vectors of a grape disease, to go look for them. Cornara spent a night reading up on *Xylella*, realized that *P. spumarius* (Figs. 1.7, 5.1, 8.1, Plate 11a–c) is a prime suspect vector, and headed for an olive grove near Gallipoli the following day. Barely out of his car, he found a *P. spumarius* individual perched on his headlight. He collected the specimen. It later tested positive for *Xylella*. He found many more specimens that day. The site was teaming with *P. spumarius*. In Italian, spittlebug is *sputacchina* (pl. *sputacchine*). As the epidemic progressed, *P. spumarius* became *la sputacchina dell'olivo*, the olive spittlebug.

Shortly afterwards, in fall 2013, Cornara set up a makeshift lab at a local agricultural cooperative and went about testing for the proportion of local *P. spumarius* carrying the bacterium and whether *P. spumarius* can transmit it to uninfected hosts. These experiments had to be done inside the infected area to avoid the possibility of accidentally spreading the disease with escaped *Xylella*-carrying vectors. Two-thirds of the November spittlebugs tested positive, and they successfully inoculated periwinkle, a known lab host for *Xylella* (Saponari et al. 2014). The first test on olives failed to confirm transmission.

The next year, Cornara and the Bari group went back to the Gallipoli area to investigate the potential role of *P. spumarius* as a vector in more depth (Cornara et al. 2017). They found three species of spittlebugs in the olive groves. *Philaenus spumarius* was again abundant. Over its whole seasonal life cycle about 50% tested positive for *Xylella*. *Neophilaenus campestris* was present, including in the olive canopy, but not as abundant, and only 1 out of 200 tested positive. A third spittlebug, *Cercopis sanguinolenta*, was rarer, never found in the olive canopy, and tested negative. When caged on infected trees in the field, both *P. spumarius* and *N. campestris* acquired *Xylella* but only *P. spumarius* subsequently transmitted it to olives. Although there were loose ends and the scale of the infection experiments was small, this work effectively established *P. spumarius* as the prime and likely only consequential vector of *Xylella* in the Salento orchards. The Bari group also demonstrated that *Xylella* is the primary and likely only cause of what was about to be formally named Olive Quick Decline Syndrome (OQDS) (Saponari et al. 2016), but this demonstration took two years because of the delayed expression of symptoms after the experimental inoculation of olive plants.

In Salento, a quirk of *P. spumarius* ecology magnified its role as vector. While both adults and nymphs prefer dicot herbs, which are the primary hosts of the spring nymphs, dry Mediterranean climate summers brown-out the herbaceous understory and drive adults into the local tree canopy, where they feed until rain returns in autumn. In most of Salento in 2013, the tree canopy was olives. The combination of an introduced bacterium, a spittlebug carrier that spends the summer moving and feeding among olive trees, and highly susceptible local olive cultivars, laid the biological basis for the catastrophe in Salento.

The Special Dangers Posed by the Unique *Xylella*-Insect Vector Relationship

Three factors combine to make *Xylella*, in the words of the EU, "one of the most dangerous plant bacteria worldwide" (European Commission 2024). One, vectors can transmit *Xylella* immediately after feeding on an infected plant. There is no latent period while it reproduces within the insect. Immature vectors lose transmission with molting, but adult vectors can transmit it for life (Severin 1950). Two, uncharacteristically for vector-borne pathogens, there is no vector-*Xylella* strain specificity. It appears that any xylem-feeding insect can transmit any variant of *Xylella*. And three, wherever *Xylella* is introduced, there are likely to be local xylem-feeding insects ready to transmit it to local plants (Almeida 2018; De La Fuente et al. 2024).

Strong Reaction to Proposed Mass Cutting of Infected Olive Trees

While the Bari group was working, the European Commission (executive arm of the EU) declared a policy of eradication of the infection in July 2014 (European Commission 2014). It required the destruction of infected olives trees and olive trees nearby. This, predictably, set off a firestorm of local controversy, controversy that did not abate when the Commission moved to a policy of containment in May 2015 (European Commission 2015). Like eradication, containment required mass destruction of trees to try to halt the spread of the infestation up the Salento Peninsula, which was proceeding at about 10 km/yr (Kottelenberg et al. 2021), in what Boscia described as a "slow tsunami" (Fasano 2020).

Boscia got a preview of the intensity of local reaction when he met with hundreds of anxious and angry farmers in the fall of 2013. They were divided. Some wanted effective measures to save the olive oil industry, even if it meant sacrificing trees in the short term. Others feared for their livelihoods and opposed any cutting, especially when, at this early stage in the epidemic, many trees were still symptomless.

A cadre of local environmental activists stepped into the breach. They were already concerned about issues like excessive use of pesticides and long-term loss of soil organic matter. Now they were convinced that the olive disease must be a result of factors like these, or perhaps other pathogens, not *Xylella*. They vehemently opposed the cutting of trees. Rumors flew that *Xylella* has been

FIGURE 9.2. Striking graffiti on a huge wall alongside the main road from Lecce to Bari, Italy reads *"XYLELLA È MAFIA,"* Xylella is Mafia, implying that the epidemic itself embodies corruption. The graffiti reflects the numerous conspiracy theories circulating early in the epidemic, some with credible roots, others bordering on science fiction. The Mafia is often present during crises, seizing economic opportunities amid disorder in discreet, often inventive ways. Yet, this graffiti may also reference theories that the epidemic was intentionally triggered by people with certain interests, whether construction investors looking to exploit land previously dedicated to olive groves, corporations like Monsanto aiming to introduce genetically modified trees, or even the Trans Adriatic Pipeline (TAP) consortium, seeking to clear land for pipeline construction. Reproduced with permission from *Fastidiosa* by Jean-Marc Caimi and Valentina Piccinni (Overlapse, London, 2022), with special thanks to the authors for providing the text for this caption.

introduced intentionally in a plot by the international agribusiness Monsanto to replace the treasured local groves with genetically modified olives. The proof: Monsanto owns a subsidiary company Alellyx, which is close to "Xylella" spelled backwards. People reported strange men in white hazmat suits wandering the countryside and suspected the shadowy influence of the Mafia (Fig. 9.2). At one point, visiting EU inspectors were accosted in an orchard by a group of tree-guarding vigilantes wearing "fake 'nature guard' uniforms—complete with hats" (Rielli 2023, 81–82). In February 2015, thousands of farmers, organized as the "Voice of the Olive Tree," demonstrated in Lecce, the administrative center of

Salento, pressing for more action on the part of the government. That same month the national government declared a state of emergency and appointed a military-police general, Giuseppe Silletti, as Extraordinary Commissioner to lead the government response to the *Xylella* crisis. A month later, thousands of environmentalists, organized as the "Olive Tree People," demonstrated in Lecce to oppose all cutting and staged a stations-of-the-*Xylella*-cross march at the tip of the peninsula.

An Outlying Infection in Oria Becomes a Center of Bitter Contention

Following EU regulations, Silletti drew up a map of the infected areas, outlining a 20-kilometre buffer zone that was mostly free of infection, where authorities were to monitor trees with particular care (Abbott 2017). The program aimed to eradicate localized outlying infections by felling trees as necessary, healthy trees as well as infected ones. It was already too late in the heart of the infected area. This put him directly in the crosshairs of both aggrieved groups. The farmers were set to lose their livelihood with no compensation on the horizon. The environmentalists were opposed to cutting any trees. The stage was set for heated confrontations.

In Oria, 35 km north of what was then the infection zone, a married couple, local agronomists, had reported an outlying *Xylella* infestation in their olive grove. Neighbors ostracized them and accused them of plotting to cut their orchard to put in a go-kart track. In April 2015, the Extraordinary Commissioner cut a handful of trees on their property. Protesters reacted violently, climbing among the branches, throwing themselves at excavator buckets, and stopping the operation. In July 2015, the Commissioner sent in a quasi-military force of 400 officers to occupy the farm, but fewer than fifty trees were actually cut. The protesters sued, and further cutting was blocked in Oria.

The Prosecutors Strike: Local Scientists Are Accused of Spreading *Xylella* and Lying

In the meantime, an extraordinary story was unfolding in Apulia, the region encompassing Salento. On 3 April 2015, the judicial police visited the Bari labs. They questioned Maria Saponari, a close colleague of Boscia and one of the leading researchers working on the *Xylella* outbreak. At the same time, Boscia was talking with prosecutors in Lecce. When the controversy over proposed cuts first emerged in Fall 2013, a handful of individuals had filed a formal complaint, and the Lecce prosecutor had opened an investigation into possible "culpable spread of plant disease." The investigation had been underway in secret for a year and a half.

How Innocent Facts Can Be Twisted to Support Baseless Conspiracy Theories: An Instance from the *Xylella* Controversy

In their indictment, the Lecce prosecutors made much of the fact that a strain of *Xylella* had been brought to Bari from California in October 2010 for a European workshop on prevention of quarantine pathogens. The controlled importation was intended to provide basic training to researchers in countries like Italy, where the pathogen was not yet present, so that in the event of an introduction they would have skills for dealing with it. The bacterial cultures had been imported under permit by my Berkeley colleague Alexander Purcell and kept carefully contained. They were a different subspecies of *Xylella* from the strain that later caused the Salento epidemic. This did not stop the prosecutors and conspiracy theorists from suggesting a sinister link, based on the coincidence between the timing of the workshop and the first appearance of OQDS in the Gallipoli area. On social media, the workshop cultures were characterized as a purposeful introduction, fueling public outrage against the scientists. A second Berkeley Colleague, Rodrigo Almeida, was personally accused of deliberately introducing the pathogen. By his account, it took a personal toll and "occupied my mind more than it should have." (Almeida 2018).

Alarmed by the police visit and the tone of the conversations, Boscia sent an email to the prosecutor's office clarifying some technical points, offering collaboration, and inviting the staff to tour infected orchards. A few days later, the prosecutors seized the researchers' computers. Then on July 14, while Boscia was in Milan lecturing on the epidemic, two prosecutors arrived at the labs and interrogated Saponari for five hours. The transcript runs to 145 pages. In what was probably a historical first, a spittlebug figured prominently in a criminal investigation, as Saponari explained that the sputacchina was an a priori vector suspect because it feeds on the xylem sap where the bacterium resides. The discussion was heated. The prosecutors did not seem to believe that Martelli could have recognized *Xylella* symptoms intuitively. At one point they indicated that they were beginning to understand the situation, but did not know if the politicians wanted to understand. They expressed some grudging respect for the work of the scientists. A few days later, the judicial police seized documents concerning the case from the Ministry of Agriculture in Rome (Abbott 2015b).

Five months later, on 18 December 2015, police officers arrived to hand Boscia, Saponari, Extraordinary Commissioner Silletti, and seven other people formal "notices of investigation." They were suspected of a long list of serious crimes, including "negligent spreading of the plant disease, presenting false information and materials to officials, environmental pollution and disfiguring natural beauty" (Abbott 2015a). The chief prosecutor held a press conference and declared that "The European Union was misled with a false representation of the *Xylella fastidiosa* emergency, based on improper data and on the non-existence of a real causal link between the bacterium and the desiccation of the olive trees" (Rielli 2023, 192). In a radio interview shortly thereafter, he said that with abundant irrigation and hoeing of the soil the olives would have recovered. The prosecution was second guessing the science. A month later, the first olives experimentally infected with the Salento strain of *Xylella* started to sicken, after a slow incubation of about two years. The prosecutor was simply wrong.

In the Aftermath of Indictments the Containment Campaign Grinds to a Halt

The day after the indictments were handed down, Extraordinary Commissioner Silletti resigned, "saying that his ability to implement his containment plan had been blocked at every turn" (Abbott 2017). In the course of his year in office, he cut down a total of 1500 trees, fewer than ordinarily uprooted for routine public works. There was a secret addendum to the warrant notices, which was quickly leaked. The prosecutor was granted preventive seizure of the olive trees destined for felling. This put an abrupt halt to the containment campaign. The commissioner appointed by the national government to oversee the effort had been forced from office. All the indictees were threatened with jail. The *sputacchine* were free to keep spreading the bacterium from a now vast and increasing reservoir of infected trees.

The result was biological disaster and deepening social tragedy. Felling of trees came to a complete halt. Even private landowners who wished to cut were barred from doing so by strict Italian laws that forbid the destruction of heritage olive trees without special permission. Testing for *Xylella* infections also came to a stop, dropping from 1,800 samples tested in September 2015 to zero in the first three months of 2016 (Abbott 2017). Substantial testing did not resume until December 2016. For a year scientists could not even monitor the spread. By 2017 approximately 6.5 million olive trees in large orchards were severely damaged, about 8.1 million adding in small orchards (Scholten et al. 2017). The immediate economic damage was estimated to be at least €1.6 billion (Fasano 2020).

Why was this fiasco permitted to unfold? The major cause appears to have been failure from the beginning to address the interests of the key local stakeholders (Almeida 2018). Saponari and her colleagues describe the circumstances well:

> . . . several critical factors may have contributed to the poor engagement of stakeholders. The lack of a structured plan to provide farmers with financial support to remove infected trees from their orchards and to provide some financial assistance to counteract the impact of the olive disease was [a] major factor that discouraged their active cooperation . . . [and] . . . lack of public support for the additional interventions required by the contingency plan (i.e., removal of the weeds in spring for vector control) was another aspect that limited the inclination of the farmers to effectively contribute to the containment plan. Farmers in the infected area . . . felt they were left alone to combat this disease, and related social and economic threats. Moreover, the prohibition in the infected area of planting host plants, including plantation of new olive orchards and the absence of any structured plan to reconstruct the agricultural landscape with alternative crops contributed further to dampen farm businesses.
>
> It should be added that difficulties were also encountered in applying containment measures by the public bodies, i.e., control of the vector population in municipalities properties, due to the shortage in public funds. This induced contradictory feelings of local farmers who still received penalties in case of noncompliance with regulatory actions aiming at controlling the vector. In addition, numerous farms in the area are managed under [an] organic regime that makes insecticide use incompatible with organic practices. (Saponari et al. 2019)

In other words, farmers were left on their own, uncompensated for the sacrifice of their olive trees, fined when they failed to plow their dying orchards to help control spittlebug nymphs, and, in some cases, compelled to use insecticides that were incompatible with their organic business models. Combine this economic ruin with brutal destruction of a cherished landscape and no wonder there was visceral opposition to the eradication and containment efforts.

This made fertile ground for the conspiracy theories (Fig. 9.2), reinforced by social media and by the natural inclination of politicians to duck responsibility and hope things will work out.

EU Pressure Leads to Reinstatement of Containment, but It Is Too Late for Salento

While the local prosecutors continued their investigation in secret, the EU ratchetted up pressure on Italy to obey the European Commission mandates. In March 2016, an EU scientific panel issued an opinion that containment is the proper scientific strategy. In June 2016, the European Court of Justice determined that the EU-mandated control measures were legal. In July 2016, with Italy under threat of economic sanctions from the EU, the Lecce prosecutor lifted the ban on cutting. In August 2016, three full years after the discovery of the epidemic, farmers were finally offered some compensation for the destruction of their trees and limited *Xylella* testing resumed. But the damage had been done and was irreversible. By spring 2017, the olive trees defended by the protestors in Oria were dying. In May 2017, the EU issued a scathing audit of Italy's failed containment efforts. By 2018, the infected area had reached 715,000 hectares, compared to 8,000 in 2013 when the epidemic first came to light.

After Five and a Half Years the Prosecution Is Finally Dropped

In May 2019, three and a half years after the indictments and five years after the beginning of the investigation, the prosecutors abruptly dropped the charges. But, in a miscarriage of justice, the dismissal degree reiterated many unfounded charges and published private emails that were twisted and quoted out of context to make it appear that the scientists knew that *Xylella* was not pathogenic but supported cutting trees anyway. This led to ongoing personal harassment of some of the scientists in question. The whole case was a fiasco for the Italian justice system, reminiscent of the trial of six seismologists following the L'Aquila earthquake, which killed about 300 people in 2009 (Duhaime-Ross 2014). They were convicted of manslaughter in 2012 for failing to warn of the catastrophe and each sentenced to six years in prison. The convictions were overturned on appeal in 2014 but not before tremendous damage to the reputation of the Italian courts. The Bari scientists never went to trial, but their harassment by the system is a stain on the reputation of the local prosecutors. In the end, it was the willfully ignorant actions of the prosecutors that exacerbated the effects of the epidemic, not the heroic but ultimately unsuccessful efforts of the scientists to keep it contained.

Starting About 2018 the Pace of *Xylella* Advance Markedly Slows

At about the time the charges were dropped, the northward advance of *Xylella* began to slow. Between 2018 and 2023 there was relatively little change in the demarcation of the infected zone, barring some small advances on the Adriatic coast (Boscia et al. 2023). The reasons for the slowdown are not entirely understood but probably include movement into a less suitable climatic area, a big decline in the reservoir of infected vectors due to cutting and mass die-off of olives, and changes in the management of orchards, with plowing of the understory to remove spittlebug nymphal habitat. At the same time, the severity of OQDS seems to have abated in the southern part of the peninsula, below Brindisi (Boscia et al. 2023). New symptoms of desiccation are milder and less frequent, and there are widespread instances of remission in mid-age surviving trees, leading, in some areas, to partial recovery (Scortichini and Ragno 2024). This has lent some support to those who argued all along that the etiology of OQDS was more complex than just the action of a killer bacterium. The reasons for these phenomena are under investigation. There is natural resistance to OQDS in some olive cultivars (La Notte et al. 2024). In the northern part of the infected area the disease is still destructive and severe (Boscia et al. 2023). In the long run, it appears that screening and selection for olives resistant to OQDS is the most likely path to replanting and restoration of the olive industry in Salento (Saponari et al. 2019; Morelli et al. 2021), a fact reflected in a 2017 EU edict permitting replanting with resistant cultivars.

Independent Spittlebug-Vectored *Xylella* Outbreaks in the Balearic Islands and the Alicante Region of Spain

In the meantime, *Philaenus spumarius* remains at the heart of concern regarding the spread and dissemination of *Xylella* to and within other regions of Europe. In 2020 the EU issued comprehensive regulations regarding the containment and management of *Xylella* (European Commission 2020). The 13,000-word document does not mention *P. spumarius* by name but uses the word vector or vectors no fewer than seventy-three times (European Commission 2020). The possibility of spread is not hypothetical. *Xylella* is already present in Spain in the Balearic Islands, where it infects grapes, almonds, and olives (Moralejo et al. 2020; Olmo et al. 2021), and in the mainland Alicante region, where it attacks almonds (Marco-Noales et al. 2021). It also occurs in scattered areas in Portugal, where, so far, it is symptomatic only in lavender (Loureiro et al. 2024), and it has been found in Corsica (Albre et al. 2020) and the South of France. Each of these cases appears to represent an independent introduction, some of them earlier than the outbreak in Salento. Excepting the outbreak on olives in Ibiza (and now Mallorca), all these outbreaks involve subspecies of *Xylella* other than the ssp. *pauca* devastating Salento. In each case *P. spumarius* is present in the environment and likely the major local vector.

Two of these local outbreaks are of serious economic consequence. In Mallorca, *Xylella*-caused almond leaf scorch disease affects 81% of almonds and was the major factor in a drop in almond orchard area from about 30,000 hectares in 2010 to 12,000 in 2019 (Olmo et al. 2021). *Xylella*-caused Pierce's disease is also widespread in Mallorcan grapes. In Ibiza, the subspecies *pauca* causes deadly dieback in olives but is less virulent than the strain in Salento. In the Alicante region of Spain, a *Xylella* infestation has led to the felling of 242,840 almond trees in an eradication campaign that has paid out €1,465,000 in compensation since 2018 (Servicio de Sanidad Vegetal 2024). In one measure of success, the local *Xylella* infection rate of *P. spumarius* dropped from 6.3% in 2017–2022 to 1.6% in 2023.

The 2020 EU regulations mandate that each member state conduct annual monitoring for the presence of *Xylella* in infected and uninfected zones (European Commission 2020). In fall 2022, researchers carrying out routine local monitoring found two *Xylella*-positive *P. spumarius* individuals in a municipality just southeast of Bari, outside the existing *Xylella* containment zone. The next fall they found thirteen. This led to intensive sampling of local plants. Five species, including almond and grape, tested positive for *X. fastidiosa* subspecies *fastidiosa* (Cornara, Boscia et al. 2024). The spittlebugs had served as effective sentinels, alerting the researchers to the presence of a new *Xylella* outbreak, a role that had been foreseen by French researchers in Corsica (Farigoule et al. 2022). Subsequent eradication efforts may have headed off a second *Xylella* epidemic in Apulia, biologically independent of the first.

The Crisis in Salento Unleashed a Proliferation of Spittlebug Studies

When the *Xylella* crisis unfolded in Salento in 2013, European spittlebug studies had been a sleepy affair for decades, centered primarily around the mysteries of the color polymorphism of *P, spumarius* and other aspects of spittlebug natural history and taxonomy. The demonstration that *P. spumarius* was the primary vector of *Xylella* in Europe led to a rapid proliferation of new studies of *P. spumarius* ecology, host plant relationships, and feeding biology, including nineteen papers on *Xylella* vectors published from 2014 to 2023 (Fig. 9.3). Bodino et al. (2023) and Serio et al. (2024) provide overviews of this contemporary research boom. Weaver and King (1954), Halkka and Halkka (1990), Yurtsever (2000), and Cornara, Bosco et al. (2018) review the older *P. spumarius* literature, which is voluminous, and Thompson et al. (2023) review and synthesize the host plant literature.

In Italy, researchers have produced thorough studies of *P. spumarius* host plants and seasonal ecology (Dongiovanni et al. 2019, 2023; Latini et al. 2019; Bodino et al. 2019, 2020; Cornara et al. 2021; Trotta et al. 2021). Studies have also been carried out in Spain (Morente et al. 2018), Portugal (Villa et al. 2020), Corsica (Albre et al. 2021), Albania (Kokiçi et al. 2024), and Greece (Antonatos et al. 2021). These studies show that *P. spumarius* is abundant and widespread in many *Xylella*-vulnerable Mediterranean areas, occurs on a wide range of host plants, and often moves in the dry summer period from the herbaceous understory to the olive canopy and native trees and shrubs (Cornara et al. 2021;

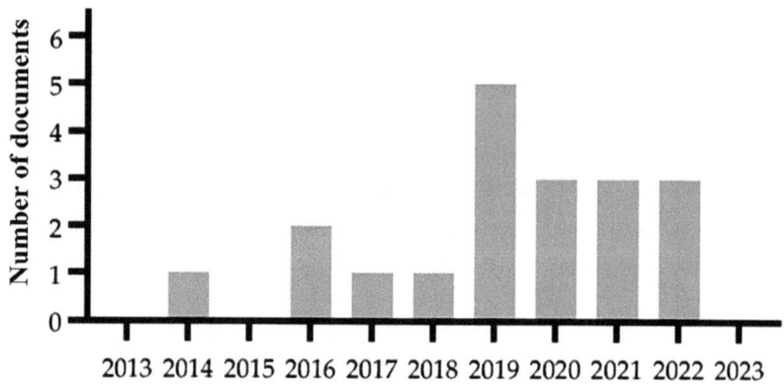

FIGURE 9.3. Number of publications per year concerning *Xylella fastidiosa* vectors and the Apulia region outbreak, 2013–2023. Reproduced with minor modification from Serio et al. (2024) under Creative Commons Attribution 4.0 International Deed (https://creativecommons.org/licenses/by/4.0/).

Thompson 2022). Its ubiquity, host versatility, and movement in local environments make it an excellent vector, reflected in the frequently high levels of *Xylella* infected *P. spumarius* individuals in outbreak areas (Bodino et al. 2023).

The *Xylella* crisis has also stimulated studies on *P. spumarius* feeding mechanisms and the relationship between feeding and the conveyance of *Xylella* from one plant host to another (Cornara et al. 2016, 2017; Cornara, Garzo et al. 2018; Cornara, Mara, Morente et al. 2018; Cornara, Zaffaroni et al. 2024). *Philaenus spumarius* acquires *Xylella* through ingestion from *Xylella*-infected host xylem vessels (Chapter 3). The bacterium propagates in the cibarium and precibarium, forming biofilms and possibly feeding in part on the chitin lining (Labroussaa et al. 2017; De La Fuente et al. 2024). It does not circulate in the body of the insect and is not passed from generation to generation. Based on analogy with sharpshooter leafhoppers, it is lost during molts, so that nymphs must reacquire it in each instar, as must adults after the last molt (Purcell and Finlay 1979). This, in addition to low mobility, makes nymphs ineffective vectors. In adults, *Xylella* persists for weeks after acquisition. Transfer from adults to plants is not highly efficient but does not need to be given the vector's large numbers and persistent presence from late spring through fall.

Spittlebug Dispersal Ability

Xylella can travel in two ways: in infected plants and via movement of vector insects. Dissemination via infected plants accounts for long distance jumps across oceans and the movement of *Xylella* from the New World to the Old. Given the diversity of known symptomless host plants (Purcell and Saunders 1999) and the long history of plant movements from the Americas to Europe, it is a bit of a mystery why *Xylella* diseases were not recognized outside the Americas until the end of the twentieth century (Purcell 1997). Dissemination via vector insects is responsible for the propagation and spread of local epidemics. Consequently, vector dispersal rates are a major parameter of models for the spread of *Xylella* in Europe and of tremendous practical consequence.

Since *P. spumarius* is the only consequential European vector, there have been several attempts to estimate its dispersal rate, leading to somewhat contradictory results. As summarized by Bodino et al. (2023), field mark-release-recapture experiments indicate that most *P. spumarius* individuals stay close to home, moving from a few dozen to a few hundred meters in the course of a season. In contrast, one laboratory tethered flight-mill experiment suggested the potential for a half-hour movement of 460 meters and a maximum five-hour movement of five kilometers (Lago, Garzo et al. 2021), an observation at variance with the historical description of *P. spumarius* as weak fliers (Weaver and King 1954). Casarin et al. (2023) measured shorter mean flight-mill distances for *P. spumarius* (102 m) but observed a maximum flight of 1.54 km in 2.5 hours. They also recorded mean flights of 623 m and a maximum flight of 6.2 km for the potential *Xylella* vector *Aphrophora salicina*, a larger insect. In mark-release-recapture experiments, they estimated mean dispersal of only 1.5 m/day in *P. spumarius* and 3.5m/day in *A. salicina*. On the other hand, a mark-release-recapture experiment on the alternative *Xylella* vector *Neophilaenus campestris* recovered individuals as far as 2.4 kilometers away after 45 days (Lago, Morente et al. 2021).

As Casarin et al. (2023) note, it only takes a small number of long-distance dispersers to effectively spread an epidemic. The measured slow dispersal of local *P. spumarius* in mark-release-recapture experiments could not drive the 10.5 km/yr rate of northward spread of the Salento epidemic (Kottelenberg et al. 2021). The *sputacchine* moved faster than localized observations predict, but in line with the evidence that *P. spumarius* is capable of long-distance dispersal. The rapid advance of the epidemic in Salento also may have been aided by spittlebug hitchhiking in or on vehicles, a hypothesis bolstered by the epidemic's apparent spread along major road networks (Bajocco et al. 2023). Spittlebugs must feed on plants to live, but they can live for hours without food if it is cool and even longer if cold. They could easily be transported long distance by cars or trucks. Extraordinary Commissioner Silletti was well aware of this. "The truth," he admitted, "is that we should take all the people, all the wild animals, all the vehicles and inspect them before letting them out, but it can't be done" (Rielli 2023, 108).

Spittlebugs do get around. Weaver and King (1954) observed quick repopulation of large experimental plots cleared of *P. spumarius* by DDT treatment in August, when only adults were present. This is indicative of major local migration. Migration on a grander scale has also been observed on occasion. Putman (1953) reported that

> a spectacular concentration of adults appeared in early July 1945, along the Lake Ontario shore . . . with a slight breeze. Attention was drawn to them by the patter, like rain, as they hit the leaves of trees. They were flying . . . from just above the ground to as high as they could be seen.

Philaenus spumarius has also been recorded from high rooftops in my own state in New York City, Albany, and Westchester County (Felt and Chamberlain 1935) and has colonized dozens of small islands in the Gulf of Finland over distances of 1 to 8 km (Halkka et al. 1971). The spittlebug record for migrating altitude seems to be held by *Clastoptera xanthocephala*, a specimen of which was taken at 2,000 feet (610 m) in aerial plankton trawls over Louisiana, using biplanes otherwise designated for crop dusting experiments against boll weevil and mosquitoes (Glick 1939).

A similar pattern of movement has been observed for tropical pasture grass spittlebugs. For the most part they stay close to home, but they can and do migrate long distances. Nilakhe and Buainain (1988) made a thorough study of movement in the Brazilian pasture pests *Deois flavopicta* (Plate 15a) and *Notozulia entreriana* (Plate 10b). They tracked dispersal by sticky traps, mark-release-recapture experiments, and lab experiments with tethered individuals. They even chased dozens of individual flying spittlebugs across fields, recording their trajectory. Most individuals in their mark-release-recapture experiments stayed within 100 m of their release point, consistent with later work by Sujii et al. (2000), but there were important exceptions. On one occasion, 11 of 2,000 spittlebugs released were recovered 750 m away only 10 minutes after they were freed.

When Nilakhe and Buainain released and chased individual *D. flavopicta*, mean flight distances until landing ranged from 14 to 220 m, and mean distances until individual bugs flew higher and out of sight ranged from 24 to 310 m. One *D. flavopicta* individual flew 910 m before it disappeared, and a *N. entreriana* individual flew 295 meters in a single flight. During the chases, *D. flavopicta* flew at an average 11 km/hr. Tethered *D. flavopicta* individuals averaged about 8 minutes per flight and made repeated flights, with a maximum flight time of 18 minutes, corresponding to estimated travel of 3.4 km. Nilakhe and Buainain also observed a mass movement of spittlebugs flying at sunset from a pasture to an adjoining rice field, many at heights exceeding 20 m and in the absence of wind. It is one of many recorded instances of spittlebugs spilling out of pastures and into rice plantings (refs. in Chapter 7).

A Spittlebug Cross-Ocean Pioneer: Krakatoa After the Eruption

In 1883 the volcanic island of Krakatoa in the strait between Java and Sumatra was almost obliterated by a massive, explosive eruption, leaving a barren landscape cleared of all terrestrial plants and animals. This presented a blank slate, permitting scientists to monitor the recolonization of the remnants of the island. One of the early immigrants was the spittlebug *Clovia conifera* (Fig. 8.3c, Plate 13d), present by 1908 (Dammerman 1948). To get there it had to have crossed at least 14 km from the nearest neighboring island or 35 or 45 km from Sumatra or Java, respectively. *Clovia conifera* is probably the most widespread and abundant spittlebug in mainland and island Southeast Asia. I have collected it in numbers in Vietnam, Sarawak, and the Philippines. Its early post-eruption advent in Krakatoa is a testament to its natural dispersal ability.

The data from *P. spumarius*, *N. campestris*, *A. salicina*, and Brazilian pasture pests are consistent. When local conditions are favorable, spittlebug dispersal is minimal. But spittlebugs are good fliers, capable of dispersing long distances when molested or when local conditions deteriorate, and some species sometimes engage in mass movements. Add to this the likelihood that spittlebugs also hitch rides with local vehicles, and one may reasonably assume that any vector containment scheme based on hypothesized limited mobility is likely to fail.

In addition to local dispersal, spittlebugs can also travel very long distances from continent to continent or to distant islands. Long-distance invasions, which have implications for long-term *Xylella* threats, are covered in Chapter 10.

European *Xylella* Vectors Other than *Philaenus spumarius*

In addition to *P. spumarius*, two other European spittlebugs have been experimentally demonstrated to transmit *Xylella*: *N. campestris* and *Philaenus italosignus* (Cavalieri et al. 2019; Nencioni et al. 2024). Neither species is believed to be epidemiologically relevant in Salento. Though *P. italosignus* adults feed on a variety of trees during the summer, the nymphs feed only on understory lilies of the genus *Asphodelus* (Drosopoulos et al. 2010), limiting both their abundance and local occurrence. *Neophilaenus campestris* is much less abundant than *P. spumarius* in Italy. Its nymphs prefer grasses, but in Mediterranean summers the adults, like those of *P. spumarius*, move to a variety of trees and shrubs (Morente et al. 2018; Cornara et al. 2021). It is relatively more abundant in Spain (Morente et al. 2018; Gallego et al. 2023). There are about two dozen other potential spittlebug vectors in Europe, listed in Cornara et al. 2019. Most, by virtue of rarity or host specialization, are unlikely to be threats as vectors, but two in particular, *Lepyronia coleoptrata* (Plate 13h) and *Aphrophora alni* (Plate 11f), are highly polyphagous and widely distributed, suggesting they merit special attention should *Xylella* spread. *Aphrophora salicina*, primarily a spittlebug of willows (*Salix* spp.) and poplars (*Populus* ssp.), has already received attention as a potential vector (Casarin et al. 2023).

There are not many xylem-feeding leafhoppers in Europe. Cornara et al. (2019) list just eight species. Only one, *Cicadella viridis*, has the geographical range, host versatility, and population numbers to make it a possible threat. However, it is a poor *Xylella* transmitter (Bodino et al. 2022) and so far seems not to have tested positive for *Xylella* in nature. Cicadas are all xylem feeders and common in many Mediterranean environments but have received little scrutiny. Fewer than 2% of the abundant species *Cicada orni* collected in infected olive orchards tested positive for *Xylella*, and the species failed to transmit *Xylella* when tested (Cornara, Marra, Tedone et al. 2020). Cicadas in *Xylella*-infested areas of Corsica tested negative (Mesmin et al. 2023). There are a couple of reports of transmission by cicadas, but they require confirmation (Cornara et al. 2019). In sum, none of the other potential vectors are likely to displace *P. spumarius* as the *Xylella* vector of consequence in Europe.

Spittlebug Vectors of *Xylella* Outside Europe

North America

Outside Europe, there are several suspected but no proven spittlebug vectors of *Xylella*. In Northern California, in the first study of insect vectors, when the infective agent was still thought to be a virus, Severin (1950) demonstrated that four spittlebug species were capable of transmitting the cause of Pierce's disease to or among grapevines experimentally: *P. spumarius*, *Aphrophora permutata* (Fig. 1.91), *Aphrophora angulata*, and *Clastoptera lineatocollis* (misidentified as *C. brunnea*). Cornara et al. (2016) have reconfirmed experimental transmission by *P. spumarius*. Closer work with *Xylella* in local vineyards later indicated that two or three xylem-feeding leafhoppers were responsible for most transmission of *Xylella* in Northen California (Purcell 2013). However, the local distribution of spittlebugs, leafhoppers, and Pierce's disease suggests that a taxon identified as *A. permutata* (or a closely related species) may play a role in some vineyards, and it has proved experimentally capable of transmission (Beal et al. 2021). In Southern California, Pierce's disease has been a severe problem, vectored by *Homalodisca vitripennis*, a xylem-feeding leafhopper introduced from the southeastern United States (Purcell 2013).

In the southeastern United States, *Xylella* causes Pecan Leaf Scorch (Sanderlin and Heyderich-Alger 2000). Two spittlebugs, *Clastoptera achatina* (Plate 5d) and *Lepyronia quadrangularis*, have experimentally transmitted *Xylella* to pecan trees, though neither tested positive for *Xylella* in the field (Sanderlin and Melanson 2010). As a pest of pecans (Chapter 8), *C. achatina* is well-suited to vector *Xylella* in pecan orchards, but it is not yet proven to do so. In New Jersey, some spittlebugs recovered from oak canopies associated with oak leaf scorch tested positive for *Xylella*, including 33% of *Aphrophora quadrinotata* individuals tested, 7% of *Clastoptera obtusa*, and about 15% of *P. spumarius*. The epidemiological role of spittlebugs in this system, if any, is unclear.

Asia

Apart from the immediate vicinity of the Mediterranean, there are two confirmed instances of epidemic *Xylella* in Asia. (Cornara et al. 2019). In Iran, *Xylella* has been found in grapevines and almond, causing almond leaf scorch and Pierce's disease (Amanifar et al. 2014). Most of the potential vectors are spittlebugs (Mozaffarian and Jafary 2023). In Taiwan, *Poophilus costalis* (Fig. 8.3f, Plate 13g) and a xylem-feeding leafhopper have been experimentally demonstrated to acquire and transmit *Xylella* (Lin and Chang 2012). Whether they underlie the local epidemic of Pierce's disease remains to be demonstrated. *Poophilus costalis* is a wide ranging and often abundant spittlebug found from West Africa to East Asia. It is a pest of several crops (Chapter 8) and extremely polyphagous, with dozens of known hosts, including a wide range of plant taxa in Taiwan (Shih et al. 2005).

The Neotropics

Xylella in the form of native *X. f*astidiosa *pauca* is the cause of four serious plant diseases in Brazil: citrus variegated chlorosis, coffee leaf scorch, plum leaf scald, and olive leaf desiccation syndrome (OLDS). Spittlebugs are important candidate vectors only for OLDS, a disease equivalent to OQDS in Salento. The other diseases are transmitted by xylem-feeding leafhoppers (Cornara et al. 2019; Froza et al. 2024). In olives, the most likely vector is an undescribed spittlebug of the genus *Clastoptera* (Froza et al. 2023). Nymphs and adults are abundant on olives, and seven of seventy-five adults tested were positive for *Xylella*.

Though spittlebugs do not seem to be *Xylella* vectors in coffee in Brazil, they have been observed in large numbers in coffee plantations in Costa Rica, where Rojas et al. (2001) found a *Clastoptera* sp. to be one of three abundant xylem-feeding insects. Castillo et al. (2020) demonstrate that Costa Rican coffee hosts a rich array of *Xylella* strains, including both *X.f. fastidiosa*, for which Central America is the center of origin, and *X.f. pauca*, for which Costa Rican coffee is the putative origin of the destructive strain wreaking havoc in Salento. Costa Rica is home to dozens of *Clastoptera* species (Thompson et al. 2020). None, to my knowledge, have been demonstrated to carry *Xylella*, but they have been there coevolving with it for a long time, suggesting that they may have played a role in its evolution and spread among native plant taxa and introduced crops like coffee.

Control Measures

The devastating *Xylella* epidemic in Salento and the smaller epidemics in the Iberian Peninsula and Balearic Islands have led to an enormous amount of research aimed at control. These have been reviewed in detail by Morelli et al. (2021), who give a systematic overview of the various approaches that have been attempted. Cornara et al. (2018) and Serio et al. (2024) also provide considerable information on control. Despite frenzied efforts to find cures, prevention, or palliatives for *Xylella* itself, many approaches have failed and none have succeeded in field tests. For now, *Xylella* remains an incurable disease with the exception of regions such as Napa Valley, California where cold winter temperatures eliminate *Xylella* from some grapevines (Purcell 1980). As a result, control efforts continue to focus on a classic suite of phytosanitary measures: monitoring and detection, culling of infected plants, suppression of vectors, and selection or screening for resistant or tolerant cultivars (Morelli et al. 2021).

This puts continued emphasis on controlling *P. spumarius*. The major practical approaches are cultivation of orchard understories to control nymphs and use of pesticides against adults in olive canopies. Tilling the understory in spring destroys most of the nymphal host plants. This is sometimes accompanied by sowing a grass understory, generally unfriendly to *P. spumarius* (though not to the possible secondary vector *N. campestris*). Many pesticides are effective against adults, particularly synthetic pyrethroids. None of these methods eliminate all adults, and even a small surviving fraction of a large initial population can continue to do considerable harm as vectors. The methods are also expensive, contributing to a 31% increase in orchard management costs from EU-mandated control practices (Cornara et al. 2019). Moreover, massive tilling of the countryside herbaceous understory and heavy use of pesticides have their own serious and incompletely explored ecological costs in terms of loss of sensitive species and biodiversity (Labadessa et al. 2024).

As a result, there has been continuing interest in identifying alternative means of control. These include the use of natural enemies, such as the wasp egg parasitoid *Ooctonus vulgatus* (Fig. 5.6) (Mesmin et al. 2020), the big-headed fly *Verrallia aucta* (Fig. 5.7) (Molinatto et al. 2020), and the assassin bug *Zelus renardii* (Fig. 5.5) (Lahbib et al. 2022). The biology of these parasites and predators is covered in more detail in Chapter 5. There has also been exploration of the use of entomopathogenic fungi and nematodes (Yousef-Yousef et al. 2023; El-Khoury et al. 2024; Bodino et al. 2024), analogous to those used to control sugarcane spittlebugs in the Neotropics (Chapter 6). In addition, the endosymbiont

Wolbachia has been suggested as a control agent (Formisano 2022). Finally, investigators have explored the possibility of disrupting vibratory communications or using pheromones or other chemical compounds as lures or deterrents (Avosani et al. 2023). None of these approaches have short-term prospects for practical application, but they underline the importance of understanding basic spittlebug biology for finding long-term solutions for control (Cornara et al. 2018; Bodino et al. 2023).

Where Might Spittlebug-Transmitted *Xylella* Appear Next?

The severe Salento outbreak of *Xylella* was based on the unique confluence of a newly introduced *Xylella* strain and a few million highly susceptible olive trees concentrated in orchards with an effective, locally abundant spittlebug vector. Where might spittlebug-vectored *Xylella* strike next? Aside from the much-discussed possibility of geographical expansion in Europe, which the EU regulations are designed to forestall, spittlebug-vectored *Xylella* might spread to new hosts or to more distant geographical areas.

Xylella poses a particular threat to new hosts because strains can mutate and recombine, producing unpredictable host shift "leaps" to phylogenetically distant plant species (Almeida and Nunney 2015; Kahn and Almeida 2022). This potential is exacerbated by the convergence of *P. spumarius* and *Xylella* to polyphagy on an astonishing breadth of host plants—*Xylella* with 712 hosts in 312 genera and 89 families (EFSA et al. 2024) and *P. spumarius* with 1,311 host plants in 631 genera and 117 families (Thompson et al. 2023). This combination of an extremely versatile and genetically malleable plant pathogen combined with an extremely polyphagous insect host makes for an especially threatening brew. In a European example, *Xylella* may pose a threat to native forest trees (Desprez-Loustau et al. 2021). It has already been detected in European chestnuts, though they are presently asymptomatic (Greco et al. 2023).

Likewise, *Xylella* has not exhausted its geographical potential. It recently appeared in the Israel-Lebanon border region. In 2022, Israeli workers reported a *Xylella* outbreak on almonds in the Hula Valley in northern Israel, immediately adjacent to Lebanon (Zecharia et al. 2022). Lebanese workers followed shortly with a report of *Xylella* in southern Lebanon, just a few kilometers from the Hula Valley (Choueiri et al. 2023). An Israeli study has identified the spittlebug *Mesoptyelus impictifrons* as the most likely vector (Tomer et al. 2024). Tests of olives and stone fruits in the West Bank area of Palestine proved negative

(Alabdallah et al. 2024), consistent with evidence that *Xylella* has not spread within Israel outside the north (Zecharia et al. 2022). The Hula Valley and adjacent Lebanon are well situated for a cross-border collaboration to contain the *Xylella* outbreak, an initiative precluded, as I write in fall 2024, by the Israel-Hezbollah conflict and forced displacement of local civilians on both sides of the border. *Mesoptyelus impictifrons* occurs north at least to southeastern Turkey (Tanyeri et al. 2023) and may be the second spittlebug vector of importance identified in the Mediterranean region. *Philaenus spumarius* does not extend to Lebanon and Israel.

Philaenus spumarius is also probably absent from North Africa. It has not been reported there recently, and my guess is that older reports confused it with *Philaenus tesselatus*, a closely related sister species (Drosopoulos and Quartau 2002; Seabra et al. 2021). Confirmation of this hypothesis would require tracking down a lot of old specimens in European museum collections. What is certain is that *P. tesselatus* is common in Mediterranean Africa, at least in Tunisia and Morocco, and a prime candidate vector should *Xylella* spread to the area (Boukhris-Bouhachem et al. 2023; Haddad et al. 2024). Further afield, there are lots of opportunities for the spread of both *Xylella* and *P. spumarius*, including Australia (Taylor 2022), New Zealand—where *P. spumarius* has already arrived (Nielsen et al. 2019)—and Argentina (Giménez-Romero et al. 2022).

Other Spittlebug-Vectored Diseases

Beyond *Xylella*, spittlebugs have been implicated in the transmission of a limited but interesting range of pathogens.

Spittlebug-Transmitted Viruses Are Known Only from Africa

In a review of arthropod transmission of viruses, Nault (1997) states that neither spittlebugs nor other xylem-feeding Auchenorrhyncha are vectors of plant viruses. However, there seem to be two exceptions. As noted in Chapter 8, *Locris rubra*, *Locris maculata*, and *Poophilus costalis* (Fig. 8.3f, Plate 13g) are all demonstrated lab vectors of rice yellow mottle virus (Koudamiloro et al. 2015), and *L. rubra* may play an important role in its transmission (Nwilene et al. 1999). *Locris maculata* has also been cited as a vector of maize streak virus (Egwuatu and Ita 1982; Dike et al. 2002).

Sumatra Disease of Cloves Caused by a Xylem-Limited Bacterial Pathogen

In addition to *Xylella*, three other bacterial pathogens are limited to xylem sap: *Erwinia trachephila*, *Ralstonia syzgii*, and *Xanthomonas albilineans* (Kahn and Almeida 2022; De La Fuente et al. 2024). As in *Xylella*, different subspecies of *R. syzgii* attack very different hosts. One subspecies, transmitted by machaerotid tube-building spittlebugs of the genus *Hindola*, causes Sumatra disease of cloves in Sumatra and Java (Eden-Green et al. 1992; Lomer et al. 1993; Safni et al. 2018). Sumatra disease kills trees over the course of six to eighteen months after symptoms appear and causes annual economic losses of up to 5 to 10% per year. This seems to be the only firmly established case other than *Xylella* in which a spittlebug serves as the vector of a deadly bacterium causing important economic losses.

Phytoplasmas Are Probably Not Transmitted, But a Couple of Cases Are Open

Cornara et al. (2018) note several cases in which *P. spumarius* has been associated with phytoplasma pathogens but say none are confirmed. In one case they do not address, Jenser et al. (1981) report that *P. spumarius* can transmit the phytoplasma causing Rubus stunt disease from raspberry to celery (it was not tested in the opposite direction). In the latest review of Rubus stunt, Linck and Reineke (2019) cite this example but state that *P. spumarius* probably does not play an important role in the natural spread of Rubus stunt disease because it does not specifically live on *Rubus*. However, *P. spumarius* feeds on twenty-one species of *Rubus* (Thompson et al. 2023), suggesting ample opportunities for transmission. This case appears to remain open. A recent report (Avramov 2022) that both *P. spumarius* and *Lepyronia coleoptrata* (Plate 13h) are vectors of "*Candidatus* Phytoplasma *solani*" requires confirmation.

Xanthomonas albilineans: A Xylem-Limited Bacterial Pathogen of Sugarcane

Very recently, the sugarcane pest *Mahanarva fimbriolata* (Plate 14g) has been implicated in the transmission of *Xanthomonas albilineans*, a xylem-inhabiting bacterium that causes leaf scald, a destructive disease of cane. Bini et al. (2023) have demonstrated experimentally that *M. fimbriolata* can acquire and transmit *X. albilineans*, both in the context of an artificial diet and among cane plants. So far, there is no direct evidence of transmission in the field, but the investigation was triggered by an outbreak of leaf scald in a controlled, disease-free nursery plot that coincided with its infestation by *M. fimbriolata*. Most interestingly, *X. albilineans* is closely related to *Xylella*. The *Xylella* lineage is phylogenetically embedded "as a minor variant lineage" in a larger lineage of *Xanthomonas* species, leading to a proposal to subsume it in an enlarged genus *Xanthomonas* (Bansal et al. 2021). In this larger lineage, *X. albilineans* clusters adjacent to *Xylella*. They share reduced genomes and are the only xylem-limited Xanthomonadaceae (Kahn and Almeida 2022). The relationship is close enough that one study used *A. albilineans* as a faster growing surrogate in searches for *bacteriophages* (bacteria that infect and replicate within bacteria) to be used against *Xylella* (Clavijo-Coppens et al. 2021). In this study, one of the phages attacking *Xylella* was identified in *P. spumarius* samples collected in a *Xylella*-infested area in Corsica, hinting that *P. spumarius* might serve as a delivery vehicle for injecting phages into the xylem stream, should phages ever prove to be practical to use against *Xylella*. If so, there would be distinct irony in using the main vector of the bacterium to attack the bacterium. Regardless of that connection, further studies of *M. fimbriolata* and *X. albilineans* may open the path to a new class of spittlebug-vectored, xylem-hosted bacteria.

Spittlebug Vectors: Prospects in a Future World of More Widely Distributed *Xylella*

Prophesy is a dangerous game for scientists. Nature inevitably finds paths unanticipated and unpredictable. Nevertheless, one may reasonably speculate that *Xylella* will eventually find its way to corners of the world now untouched, and it may already have spread to other areas undetected. Areas at risk include large portions of temperate and tropical Asia and Africa as well as New Zealand and parts of Australia. Each has local spittlebugs, often abundant, often highly polyphagous, ready to step in as vectors.

The range of *P. spumarius* in Eurasia, for example, extends well past the present areas of *Xylella* introduction in Southern Europe, reaching from the British Isles through Northern and Eastern Europe, Russia, Turkey, and Iran to Kirgizstan, where it is abundant, and to temperate East Asia, where its distribution is poorly documented. *Philaenus spumarius* is already documented as a *Xylella* vector in Italy and California. There is no reason it should not be an effective *Xylella* vector over vast portions of Eurasia if, as appears inevitable, infection spreads as far as climate and flora permit. In the North African Maghreb, *P. tesselatus* looks poised to serve the same role. In warmer parts of Africa and Asia, *P. costalis* (Fig. 8.3f, Plate 13g), another highly polyphagous, demonstrated *Xylella* vector, is extremely wide-ranging and often abundant, making it a prime candidate for transmission in many presently uninfected areas. There are many other candidate spittlebug species ready to audition. In some areas, as presently in California and Brazil, xylem-feeding leafhoppers may turn out to be more important epidemiologically, but I would be mightily surprised were spittlebugs not to play an important role in future epidemics. This is all, of course, subject to the vagaries of climate change and the peregrinations of spittlebug vectors, which is the subject of the next and last chapter.

10

SPITTLEBUGS IN A CHANGING WORLD

Shifting Risks and Threats

In the preceding chapters we have gotten an overview of spittlebugs and their past and present roles in natural and agricultural settings. This chapter looks toward the future. Human activity drives ceaseless environmental change, either directly, by transforming natural habitats and cropland, or indirectly, by stimulating climate change through greenhouse gas emissions. During the past century, changes in Neotropical sugarcane and pasture grass cultivation led to enormous increases in the importance of spittlebug pests of those crops. Likewise, in this century, the introduction of Xylella fastidiosa *turned Italian populations of* Philaenus spumarius *from abundant but relatively inconsequential insects into fearsome vectors of a destructive disease. What changes might we anticipate in the distribution and importance of spittlebug pests going forward? How might spittlebugs fare in the face of global warming? And, in this context, what developments hinder and help the study of spittlebugs?*

Spittlebugs Were Among the First Insects Studied for Potential Effects of Climate Change

When climate change studies first took off widely in the 1990s, British ecologist John Whittaker (see sidebar: John Whittaker (1939–2024): Pioneer of Spittlebug Climate Change Studies) was well positioned to make a mark. In 1971, he demonstrated that toward the upper margin of its altitudinal range, *Neophilaenus lineatus* (Plate 11d) population size is determined by nonbiological factors like weather and temperature (Whittaker 1971b). In the mid-1990s, Whittaker and his colleagues built on this work to examine the likely effects of long-term global warming on spittlebug population size and distribution.

Using 10-year data from an 880 m altitudinal transect in Scotland, Whittaker and Tribe (1996) demonstrated a strong relationship between spring temperatures and the maximum altitude at which *N. lineatus* nymphs occurred in a given year (Fig. 10.1). They calculated that a global mean annual temperature increase of 2 to 3°C would permit populations to expand 300 m uphill. It would also accelerate egg hatching and shift nymphal development two to three weeks earlier in the season. In a subsequent paper they extended the analysis to long-term population density at Moor House in the Pennine Mountains in England (Whittaker and Tribe 1998). Based on 1961 to 1992

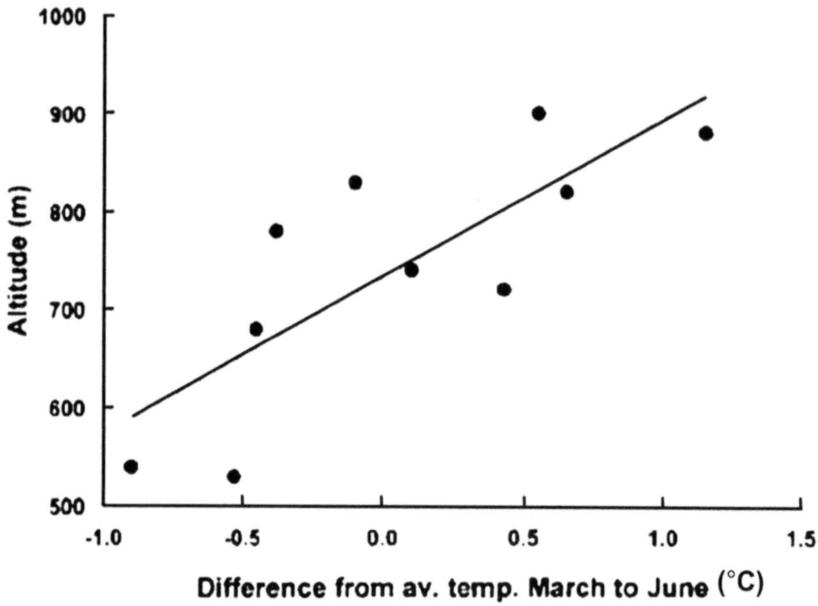

FIGURE 10.1. Relationship between the maximum mountainside altitude at which *Neophilaenus lineatus* nymphs occurred and average spring temperature during the same growing season. Temperature is expressed as deviation from the long-term average. Negative deviations represent colder springs, positive deviations warmer springs. There is a strong positive relationship. The warmer the spring, the higher nymphs occurred on the altitudinal gradient. Reproduced from Whittaker and Tribe (1996) under Creative Commons Attribution 4.0 International License (https://creativecommons.org/licenses/by/4.0/.

meteorological observations, population density there correlated best with mean minimum September temperature, a reflection of the fact that autumn frosts kill egg-laying females, truncating the reproductive season (Fielding et al. 1999). Combined with the previous year's population size, September temperature explained 75% of the annual variation in population size and predicted that a 1°C mean temperature increase would increase N. lineatus population density by 50%.

Whittaker and Tribe tested this prediction experimentally by covering eight circular vegetation plots with *cloches*, plastic covers that raised internal temperature by a mean of 1°C over ambient for the May through September observation period. *Neophilaenus lineatus* population densities rose 157%, substantially more than predicted, which, they suggest, might be attributable to protection from wind and restricted dispersal of adults. In a separate experiment with *P. spumarius* to test the effects of winter warming, electric heating cables (roughly equivalent to an electric blanket) raised the November through April temperature of the soil surface in vegetation plots 3°C over ambient. This sped up the date of egg hatch but did not influence the subsequent rate of nymphal development (Masters et al. 1998) and had no appreciable effect on population density (Whittaker 2001).

While elevated CO_2 causes long-term temperature increase, it might also have direct effects on organisms. Brooks and Whittaker (1999) investigated the consequences of increased CO_2 concentrations on *N. lineatus* survival and development (Fonseca et al. 2023). They excavated blocks of turf, with the surface vegetation intact, from the Moor House site and installed them in solar domes in Oxfordshire. Two domes had ambient CO_2 levels, which stood at about 350 ppm at the time of the experiment in 1993 to 1994. Another two domes had levels of about 600 ppm, simulating expected increases in CO_2 levels. Each solardome had six replicate turf blocks, which contained naturally deposited *N. lineatus eggs*. The solardomes were sealed enclosures with inflowing air cooled by chilling units. This countered the effects of solar heating and maintained internal temperatures comparable to the site of origin. After the eggs hatched, nymphs of that generation and the one succeeding were monitored in the domes, corresponding to two generations. Elevated CO_2 reduced nymphal survival by 20% and 27% in the first and second generations, respectively, and in some instances delayed nymphal development. Brooks and Whitaker suggest that these effects may have been due to a decrease in host vegetation transpiration by about half during daytime hours. Lower transpiration leads to decreased xylem sap flow, potentially reducing the nutrient flow available to nymphs (see Chapter 3). A parallel experiment with *P. spumarius* showed no significant effect of elevated CO_2, although nymphal survival was lower than in controls (Whittaker 2001).

Overall, it appears that temperature increases from global warming will extend the British elevational range of *N. lineatus* upward and increase its numbers at higher elevations, but it will not have much effect on its geographical range (Fielding et al. 1999), whereas elevated CO_2 will have little consequential direct effect on either *N. lineatus* or *P. spumarius* (Whittaker 1999).

John Whittaker (1939–2024): Pioneer of Spittlebug Climate Change Studies

Midway through my PhD studies, in summer 1973, John Whittaker (Plate 16a) dropped by to see me in Chicago. He was flying back to England after a trip to Alaska. I had published my first two papers on *Philaenus spumarius* the spring before (Thompson 1973; Thompson and Halkka 1973). He already had a decade worth of publications on spittlebugs under his belt and swooped in to see the new fellow in *Philaenus* studies. He stayed with us, we talked, and I showed off a few Chicago area field sites, including alfalfa plantings. He marveled at the taxonomic diversity of local Auchenorrhyncha, and I think we discussed the possibility that the preference of *P. spumarius* for nitrogen-fixing legumes like alfalfa might be linked to xylem feeding. This kernel of insight bore fruit two decades later in papers analyzing the biology of xylem feeding (Press and Whittaker 1993) and a pattern of spittlebug preferences for nitrogen-fixing plants (Thompson 1994).

Whittaker had begun PhD fieldwork in 1960 at the Moor House National Nature Reserve, a montane site said to have "probably the harshest climate in England" (Whittaker 2001) and where both he and the spittlebugs "proved to be at the edge of their ranges." (John B. Whittaker, personal comm. 2023). There he had the good fortune to interact occasionally with Charles Elton, perhaps the best-known animal ecologist of the period. Elton later hired Whittaker to work at Oxford University's Bureau of Animal Population, permitting him to make a parallel study of local lowland spittlebug populations. Comparative studies in these two habitats and a serendipitous 1966 move to Lancaster University, where work was already underway on the effects of greenhouse gases on plants, set the stage for his pioneering 1990s work on spittlebugs and climate change. In 2001 he was elected President of the British Ecological Society and delivered a presidential address titled "Insects and Plants in a Changing Atmosphere" (Whittaker 2001).

Climate Warming May Be a Cause of *Philaenus spumarius* Decline on the California Coast

Half a world away from Britain, in 1983, Richard Karban initiated a continuing series of investigations that implicate global warming and climate change in declining numbers of *P. spumarius* on the California coast. Karban, an ecologist at the University of California, Davis, tracked *P. spumarius* nymphal populations on a single, abundant native host, the sea daisy *Erigeron glaucus*. He and Sharon Strauss observed that between 1988 and 2001, the range of *P. spumarius* seemed to shift northward, which they attributed to lower humidity and rising temperature due to a warming climate (Karban and Strauss 2004). Extending these studies to 2018, Karban and Mikaela Huntsinger concluded that rather than shifting northward, *P. spumarius* was experiencing a general and very marked decline along its whole coastal range (Karban and Huntsinger 2018). More recently, they have extended their population surveys and identified protected coastal refugia where nymphal spittlebugs occur on plants other than sea daisy, demonstrating that *P. spumarius* is still present in those locales despite its decline on sea daisy (Karban and Huntzinger 2021).

Their results are consistent with an unpublished survey I conducted of coastal *P. spumarius* adults in 2018. South of Carmel I found only sparse populations in streamside refugia extending partway down Big Sur. However, north of San Francisco I found robust populations at Stinson Beach and sites close to Manchester and Beatrice, California, with population levels comparable to those of populations I had sampled at or close to the same localities in 1970 (Thompson and Halkka 1973). The upshot is that while there is strong evidence that *P. spumarius* populations are in decline in some coastal California habitats, perhaps linked to climate change, the area-wide population dynamics are complex and yet to be fully explored.

European *Philaenus spumarius* Populations Show Correlations with Climate That Portend Future Change

Climate warming may reduce local *P. spumarius* populations in at least one area of Finland. Halkka et al. (2006) studied the relationship between the North Atlantic Oscillation (NAO) index and *P. spumarius* population size on three small islands in the Baltic Sea in southern Finland. The NAO index is based on the surface sea-level pressure difference between the Atlantic

Subtropical (Azores) High and the Subpolar Low. When the index is positive, winter temperatures in Northern Europe tend to be warmer, and when negative, cooler. Using spittlebug population size data for 1970 through 2005 and survival data for 1969 through 1978, they correlated warmer winters with smaller populations. Apparently, warmer winters had less snow, increasing the probability of desiccation of host plants the following spring and causing greater mortality in late instar nymphs. This implies that longer-term warming might cause the decrease of these and similarly situated *P. spumarius* populations in Northern Europe.

Mathematical modeling suggests that several parts of Mediterranean Europe will also become less suitable for *P. spumarius* with warming (Godefroid et al. 2021). As in Finland, climate parameters related to desiccation are important components of this prediction. Based on projected changes, the Iberian Peninsula, coastal Turkey, central and southern France, and most of the Italian lowlands will become less hospitable to *P. spumarius* by 2040–2060. In some more northerly European areas, climate may become more suitable. The possible contradiction between this forecast and the Finnish projection may be a matter of local geography. On the small Finnish islands investigated, soils are thin and especially susceptible to drought.

Philaenus spumarius Color Polymorphism Might Be a Barometer of Climate Warming

Cooler climates often favor darker (melanic) color forms. These forms heat more quickly and reach higher equilibrium temperatures in sunlight. This phenomenon, called thermal melanism, gives melanic forms an advantage in cooler weather (Clusella Trullas et al. 2007). As noted in Chapter 5, P. *spumarius* exhibits a striking color polymorphism, including several dark forms (Figs. 5.1, 8.1). Consistent with the hypothesis that they have been selected for thermal melanism, these dark forms occur at higher frequency in northern localities in both North America and Europe (Thompson 1984b, 1988). As a corollary, warming climate should lead to a reduction in the frequency of *P. spumarius* melanic color forms in cooler northern localities.

Ruth Moscovitch and I tested this corollary in northern Minnesota. This is one of the coldest areas in the North American distribution of *P. spumarius* and the area that exhibited the highest frequency of melanic color forms in a continent-wide survey of *P. spumarius* color polymorphism (Thompson 1984b). Between 1974 and 2021 the mean September temperature in the area rose 2.7°C, while the frequency of melanic forms (LOP, LCE, FLA, LAT, Fig. 5.1) declined by approximately one third, from 22.1 to 14.2% in females and from 12.7 to 7.3% in males (Thompson and Moscovitch 2022). In addition, the proportions of the local nonmelanic color forms TYP and POP (Fig. 5.1, 8.1) shifted toward greater predominance of TYP, the darker of the two. We chose September temperature to measure change because, as in parts of Britain (Fielding et al. 1999), it is the season when adult females must lay eggs before being killed by frost. The results are consistent with relaxation of local selection for thermal melanism as a result of a warming climate. If they can be generalized to other areas, the *P. spumarius* color polymorphism might serve as a barometer of climate change, at least in colder climates.

The Perils of a Five-decade Experiment in Uncontrolled Locales

When Ruth and I returned to northern Minnesota in 2021, we could locate the exact location of only two of our five original 1974 sites with certainty (Thompson and Moscovitch 2022). One was at a crossroads. The local vegetation had changed, but the site remained undeveloped. Three sites had changed beyond recognition due to development or ecological succession (in the early 1970s we had no GPS to record site locations). The fifth site, a local monument to the first iron mine in the area, had been relandscaped in spittlebug-inimical fashion. What had been an unkempt meadow in 1974 had turned into a manicured lawn in 2021. An Illinois comparison site that had been an alfalfa field in 1971 had become a housing subdivision decades before we returned in 2021. In each case we found suitable nearby substitute sites, but it was a lesson in the impermanence of human influenced habitats. A chance encounter with a retired biologist who was inline skating down the road in the middle of nowhere at one of our substitute sites led to two articles on our work in the local newspaper, the Tower, Minnesota *Timberjay*, a high point in local spittlebug news (Helmberger 2021, 2023).

Climate Change Will Influence Spittlebug Spread of *Xylella*

Saifi et al. (2024) and De La Fuente et al. (2024) review prospective *X. fastidiosa* risks in light of projected climate change. The strain of *Xylella* causing Pierce's disease presently seems to pose no major danger to existing grape growing areas outside North America. The zones at risk are predicted to expand with climate change by 2050, but not dramatically. Yoon and Lee (2023) and Bosso et al. (2016) suggest that *P. spumarius* is ecologically flexible enough to be irrelevant to predictive models, but Godefroid et al. (2021) make a persuasive case that *P. spumarius* climate tolerances are an important component of models assessing *Xylella* threats in Europe. Based in part on these tolerances, several studies predict lower future vulnerability to *Xylella* in large parts of Mediterranean Europe (Godefroid et al. 2021, 2022; Giménez-Romero et al. 2022).

In one local study, Farigoule et al. (2022) tracked the presence of *X. fastidiosa* in *P. spumarius* samples in Corsica using the spittlebugs as sentinel organisms. Higher rates of infection correlated with warmer local climate, leading them to predict that, all other things equal, the frequency of *X. fastidiosa* in *P. spumarius* is likely to increase with increasing ambient temperature. This would put areas experiencing milder winters and warmer springs and falls as a result of climate change at greater risk of outbreaks. However, warming would also drive the vector toward more elevated areas, shifting the local geography of risk.

The Ranges of South American Pasture and Sugarcane Pests May Shift Under Climate Change

Schöbel and Carvalho (2019) and Hernández et al. (2021) have modeled the potential ranges of South American sugarcane and pasture pest spittlebugs under both present climatic conditions and those predicted under climate change. Several pest species appear to have the potential to expand their ranges under current conditions. These include the cane pests *Mahanarva fimbriolata*, which could expand south; *M. posticata*, which could expand south and west (Schöbel and Carvalho 2019); as well as the pasture and/or cane pests *Aeneolamia lepidior*, *A. reducta*, *A. varia*, *Prosapia simulans*, *Zulia carbonaria*, and *Z. pubescens*, all of which appear to have the potential to expand their ranges into contiguous areas (Hernández et al. 2021).

Mahanarva fimbriolata also has the potential to invade trans-Andean Ecuador and Colombia, as well as Mexico's Gulf Coast and Baja California, if it were to be introduced. With the exception of Baja California, these regions include extensive areas of sugarcane cultivation, putting them at risk. *Mahanarva posticata* likewise appears to have the potential, if introduced, to invade trans-Andean Ecuador, Guatemala, and parts of western Mexico (Schöbel and Carvalho 2020). Long-distance introductions have been uncommon among spittlebugs, but far from unprecedented (Table 10.1, and see section on Long Distance Invasions, this chapter).

With climate warming, the present Brazilian ranges of *M. fimbriolata*, *M. liturata*, *M. posticata*, and *M. spectabilis* would all become less hospitable, though opportunities to colonize other Neotropical areas, some requiring long-distance introductions, would open up (Schöbel and Carvalho 2020). The modeling results for *M. spectabilis* are roughly consistent with experimental measurements of *M. spectabilis* nymphal survival, adult longevity, and egg production at experimental temperatures ranging from 16 to 32°C (Fonseca et al. 2016). The recent discovery (see Chapter 6) that *M. fimbriolata*, *M. spectabilis* and a newly discovered sibling species constitute a previously unrecognized species complex throws the predictions for *M. fimbriolata* and *M. spectabilis* into doubt, pending reanalysis of their respective current distributions.

Aeneolamia lepidior and *A. reducta* seem poised to expand into large areas of Amazonia (Hernández et al. 2021). The recent extension of *A. reducta*'s range to Roraima state in Brazil (Fidelis et al. 2021) may be a harbinger of such expansion. *Prosapia simulans* stands to lose ground in its present range in Central America, Colombia, and Venezuela, but pick up suitable climates further south in South America. *Aeneolamia varia* stands to lose favorable habitat, suggesting it will contract in range and/or numbers with climate warming (Hernández et al. 2021). In addition, climate change could also, unpredictably, favor the recruitment of new pests from among the many grass-feeding Neotropical Cercopidae potentially available for the role.

The Limitations of Climate Models

Modeling exercises have to be taken with a grain of salt. Models are only as good as their assumptions and underlying data, which are both imperfect. They inevitably simplify complex climatic and biological realities. They also implicitly assume that spittlebug evolution is static in the face of a changing environments (Fonseca et al. 2016). We have already seen evidence that *P. spumarius* evolved in response to climate change in a few dozen generations (Thompson and Moscovitch 2022), and *A. varia* evolved even more quickly in response to pesticide pressure in in Trinidad (Chapter 6). Also, changing sugarcane and pasture grass distributions will lead to new grass/spittlebug species combinations in local habitats with unknown practical consequences for agriculture. We know only that the spittlebugs will evolve in response to changing temperature, humidity, and planting regimes as the earth warms, with unpredictable results.

Climate Change Is Implicated in the Range Expansion of Two Old World Spittlebug Pests

In recent years, the rice spittlebug *C. versicolor* has expanded northward, apparently in response to warming climate, a contributing factor to its emergence as an important pest in China (Chen and Liang 2012). In Europe, the pine pest *Haematoloma dorsata* has also expanded northward in the past few decades, also probably due to climate warming. In the Netherlands, its increase as a pest has been reinforced by understory proliferation of a nymphal host, the grass *Deschampsia flexuosa*, its spread apparently driven by increased anthropogenic nitrogen deposition (Moraal 1996). Both these pests are discussed in more detail in Chapter 8.

Long-Distance Invasions

Local dispersal, an important aspect of spittlebugs' threat as regional pests, is covered in detail in Chapter 9. Long-distance dispersal poses broader geographical threats to crops and forestry. For example, following its introduction to North America, *P. spumarius* became a major component of the insect fauna in disturbed habitats—such as meadows, roadside herbage, and successional old fields—and eventually became an important pest of forage legumes and strawberries (Chapter 8). *Prosapia simulans*, already a pasture pest in Mexico and Central America, has become a pest in Colombia after its recent introduction into South America, and *Prosapia bicincta* has been cutting a swath of destruction through Hawaiian pastures after recent introduction there (Chapter 7). When introduced to Bermuda, *Clastoptera undulata* caused damage to *Casuarina* plantings (Bennett and Hughes 1963; Cock 1985). It also attained nuisance levels on *Casuarina* in Florida when introduced there in the same period (Mead and Bennett 1987). Climate change may increase the potential for long distance dispersal by turning formerly unsuitable habitats into areas amenable to invasion.

Overall, long distance spittlebug introductions are uncommon events. Both nymphs and adults must feed almost constantly on living plants, making them poor candidates for transoceanic dispersal. When spittlebugs do invade, it is most likely to occur via dry season or winter diapause eggs embedded in soil or plant debris; this, for example, the probable mode of dispersal of *P. bicincta* to Hawaii (Wilson et al. 2023). There are eighteen known instances of long-distance introductions involving eleven spittlebug species, which are summarized in

TABLE 10.1. Long-distance introductions of spittlebugs

SPECIES	ORIGIN	INTRODUCED TO	TIME OF INTRODUCTION	REFERENCE
Aphrophora alni	Palearctic	Eastern North America	Before 1927	Hamilton 1982
Aphrophora salicina	Palearctic	Eastern North America	Before 1921	Hamilton 1982
Bathylus alba	Australia	New Zealand	Before 2004	Ashcroft & George 2004
Clastoptera querci	Cuba?	Florida	Before 2012	Thompson et al. 2020
Clastoptera undulata	Caribbean Islands	Bermuda	Before 1959	Bennett & Hughes 1963
Clastoptera undulata	Caribbean Islands	Florida	Before 1953	Mead & Bennett 1987
Clastoptera xanthocephala	SE North America	Hawaii	Before 1997	Wilson et al. 2023
Hindoloides bipunctata	Japan?	Korea	Before 2018	Lee et al. 2018
Lepyronia coleoptera	Palearctic	E North America	Before 1940	Hoebeke & Hamilton 1983
Neophilaenus lineatus	Europe	E&W North America	1800s?	Hamilton 1982
Philaenus spumarius	Palearctic	E&W North America	1800s?	Seabra et al. 2021
Philaenus spumarius	Palearctic	Azores	Unknown	Seabra et al. 2021
Philaenus spumarius	Palearctic	Hawaii	Before 1945	Davis & Mitchell 1946
Philaenus spumarius	Palearctic	New Zealand	Before 1960	Archibald et al. 1979
Philaenus spumarius	Palearctic	Iceland	1970s?	Endrestøl 2013
Philaenus spumarius	Palearctic	Réunion Island	Before 1998	Quilici et al. 1998[a]
Prosapia bicincta	SE North America	Hawaii	Before 2016	Wilson et al. 2023
Prosapia simulans	Mexico/Central America	N South America	Before 1980	Peck et al. 2001

[a] This introduction seems not to have persisted.

Table 10.1. Six of the eighteen invasions involve *P. spumarius*, which clearly has a proclivity for distance travel. This makes it an even more dangerous threat for the introduction of *Xylella* into novel environments. Areas at particular risk include Australia (Taylor 2022), New Zealand—which already has large, introduced populations of *P. spumarius* (Nielsen et al. 2019)—and grape-growing areas of Argentina (Giménez-Romero et al. 2022).

Long-term Ecological Effects of Higher CO_2 Levels

As noted in Chapter 3, all important spittlebug grass pests feed on C4 grasses, which include sugarcane. In warm environments with adequate water, C4 plants have more efficient photosynthesis than C3 plants (see sidebar: C4 Photosynthesis and C4 Plants, Chapter 3). This is what makes C4 crops, such as sugarcane, maize, and the extensively planted C4 pasture grasses (Table 7.1), so productive. C4 grasses have been abundant for only the last six to eight million years. The New World grass-feeding Cercopidae have apparently evolved in tandem (Thompson 2004; Paladini et al. 2015; and see Fig. 2.11).

C4 grasses evolved as an adaptation to relatively low levels of atmospheric CO_2. As levels rise, they lose their photosynthetic advantage (Sage 2004). Taken as an isolated factor, increasing CO_2 concentrations should favor C3 grasses, disadvantaging C4 grass-feeding spittlebugs in natural habitats. However, the spread of warmer habitats should favor C4 grasses, so long as these habitats are not too dry. Depending on how these competing factors play out, spittlebugs may gain (Sage and Kubien 2003) or lose (Collatz et al. 1998) suitable C4 grassland habitat on a large scale. Right now, the long-term ecological effects of increasing CO_2 levels on spittlebug-hosting natural grasses are unpredictable.

In an ironic turn, planted pastures in parts of the Neotropics may themselves help mitigate the rise of atmospheric CO_2. Replacement of Orinoco area natural savannah in Venezuela with *Urochloa decumbens* led to large increases in root-sequestered carbon, primarily because *U. decumbens* stays green and maintains photosynthesis in the dry season (San José et al. 2014), one of its great attractions to spittlebugs in the Brazilian Cerrados (Chapter 7). If this phenomenon extends to broader areas, some Neotropical pasture plantings may actually help sequester CO_2 and reduce atmospheric levels, though with inevitable offsets from, for example, climate-warming methane produced by belching cattle.

Engineered Nitrogen-fixing Grasses Could Be at Greater Risk of Spittlebug Attack

Whatever the long-term interactions among C4 grasses, CO_2 levels, and CO_2-induced global warming, C4 crop species are a tempting target for genetically engineered nitrogen fixation. Though the atmosphere is 80% nitrogen, this nitrogen is tightly bound in N_2 form, unassimilable by plants and animals. However, some plants have formed symbiotic partnerships with bacteria that fix atmospheric nitrogen in chemically accessible form. As we saw in Chapter 3, transfer of this nitrogen to plants ultimately provides a source of xylem-borne amino acids that makes nitrogen-fixing plants particularly attractive spittlebug hosts.

Engineering new nitrogen-fixing crops is one of agricultural biotechnology's most anticipated ends. If achieved it could lead to 1) large cost savings by reducing the need for chemical nitrogen fertilizers; 2) lower production of these fertilizers, manufacture of which is presently the source of about 5% of global greenhouse gas emissions; and 3) less pollution of waterways from excess fertilizer runoff (Mus et al. 2016; Guo et al. 2023; Gao and Serrenho 2023). However, biologically fixed nitrogen is not free. Splitting the molecular bonds in atmospheric N_2 is energetically expensive, potentially cutting into crop yields by diverting photosynthetic energy for that purpose. As a result, the most promising candidate crops for genetically engineered nitrogen fixation are C4 grasses, which, by virtue of their highly efficient photosynthesis, are best equipped to provide the extra photosynthetic energy needed to power nitrogen fixation.

As noted in Chapter 3, many C4 grasses already associate with naturally occurring nitrogen-fixing bacteria, which provide a substantial portion of their nitrogen needs (Thompson 2004; Chalk 2016; Zhang et al. 2022). These associative nitrogen-fixing symbioses provide a promising route to further enhance nitrogen fixation; this is done either by bolstering the performance of existing bacterial partners or by introducing or engineering new ones, approaches being applied to sugarcane (Carvalho et al. 2022), maize (Carvalho et al. 2023), and pasture grasses (Silva et al. 2023). However, there is a major potential downside. Sugarcane and pasture grasses are already subject to destructive attack by spittlebugs (Chapters 6 and 7), likely due in part to the boost that associative nitrogen fixation gives to production of the xylem-borne amino acids that sustain spittlebugs (Thompson 2004; see Chapter 3). There is substantial risk that new commercial cultivars with purposefully enhanced levels of nitrogen fixation will attract and support larger spittlebug populations, intensifying damage. As a result, sugarcane and grass breeders may need to redouble parallel efforts to incorporate resistance to spittlebugs, which is already the most promising path to combatting spittlebug damage to pasture grasses (Chapter 7).

Threats to Spittlebugs and Their Study

Spittlebugs and their study are subject to several threats. Insects in general seem to be in decline. Spittlebugs in natural or seminatural habitats are not immune to this trend. Going forward there may be fewer to collect. Independently of what is going on in nature, there are increasing bureaucratic barriers to international collecting. There is also a chronic shortage, likely to grow, of experts in spittlebug taxonomy. Even if we manage to maintain and expand insect collections, there may not be sufficient taxonomists to unlock their practical and scientific value.

No "Insect Apocalypse" for Spittlebugs, but Some Local Declines

There has been considerable coverage in the press and scientific literature of reported sharp declines in insect numbers over recent decades (Sánchez-Bayo and Wyckhuys 2019; Wagner 2020), with causes essentially unknown but habitat loss suspected as a primary driver. No data suggests a general spittlebug apocalypse, least of all for the important pasture and sugarcane pest species, but there has been some evidence of local decline.

For a few North American species, I can offer anecdotal observations from a locality I have monitored in Wonalancet, New Hampshire since the mid-1970s. The site is a pristine mountain valley, elevation about 350 m, where the only agriculture since at least the mid-twentieth century has been cultivation of mixed grass-legume hay and some grazing by farm animals, mostly sheep, with minimal application of agricultural chemicals. When I started collecting in the 1970s, *Philaenus spumarius* was by far the most common spittlebug in the valley, followed by *Neophilaenus lineatus* and *Lepyronia quadrangularis*. Both *P. spumarius* and *N. lineatus* are long-established Eurasian introductions. By the 2010s, *P. spumarius* was in decline (Seabra et al. 2021) in both a large local hayfield, mowed twice per summer, and an old field turned meadow, mowed annually. *Lepyronia quadrangularis*, a native species of woodland margins, also declined in the same period.

The most likely culprit in these declines was the invasive Eurasian spittlebug *Lepyronia coleoptrata* (Plate 13h) (Hoebeke and Hamilton 1983), which I first collected locally in 1995, and which subsequently became the most abundant spittlebug in both the hayfield and the meadow. *Philaenus spumarius* is now rare in the meadow and severely reduced in numbers in the hayfield, though it bounced back in the hayfield in 2024 after several years of very low numbers. It is tempting to attribute its long-term decline to competitive displacement by the more recent invader. Whatever the cause, this local decline in *P. spumarius* fits a broader pattern in which most of the observed broad decline in insect abundance can be explained by declines in formerly abundant species (van Klink et al. 2024)

The native spittlebug *Aphrophora quadrinotata*, which was formerly frequent in the valley but never especially abundant, has also declined to the point that I have not collected it in years. It is tempting to attribute this decline to competition from another invasive Eurasian spittlebug, *Aphrophora alni*, which first appeared in local collections in 2003 and occupies a similar ecological niche to *A quadrinotata*. Another native species, *Aphrophora gelida*, appears to have vanished locally around 2021. As recently as the mid-2010s, *A. gelida* nymphs were abundant on young red spruce (*Picea rubens*) in a hilltop forest. In spring 2020, with time on my hands from the COVID pandemic, I decided to follow its life history for a season, but it had almost vanished. Despite diligent searching over a large stand of young spruce, I found only three nymphs. The following year, and in the years since, I have found none. *Aphrophora gelida* is a spittlebug of the cool boreal zone (Hamilton 1982), suggesting that its apparent local extinction might be due to climate warming.

The best documentation for long-term decline of a spittlebug as a presumptive result of climate warming comes from the *P. spumarius* studies by Karban and his colleagues in California, summarized previously. In eastern Germany, populations of *Neophilaenus minor*, a spittlebug of dry grasslands, exhibited severe decline over a forty-year period ending in 2010, but this is most likely due to factors other than climate change (Schuch et al. 2012). In Australia, prolonged drought led to a marked decrease in the availability of spittlebug specimens for research (see sidebar on Alan Marshall, Chapter 4). New Zealand has listed four species of Aphrophoridae as "at risk," by virtue of being naturally uncommon, but none as "threatened" (Stringer et al. 2012). Michigan has declared *Prosapia ignipectus* an insect of "special concern" due to vulnerability to habitat destruction (Cuthrell 1999). In my estimation, the biggest threats to spittlebug diversity come from expansion of agriculture into natural habitats and long-term desiccation due to warming climates.

Spittlebug Taxonomists Are Faring Worse than Spittlebugs

While most spittlebugs seem not to have experienced recent decline in numbers, the same cannot be said for spittlebug taxonomists. These are the scientists who name and scientifically characterize spittlebugs so that we can make accurate identifications. Taxonomists describe new species, plot their geographical distributions, and determine their relationships to other species. This work is critical to the study of insect pests, habitat conservation, and sophisticated interpretation of biochemical and molecular studies conducted on insects. If you cannot accurately identify the species you are working with, you are shooting in the dark. Taxonomists generally work through museums like my own or in colleges and universities. Their work provides a scientific framework on which all other biological, agricultural, and conservation work depends.

Spittlebug taxonomists have never been abundant. Presently they are few and becoming fewer. Including scientists who also work on groups other than spittlebugs or in disciplines other than taxonomy, the New World count is, very approximately, one in Canada, one in the United States (if I qualify; it is not my formal training or the core of my work), one in Mexico, and three in Brazil. In the Old World, there are, roughly, two spittlebug taxonomists in France, one in the UK, one in China, one in Japan, one or two in Turkey, one in Iran, and two in the Philippines. Only six or seven of these individuals concentrate their efforts on spittlebug taxonomy. Several are formally retired. Among the younger taxonomists, a relative term, only three, to my knowledge, are committed to spittlebug taxonomy as the ongoing focus of their work, a pretty sparse cadre given the three or four thousand species yet to be described and the increasing importance of *Xylella* vectors and other spittlebug pests on the world stage.

Bureaucratic Regulation of International Collecting Threatens Ongoing Work

Specimen-based investigations are one of the key underpinnings of modern science. This gives museum collections tremendous scientific value. Collections like those of the American Museum of Natural History and the Smithsonian house key reference specimens (type specimens), which are used to confirm the identity of insect species, as well as diversified collections that record the geographic distribution, variation, and historical occurrence of hundreds of thousands of species. They also serve as a reservoir of unanalyzed specimens—many of which will prove to be new to science,—and of specimen-associated DNA that supports a tremendous variety of studies on the biology and evolution of organisms represented in the collections. These collections have been built for 200-plus years through the collective effort of travelling professional and amateur scientists who have secured specimens from around the world, in many cases from habitats that have since been destroyed by urbanization, deforestation, and the extension of intensive agriculture. They are an invaluable and irreplaceable asset for modern science.

Unfortunately, opportunities to augment and enrich these collections are rapidly contracting. Many countries outside Europe and the United States have instituted strong legal restrictions on collecting organisms, including insects, without formal permission. The details vary from country to country, but the net effect is to make it difficult or nearly impossible to secure permits to collect and export local specimens. These restrictions are well-intentioned. They aim to preserve the country's natural patrimony, often including the DNA of its flora and fauna, against exploitation or theft, and to conserve wildlife. Plants of commercial value; elephants; rhinoceroses; gaudy birds; big, beautiful butterflies; and DNA sequences that might enrich pharmaceutical companies are the sorts of things national legislators have in mind. Unfortunately, the growing regulatory net snares ordinary noncommercial insects, including spittlebugs. Most international collecting venues now require months of clearing bureaucratic hurdles in order to receive permission to collect, and many require separate export permits to take the specimens out of the country. None of this has brought professional international collecting to a halt, but it has made it more difficult and costly. It has also effectively eliminated contributions from amateurs, who, historically, have been a source of scientifically important specimens.

Collections are also caught in a second web of regulation. There is a long, unhappy history of museum collections built on cultural looting, some of it by state actors (particularly in the colonial era) and some of it by private collectors wittingly or unwittingly participating in a market for historical artifacts built on theft and deception. For years this was papered over, but as a result of public scandals and aggressive legal action by countries of origin to recover stolen or illicitly exported treasures, museums have become exquisitely sensitive to issues of provenance. For cultural artifacts this is a laudable change, but for ordinary entomological collections it has elements of disaster. Every incoming international specimen, even the most ordinary bug, must be treated as though it might have been pillaged from King Tut's tomb or looted from the National Museum of Iraq. This has led to requirements for paperwork establishing that the specimens have been collected and exported legally, documentation that, depending on circumstances, may be difficult or impossible to obtain, even in the most innocent circumstances. Combined with the difficulties of getting collection permits in the first place, the net effect is to discourage collection and import of specimens and to burden museum staff with paperwork that has nothing to do with the conservation and study of collections that are central to their work.

The old system of freewheeling collecting had many faults. Like wealth and other things of value, most specimens flowed from colonies or ex-colonies to the metropole, depriving the source areas of reference collections that could be of tremendous local use. This required and still requires many scientists from the Global South to make expensive visits (assuming visas are available) to Europe and North America to examine specimens from their own countries. No one with a sense of justice and equity would want to perpetuate this circumstance, made even less tenable by the rise in many countries of origin of well-maintained local collections. But the answer, I submit, lies not in hobbling the import of specimens to countries like the United States, but in better mechanisms—legal, professional, and cultural—to assure that collected specimens are eventually returned, where desired, to their home countries, with mutually agreed upon processes for some dispersal of materials among collections to insure the survival of representative specimens in the face of local mishaps or disasters at either end of the transaction (for example, the destruction by fire of the Brazilian National Museum insect collection in 2018).

Citizen Science Photography: A Major Positive Development in Entomology

Though amateurs are now virtually excluded from licit international collecting, advances in digital photography have led to an important, compensating, positive development. Universal access to excellent cell phone and digital cameras has ushered in a golden age of citizen scientist nature photography. Websites like iNaturalist, which originated from the California Academy of Sciences in San Francisco, permit amateurs to post photos and seek identifications from fellow amateurs and experts (Kittelberger et al. 2021). Collectively, this dedicated cadre of photographers makes major contributions to science and is the source of many of the color photos in this volume.

For spittlebugs that can be identified by photograph, this provides invaluable information on geographical distribution and seasonality, sometimes complementing information from museum collections and sometimes simply unavailable elsewhere. The downside is that many spittlebugs are impossible to determine to species, or in some cases even to genus, by photographs alone. In that case, the restrictions on international transfer of specimens outlined previously often preclude shipment of specimens for verification. Most citizen scientist photographers do not collect specimens, but in an ideal world they would be welcome and encouraged to collect and ship interesting bugs for species verification, with appropriate local limitations on collecting threatened or commercially exploited species and collecting in national parks, nature reserves, and other protected areas.

Citizen Science in Action: The Identification of an Invasive Spittlebug in Southern California

Starting in 2019, iNaturalist photographers in San Diego and Orange Counties began to record sightings of a large, conspicuous, predominantly black spittlebug, clearly family Cercopidae. By 2021 records spread to Los Angeles County. This was the first ever California occurrence of a species in the large family Cercopidae. One citizen scientist photographer graciously forwarded a specimen collected in her backyard, from which I tentatively determined the species to be *Prosapia ignifera*, a spittlebug hitherto known only from central Mexico, at least 1,500 km from the California border. This precipitated a field trip, in which I collected *P. ignifera* at one location in Los Angeles County and two in San Diego County, confirming its widespread occurrence in California and its abundance in a jewel of a seaside park in La Jolla, where it occurred in rough grass on the edges of a manicured outdoor event area for wedding photography. I found all three sites through the geolocation information associated with iNaturalist photos.

Citizen scientist photographers can also contribute to large, professionally coordinated scientific endeavors. In Britain, they played a key role in a campaign to document nymphal host plants of the *X. fastidiosa* vector *P. spumarius* (Chapter 9) in anticipation of a possible future *X. fastidiosa* introduction. The BRIGIT project (2019–2021) solicited photographs of nymphal spittles with identified hosts. It received thousands of entries, primarily through a dedicated internet portal and mostly from amateurs. After winnowing by professional scientists, the BRIGIT host list stood at 628 species, no fewer than 358 of which were records new to science (Thompson et al. 2023). The British total, moreover, was 48% of the documented total of 1,311 host species for the world. This by any measure is a striking testimonial to the power of citizen science crowd sourcing.

Epilogue

We have now come full circle. Modern spittlebug studies began at the hands of Charles De Geer, eighteenth century Sweden's version of a twenty-first century billionaire and one of the few people of the time who could afford a good microscope (Chapter 1). It grew to encompass a small cadre of nineteenth century naturalists and taxonomists. Some were professionals such as E. S. Morse, some engaged amateurs, such as David Livingstone. In the first half of the twentieth century, the number of people working on spittlebugs markedly increased. Spittlebug taxonomy took a leap with major contributions by Distant, Lallemand, Jacobi, Schmidt, and Doering, while the serious study of spittlebugs for agricultural applications was established in Trinidad in the face of the *Aeneolamia varia* threat to sugarcane (Chapter 6). Mid-twentieth century witnessed an expansion of interest in *Philaenus spumarius* as a pest of alfalfa in the American Midwest (Chapter 8).

The Trinidad work stimulated several taxonomic papers on Cercopidae by Fennah, which culminated at the turn of the twenty-first century with a major taxonomic treatise on New World Cercopidae by Carvalho and Webb (2005). In the meantime, a contingent of late twentieth and early twenty-first-century agronomists took up where the American and Trinidadian agriculturalists left off, conducting extensive investigations on the biology, ecology, and agronomics of Neotropical pasture and sugarcane spittlebugs (Chapters 6 and 7). While they worked, spittlebug taxonomy experienced a renaissance with the publication of the first DNA-based spittlebug trees of life (Chapter 2). This coincided with a proliferation of studies unleashed by the realization that spittlebugs are the key vectors of *Xylella fastidiosa* in Europe (Chapter 9).

As a result of the *Xylella* crisis and the funding it engendered, there are probably more scientific professionals engaged directly or indirectly in spittlebug studies now than at any other time in history. This has not moved the needle on taxonomists, but the volume of studies related to basic biology and vector control has increased immensely (Boscia et al. 2023; Thompson et al. 2023; Serio et al. 2024). In parallel, the study of natural history has been democratized. With the help of digital photography and the internet, spittlebug studies now embrace the creativity and effort of thousands of ordinary people intrigued by science and nature, a worldwide phenomenon. In my youth, now some time ago, I worked with others in a movement called Science for the People. One of the movement's aims was to transform science from an elite, sequestered institution, to an instrument of realization of the knowledge, power, and initiative of common citizens. In its own modest way, the study of spittlebugs is heading in that direction.

Acknowledgments

The dedication to my colleagues is heartfelt. I did not train as an entomologist. They welcomed me into the field and gave assistance and encouragement at every turn. Without them, I could not have done nearly as much or as well. Here are some of the many professional colleagues and students to whom I am indebted (those marked with † are no longer with us). Entries are alphabetical by region. My apologies to anyone I have inadvertently omitted:

United States and Canada: Rodrigo Almeida, Ameilia Adams, Charles Bartlett, Dylan Beal, Adam Bell, James Carpenter, Brandon Cooper, Monica Cooper, Jason Cryan, Christopher Dietrich, Lewis Dietz, Dmitri Dimitrov, Scott Ellis, Francis Evans†, Donald Farish†, Stephen Green, David Grimaldi, Susan Halbert, K. G. Andrew Hamilton, James Heath†, Solomon Hendrix, G. Evelyn Hutchinson†, Christine Johnson, Kyle Kittelberger, Manpreet Kohli, Monte Lloyd†, Michelle Mayfield, Stuart McKamey, Janice Konigsford, Norhashimah Mohd-Saleh, Thomas Moore†, Harry Nelson†, Al Newton, Lois O'Brien, Daniel Peck, Agnieszka Pierwola, Alexander Purcell, Mark Rothschild, Ruth Salas, Randall Schuh, Chungkun Shih. Margaret Thayer, Lydia Thompson, Ethan Tolman, Julie Urban, Leigh Van Valen†, Lucia Varela, K. Venkatachalam, Jo Ann White.

Latin America and the Caribbean: Alexander Auad, Lorelí Carranza, Gervasio Carvalho, Ulises Castro-Valderrama, Miriam Fernandez, Alvaro Foieri, Joyce Froza, Carolina Godoy, Paul Hanson, Ruth León, Humberto Lezama, João Lopes, Jean Michel Maes, John Miles†, Kenji Nishida, Andressa Paladini, Carlos Sáenz, Christian Schöbel, Daniela Takiya, José Raul Valério.

Europe: Sabina Avosani, John Badmin, Robert Biedermann, Daniele Cornara, Jürgen Deckert, Ersin Demirel, Sakis Drosopoulos†, Anti Halkka, Olli Halkka†, Claire Harkin, Werner Holzinger, Elisabeth Huber, Gernot Kunz, Maxime Le Cesne, Holger Löcker, Herbert Nickel, José Quartau, Roman Rakitov, Pio Roversi, Sofia Seabra, Adeline Soulier, Alan Stewart, Rukiye Tanyeri, Vojtek Novotny, Mick Webb, John Whitaker†, Michael Wilson, Selcuk Yurtsever, Ünal Zeybekoğlu.

Middle East, Asia, Africa, Australia, and New Zealand: Hani Abdul-Nour†, Magda Botha, Clare Butcher Morales†, Elorde Crispolon, Fariba Mozaffarian, Murray Fletcher, Misamo Hayashi, Ai-Ping Liang, Melinda Moir, Keiji Morishima†, Ngô Trần Phương Lam, Hong Thai Pham, Hsien-Tzung Shih, Michael Stiller, Sheryl Yap, Yalin Zhang.

I am indebted to the diligent staff of the Organization for Tropical Studies Las Cruces Biological Station, where I have done periodic fieldwork since 1993, including Emilce Ramírez, Gail Hewson Hull, and Luis Diego Gómez†. Gail personally rescued a semester's worth of lab and field notes, as well as many specimens, when the station burned the night before Thanksgiving, in November 1994, toward the end of a four-month sabbatical stay. George Patterson, station benefactor and our landlord for four months, lent a field vehicle to help recollect specimens lost in the fire.

I am also indebted to my hosts during my visit to Trinidad in December 2003, including Moses Kairo, Bibi Ali, Balmattee Sukha, Cynthra Persad, Graham White, and Ayub Khan; to my DIECA hosts in Costa Rica in 1994, particularly Carlos Sáenz; and to my CIAT hosts in Cali in 1997, Daniel Peck, Cesar Cardona†, and John Miles†. My work in Vietnam was facilitated by Hong Thai Pham in 2014 and 2019 and by Lê Văn Hương and Lê Văn Sơn in 2019. In 2019, Ngô Trần Phương Lam provided able cultural and linguistic interpretation as well as field assistance. In 2024, Ulises Castro-Valderrama and Jesús Romero-Nápoles hosted a field expedition in Mexico, and Cheryl Yapp hosted an expedition in the Philippines.

At critical moments, Janet Sprent and John Raven†, both of the University of Dundee, Scotland, provided encouragement and helpful counsel as I struggled to sort through the unfamiliar worlds of nitrogen fixation and xylem feeding. In addition to John Raven, colleagues who read and commented on drafts of parts of this book include Rodrigo Almeida, Christopher Dietrich, Alan Marshall, Daniel Peck, Alexander Purcell, Chungkun Shih, and José Raul Valério. Paul Hanson and Alan Stewart read through and commented critically on the entire manuscript. Daniele Rielli checked my English translation of selections from his book and gave permission to reproduce them. Lorenzo Prendini identified the spider predators in Plate 9. Cornell University Press representatives Kitty Liu and India Miraglia guided me through the complex process of turning a proposal into a submitted manuscript.

As a college administrator, I have noted that faculty members occasionally neglect to credit their home institutions, which pay their salaries and thus, indirectly but crucially, support their scholarly endeavors. Over the years, Roosevelt University, The City Colleges of Chicago, Kean University, and Metropolitan College of New York kept me fed and the house warm while I pursued this work. Their libraries have also supported me, particularly the library of Roosevelt University, which over many years patiently sought out interlibrary loans of the most obscure articles imaginable, with special thanks to head librarian Mary Beth Riedner, who for a period when I was provost there, reported to me directly.

The Field Museum, Chicago, and the American Museum of Natural History, New York, provided space, resources, colleagues, and appointments in residence that underpinned this work from the 1990s onward. Natural history museums are critical repositories of specimens and beehives of scientific activity. They make work like this possible.

I owe special tribute to my wife, Ruth Moscovitch, who has tolerated and abetted this obsession, from the early *Philaenus spumarius* collecting trips in the 1970s to the completion of this book. She read the full manuscript with a critical eye. We in turn owe thanks to friends and family, entomologists and lay people, who have fed and housed us on our travels over the years and put up with odd things in their ovens and refrigerators.

Literature Cited

Abate, Tsedeke. 1988. *Insect and Mite Pests of Horticultural and Miscellaneous Plants in Ethiopia*. Addis Ababa, Ethiopia: Institute of Agricultural Research.

Abbott, Alison. 2017. "Italy Rebuked for Failure to Prevent Olive-Tree Tragedy." *Nature* 546 (7657): 193–94. https://doi.org/10.1038/546193a.

Abbott, Alison. 2015a. Scientists Blamed for Olive-Tree Ruin." *Nature* 522 (7554): 13–14. https://doi.org/10.1038/nature.2015.17651.

Abbott, Alison. 2015b. "Italian Scientists under Investigation after Olive-Tree Deaths." Nature. https://doi.org/10.1038/nature.2015.19078.

Abdullah, Md., and Md. Abdul Mannan. 2010. "Influence of Climatic Changes on the Abundance of Major Insect Pests of Sugarcane." In *Climate Change and Food Security in South Asia*, edited by Rattan Lal, Mannava V. K. Sivakumar, S. M. A. Faiz, A. H. M. Mustafizur Rahman, and Khandakar R. Islam. Springer Science & Business Media. https://doi.org/10.1007/978-90-481-9516-9.

Aguirre, Lina M., César Cardona, John W. Miles, and Guillermo Sotelo. 2013. "Characterization of Resistance to Adult Spittlebugs (Hemiptera: Cercopidae) in *Brachiaria* spp." *Journal of Economic Entomology* 106 (4): 1871–77. https://doi.org/10.1603/EC11189.

Ahmed, Hamiz Uddin, Md. Abdul Latif, Md. Fazlul Huq et al. 2019. *Final Report on Developing Pest List of Plants and Plant Products in Bangladesh*. Dhaka, Bangladesh: Department of Agricultural Extension.

Ajayi, O., and F. A. Oboite. 1999. "Importance of Spittle Bugs, *Locris rubens* (Erichson) and *Poophilus costalis* (Walker) on Sorghum in West and Central Africa, with Emphasis on Nigeria." *Annals of Applied Biology* 136 (1): 9–14. https://doi.org/10.1111/j.1744-7348.2000.tb00002.x.

Ajayi, O., and F. A. Oboite. 2001. "Control of *Locris rubens* (Erichson) (Homoptera: Cercopidae) on Sorghum in Nigeria." *Journal of Sustainable Agriculture* 18 (1): 85–97. https://doi.org/10.1300/J064v18n01_07.

Akiyama, Masayuki, and Kazuma Matsumoto. 1986. "Biology of the Pine Spittlebug, *Aphrophora flavipes* (Homoptera: Cercopidae) in the Nymphal Stage." *Japanese Journal of Applied Entomology and Zoology* 30 (2): 136–41. https://doi.org/10.1303/jjaez.30.136.

Akre, R.D., W.B. Garnett, J.F. Mac Donald, A. Greene, and P. Landolt. 1976. "Behavior and Colony Development of *Vespula pensylvanica* and *V. atropilosa* (Hymenoptera: Vespidae)." *Journal of the Kansas Entomological Society* 49 (1): 63–84.

Alabdallah, Osama, Raed Alkowni, Jehad Radwan et al. 2024. "The First Surveillance Report for *Xylella fastidiosa* in Olive and Stone Fruit Orchards in Palestine." *EPPO Bulletin* 54 (2): 236–42. https://doi.org/10.1111/epp.13023.

Albre, Jérôme, José María García Carrasco, and Marc Gibernau. 2021. "Ecology of the Meadow Spittlebug *Philaenus spumarius* in the Ajaccio Region (Corsica)–I: Spring." *Bulletin of Entomological Research* 111 (2): 246–56. https://doi.org/10.1017/S0007485320000711.

Allard, G. B. 1987. "Prospects for the Biocontrol of the Sugarcane Froghopper with Particular Reference to Trinidad." *Biocontrol News and Information* 8 (2): 105–15.

Allard, G. B., C. A. Chase, J. B. Heale, J. E. Isaac, and C. Prior. 1990. "Field Evaluation of *Metarhizium anisopliae* (Deuteromycotina: Hyphomycetes) as a Mycoinsecticide for Control of Sugarcane Froghopper, *Aeneolamia varia saccharina* (Hemiptera: Cercopidae)." *Journal of Invertebrate Pathology* 55 (1): 41–46. https://doi.org/10.1016/0022-2011(90)90030-A.

Almeida, Rodrigo P. P. 2016. "Can Apulia's Olive Trees Be Saved?" *Science* 353 (6297): 346–48. https://doi.org/10.1126/science.aaf9710.

Almeida, Rodrigo P. P. 2018. "Emerging Plant Disease Epidemics: Biological Research Is Key but Not Enough." *PLOS Biology* 16 (8): e2007020. https://doi.org/10.1371/journal.pbio.2007020.

Almeida, Rodrigo P. P., and Leonard Nunney. 2015. "How Do Plant Diseases Caused by *Xylella fastidiosa* Emerge?" *Plant Disease* 99 (11): 1457–67. https://doi.org/10.1094/PDIS-02-15-0159-FE.

Altieri, M.A., and W.H. Whitcomb. 1979. "Predaceous Arthropods Associated with Mexican Tea in North Florida." *Florida Entomologist* 62 (3): 175–82. https://doi.org/10.2307/3494055.

Alvarenga, Roberta, Alexander Machado Auad, Jair Campos Moraes, Sandra Elisa Barbosa da Silva, and Brunno Santos Rodrigues. 2019. "Tolerance to Nymphs and Adults of *Mahanarva spectabilis* (Hemiptera: Cercopidae) by Forage Plants in Fertilized Soils." *Pest Management Science* 75 (8): 2242–50. https://doi.org/10.1002/ps.5361.

Alves, Roberto Teixeira, and Gervasio Silva Carvalho. 2014. "First Record of the Sugarcane Froghoppers *Mahanarva spectabilis* (Distant) and *Mahanarva liturata* (Le Peletier & Serville) Infesting Sugarcane Plantations in the Region of Goianésia (GO), Brazil." *Arquivos do Instituto Biológico* 81 (1): 83–85. https://doi.org/10.1590/S1808-16572014000100016.

Amanifar, Naser, Mohsen Taghavi, Karamat Izadpanah, and Ghobad Babaei. 2014. "Isolation and Pathogenicity of *Xylella fastidiosa* from Grapevine and Almond in Iran." *Phytopathologia Mediterranea* 53 (2): 318–27.

Ananthasubramanian, K. S., and S. J. William. 1989. "Notes on Immature Stages of *Carabunia orientalis* Subba Rao (Hymenoptera: Encyrtidae) Parasitic on the Nymphs of *Clovia puncta* Walker (Homoptera: Cercopidae)." *Hexapoda* 1: 123–25.

Andersen, Peter C., Brent V. Brodbeck, and Russell F. Mizell III. 1989. "Metabolism of Amino Acids, Organic Acids and Sugars Extracted from the Xylem Fluid of Four Host Plants by Adult *Homalodisca coagulata*." *Entomologia Experimentalis et Applicata* 50 (2): 149–59. https://doi.org/10.1111/j.1570-7458.1989.tb02384.x.

Andersen, Peter C., Brent V. Brodbeck, and Russel F. Mizell III. 1992. "Feeding by the Leafhopper, Homalodisca Coagulata, in Relation to Xylem Fluid Chemistry and Tension." *Journal of Insect Physiology* 38 (8): 611–22. https://doi.org/10.1016/0022-1910(92)90113-R.

Andersen, Peter C., Brent V. Brodbeck, and Russel F. Mizell III. 2003. "Plant and Insect Characteristics in Response to Increasing Density of *Homalodisca coagulata* on Three Host Species: A Quantification of Assimilate Extraction." *Entomologia Experimentalis et Applicata* 107 (1): 57–68. https://doi.org/10.1046/j.1570-7458.2003.00036.x.

Andrews, Mitchell, and John A. Raven. 2022. "Root or Shoot Nitrate Assimilation in Terrestrial Vascular Plants–Does It Matter?" *Plant and Soil* 476 (1): 31–62. https://doi.org/10.1007/s11104-021-05164-9.

Angeles, Guillermo, Barbara Bond, John S. Boyer et al. 2004. "The Cohesion-Tension Theory." *New Phytologist* 163 (3): 451–52. https://doi.org/10.1111/j.1469-8137.2004.01142.x.

Ankrah, Nana Y. D., Rebecca A. Wilkes, Freya Q. Zhang, Ludmilla Aristilde, and Angela E. Douglas. 2019. "The Metabolome of Associations between Xylem-Feeding Insects and Their Bacterial Symbionts." *Journal of Chemical Ecology* 46 (8): 735–44. https://doi.org/10.1007/s10886-019-01136-7.

Ankrah, Nana Y. D., Rebecca A. Wilkes, Freya Q. Zhang et al. 2020. "Syntrophic Splitting of Central Carbon Metabolism in Host Cells Bearing Functionally Different Symbiotic Bacteria." *The ISME Journal* 14 (8): 1982–93. https://doi.org/10.1038/s41396-020-0661-z.

Antonatos, S., D.P. Papachristos, K. Varikou, P. Vahamidis, A. Kapranas, and P. Milonas. 2021. "Seasonal Appearance, Abundance, and Host Preference of *Philaenus spumarius* and *Neophilaenus campestris* (Hemiptera: Aphrophoridae) in Olive Groves in Greece." *Environmental Entomology* 50 (6): 1474–82. https://doi.org/10.1093/ee/nvab093.

Arango Guillermo S., and Mario Calderón. 1981. "Biología y hábitos de *Zulia colombiana* (Lallemand) plaga del pasto *Brachiaria* spp." *Revista Colombiana de Entomología* 7 (1 and 2): 3–11.

Archibald, R. D., J. M. Cox, and L. L. Deitz. 1979. "New Records of Plant Pests in New Zealand: III. Six Species of Homoptera." *New Zealand Journal of Agricultural Research* 22 (1): 201–7. https://doi.org/10.1080/00288233.1979.10420862.

Armendáriz-Toledano, Francisco, Misael Adrián López-Posadas, Vinton Thompson et al. 2022. "Overview of Spittlebugs of the Family Cercopidae (Hemiptera: Auchenorrhyncha) from Mexico, with Keys to Genera and Species." *Revista Mexicana de Biodiversidad* 93 (2022): e934030. https://doi.org/10.22201/ib.20078706e.2022.93.4030.

Armendáriz-Toledano, Francisco, Misael Adrián López-Posadas, Youssef Utrera-Vélez, Jesús Romero Nápoles, and Ulises Castro-Valderrama. 2023. "More than 80 Years without New Taxa: Analysis of Morphological Variation among Members of Mexican *Aeneolamia* Fennah (Hemiptera, Cercopidae) Support a New Species in the Genus." *ZooKeys* 1139: 71–106.

Ashburner, Michael. 1981. "Entomophagous and Other Bizarre Drosophilidae." In *The Genetics and Biology of Drosophila* 3a, edited by M. Ashburner, H. L. Carson, and J. N. Thompson. Academic Press.

Ashcroft, T., and S. George. 2004. *New to New Zealand:* Bathyllus albicinctus *(Hemiptera: Aphrophoridae) Whangarei. Initial Investigation & Technical Report/Entomology.* 2004-049-261: 1–7. New Zealand Ministry of Agriculture and Forestry.

Asia and Pacific Plant Protection Commission. 1987. *Technical Document No. 135: Insect Pests of Economic Significance Affecting Major Crops of the Countries in Asia and the Pacific Region.* Food and Agriculture Organization of the United Nations. https://swfrec.ifas.ufl.edu/hlb/database/pdf/00001757.pdf.

Atencio, Randy, François-Régis Goebel, and Roberto J. Miranda. 2019. "Entomofauna Associated with Sugarcane in Panama." *Sugar Tech* 21 (4): 605–18.

Auad, Alexander Machado, Amanda Daniela Simões, Antônio Vander Pereira et al. 2007. "Seleção de genótipos de capim-elefante quanto à resistência à cigarrinha-das-pastagens." *Pesquisa Agropecuária Brasileira* 42 (8): 1077–81. https://doi.org/10.1590/S0100-204X2007000800003.

Auty, R. M. 2018. "Natural Resources and Small Island Economies: Mauritius and Trinidad and Tobago." *The Journal of Development Studies* 53(2): 264–277. https://doi.org/10.1080/00220388.2016.1160063.

Avosani, Sabina, Elissa Daher, Pietro Franceschi, Marco Ciolli, Vincenzo Verrastro, and Valerio Mazzoni. 2020. "Vibrational Communication and Mating Behavior of the Meadow Spittlebug *Philaenus spumarius*." *Entomologia Generalis* 40 (3): 307–21. https://doi.org/10.1127/entomologia/2020/0983.

Avosani, S., R. Nieri, V. Mazzoni et al. 2023. "Intruding into a Conversation: How Behavioral Manipulation Could Support Management of *Xylella fastidiosa* and Its Insect Vectors." *Journal of Pest Science* 97 (1): 17–33. https://doi.org/10.1007/s10340-023-01631-7.

Avramov, Zhelyu. 2022. "Study of Sudden Decline of Lavender in Bulgaria Caused by '*Candidatus phytoplasma solani*'." *Bulgarian Journal of Crop Science* 59 (1): 25–37.

Babos, Rezső. 1983. "A Fonófűz Termesztése, Feldolgozása." *Az Erdő* 32: 538–43.

Backus, Elaine A., and Wah Keat Lee. 2011. "Glassy-Winged Sharpshooter Feeding Does Not Cause Air Embolisms in Xylem of Well-Watered Plants." In *Proceedings of the 2011 Pierce's Disease Research Symposium*, edited by Thomas Esser. California Department of Food and Agriculture.

Badmin, John. 2007. "Spittle Sharing Among Froghopper Species." *British Journal of Entomology and Natural History* 20: 152–53.

Baerg, W. J. 1920. "An Unusual Case of Parasitism on *Clastoptera obtusa* Say (Hemip., Cercopidae; Dip., Drosophilidae)." *Entomological News* 31: 20–21.

Bailey, Peter T. 2007. *Pests of Field Crops and Pastures: Identification and Control.* CSIRO Publishing.

Bajocco, S., E. Raparelli, and S. Bregaglio. 2023. "Assessing the Driving Role of the Anthropogenic Landscape on the Distribution of the *Xylella Fastidiosa*-Driven 'Olive Quick Decline Syndrome' in Apulia (Italy)." *Science of The Total Environment* 896: 165231. https://doi.org/10.1016/j.scitotenv.2023.165231.

Balaji R. M., B. Gundappa, Manish Mani Tripathi, and S. Rajan. 2018. "Pests of Jackfruit." In *Pests and Their Management*, edited by Omkar. Springer. https://doi.org/10.1007/978-981-10-8687-8_18.

Baldani, José Ivo, Natália Dos Santos Ferreira, Stefan Shwab et al. 2024. "*Nitrospirillum viridazoti* sp. nov., an Efficient Nitrogen-Fixing Species Isolated from Grasses." *Current Microbiology* 81 (6): 1–12.

Balfas, Rodiah, Marsusi Nuhardiyati, and C. J. Lomer. 1990. "Distribution and Parasitism of *Acmopolynema* (Hymenoptera: Machaerotidae) on Clove and *Xanthostemon chrysanthus*." *Industrial Crops Research Journal* 3 (1): 6–10.

Ball, E. D. 1901. "The Food Habits of Some *Aphrophora* Larvae." *Ohio Naturalist* 1 (8): 122–24.

Balu, A., R. Rajarishi, B. Deeparaj, and S. Durairaj. 1997. "Curling and Crinkling of Teak Leaves." *Indian Forester* 123 (8): 775–77. https://doi.org/10.36808/if/1997/v123i8/6070.

Balzani, P., A. Nencioni, M. Grillini et al. 2023. "Spittlebug Invisibility Cloak: Experimental Tests on the Antipredatory Effect of the Froth of *Philaenus spumarius*." *Bulletin of Insectology* 76: 21–28.

Bansal, Kanika, Sanjeet Kumar, Amandeep Kaur, Anu Singh, and Prabhu B. Patil. 2021. "Deep Phylo-Taxono Genomics Reveals *Xylella* as a Variant Lineage of Plant Associated *Xanthomonas* and Supports Their Taxonomic Reunification along with *Stenotrophomonas* and *Pseudoxanthomonas*." *Genomics* 113 (6): 3989–4003. https://doi.org/10.1016/j.ygeno.2021.09.021.

Barber, G. W., and W. O. Ellis. 1922. "Eggs of Three Cercopidæ." *Psyche* 29 (1): 1–3. https://doi.org/10.1155/1922/71368.

Barrientos, A. 1983. "Different Methods to Control Spittlebug (*Monecphora bicinta fraterna* (Uhler)) in *Cynodon nlemfuensis*." *Cuban Journal of Agricultural Science* 17 (2): 223–28.

Barrientos, A., R. S. Herrera, N. Dorta, and C. Mora. 1991. "Evaluation of Yield and Quality Losses in *Cynodon dactylon* Cv. Coast Cross 1 from *Monecphora bicincta fraterna* (Uhler)." *Cuban Journal of Agricultural Science* 22 (3): 333–37.

Bartlett, Charles R., Lewis L. Deitz, Dmitry A. Dmitriev, Allen F. Sanborn, Adeline Soulier-Perkins, and Matthew S. Wallace. 2018. "The Diversity of the True Hoppers (Hemiptera: Auchenorrhyncha)." In *Insect Biodiversity: Science and Society*, edited by Robert G. Foottit and Peter H. Adler. John Wiley & Sons, Ltd. https://doi.org/10.1002/9781118945582.ch19.

Bass, Katherine A., Elizabeth A. John, Naomi C. Ewald, and Sue E. Hartley. 2010. "Insect Herbivore Mortality Is Increased by Competition with a Hemiparasitic Plant." *Functional Ecology* 24 (6): 1228–33. https://doi.org/10.1111/j.1365-2435.2010.01743.x.

Basu, Paramita, A. K. Sanyal, and D. Bhattacharya. 2010. "Studies on Insect Pests of Timber Yielding Tree Species in a Tropical Moist Deciduous Forest (Bethuadahari: West Bengal)." *Records of the Zoological Survey of India* 110 (1): 1–13.

Bates, Marston. 1942. "Insectos nocivos." *Revista Facultad Nacional de Agronomía Medellín* 5 (18): 333–401.

Batista, Elder Simões de Paula., Alexander Machado Auad, Vanessa Andaló, and Caio Márcio de Oliveira Monteiro. 2014. "Virulence of Entomopathogenic Nematodes (Rhabditida: Steinernematidae, Heterorhabditidae) to Spittlebug *Mahanarva spectabilis* (Hemiptera: Cercopidae)." *Arquivos do Instituto Biológico* 81: 145–49.

Bautista-Gálvez, A., and N. González-Cortes. 2005. "Tres dosis de *Metarhizium anisopliae* sobre la mosca pinta (*Aeneolamia* Spp.) en caña de azúcar en la región de Los Ríos, Estado de Tabasco." *Universidad y Ciencia* 21 (41): 37–40.

Bautista-Gálvez, A., J. F. Barrera, E. P. de la Cruz, S. Salgado-García, J. Gómez-Ruiz, and J. F. Gomez-Leyva. 2012. "Genetic Characterisation of *Metarhizium anisopliae* (Metchnikoff) Sorokin Isolates from Sugarcane Fields and Their Pathogenicity against *Aeneolamia postica* (Walker)(Hemiptera: Cercopidae)." *Universidad y Ciencia* 28 (3): 217–29.

Beal, Dylan J., Amielia G. Adams, Monica L. Cooper et al. 2021. "Assessment of Nymphal Ecology and Adult *Xylella fastidiosa* Transmission Biology of *Aphrophora* nr. *permutata* (Hemiptera: Aphrophoridae) in California Vineyards." *Environmental Entomology* 50 (6): 1446–54. https://doi.org/10.1093/ee/nvab081.

Beal, F. E. L. 1918. *Food Habits of the Swallows: A Family of Valuable Native Birds.* US Department of Agriculture.

Beckett, Kephra I. S., Anne B. Robertson, and Philip G. D. Matthews. 2019. "Studies on Gas Exchange in the Meadow Spittlebug, *Philaenus spumarius*: The Metabolic Cost of Feeding on, and Living in, Xylem Sap." *Journal of Experimental Biology* 222 (3): jeb191973. https://doi.org/10.1242/jeb.191973.

Begnami, Isabela dos Santos, Alexandre Hild Aono, Diego da Silva Graciano et al. Vigna. 2024. "Elucidating Molecular Responses to Spittlebug Attack in *Paspalum regnellii*." *Plant Molecular Biology Reporter* 43: 307–323. https://doi.org/10.1007/s11105-024-01487-w.

Behie, Scott W., and Michael J. Bidochka. 2014. "Ubiquity of Insect-Derived Nitrogen Transfer to Plants by Endophytic Insect-Pathogenic Fungi: An Additional Branch of the Soil Nitrogen Cycle." *Applied and Environmental Microbiology* 80 (5): 1553–60. https://doi.org/10.1128/AEM.03338-13.

Beirne, B. P. 1972. *Pest Insects of Annual Crop Plants in Canada: IV. Hemiptera-Homoptera, V. Orthoptera, VI. Other Groups.* Entomological Society of Canada.

Bell, Adam Jorgensen. "A Phylogenetic Investigation of the Spittlebug Superfamily Cercopoidea (Insecta: Hemiptera: Auchenorrhyncha) with Emphasis on the Family Machaerotidae." PhD diss., University at Albany, 2016. https://scholarsarchive.library.albany.edu/legacy-etd/1563.

Bell, Adam J., Gavin J. Svenson, and Jason R. Cryan. 2014. "The Phylogeny and Revised Classification of Machaerotidae, the Tube-Making Spittlebugs (Hemiptera: Auchenorrhyncha: Cercopoidea)." *Systematic Entomology* 39 (3): 474–85. https://doi.org/10.1111/syen.12066.

Bell-Roberts, Louise, Angela E. Douglas, and Gijsbert D. A. Werner. 2019. "Match and Mismatch between Dietary Switches and Microbial Partners in Plant Sap-Feeding Insects." *Proceedings of the Royal Society B: Biological Sciences* 286 (1902): 20190065. https://doi.org/10.1098/rspb.2019.0065.

Benhadi-Marín, Jacinto, María Villa, Luís F. Pereira, I. Rodrigues, M. Morente, P. Baptista, and J.A. Pereira. 2020. "A Guild-Based Protocol to Target Potential Natural Enemies of *Philaenus spumarius* (Hemiptera: Aphrophoridae), a Vector of *Xylella fastidiosa* (Xanthomonadaceae): A Case Study with Spiders in the Olive Grove." *Insects* 11 (2): 100.

Bennett, Frederick D. 1965. "Observations on the Role of *Clastopteromyia* spp. Inhabitants of Spittle-Masses of *Clastoptera* spp. in the West Indies." *Technical Bulletin. Commonwealth Institute of Biological Control* 5: 98–100.

Bennett, Frederick D., and I. W. Hughes. 1963. "Studies on the Life History and Biological Control of the Spittlebug, *Clastoptera undulata* Uhler (Cercopidae: Hemiptera)." *Entomophaga* 8 (1): 49–66. https://doi.org/10.1007/BF02381338.

Bentley, Jeffery W., and G. Rodríguez. 2001. "Honduran Folk Entomology." *Current Anthropology* 42 (2): 285–300. https://doi.org/10.1086/320010.

Bergman, Elisabeth A., Emma L. Green, and Philip G. D. Matthews. 2021. "The Cibarial Pump of the Xylem-Feeding Froghopper *Philaenus spumarius* Produces Negative Pressures Exceeding 1 MPa." *Proceedings of the Royal Society B* 288 (1954): 20210731.

Bergman, Elisabeth A., and Philip G. D. Matthews. 2024. "Born Suckers: The Cibarial Pump of *Philaenus spumarius* Scales across Ontogeny to Ensure Functional Equivalence." *Journal of Experimental Biology* 227 (17): jeb247966.

Berisha, B., Y. D. Chen, G. Y. Zhang, B. Y. Xu, and T. A. Chen. 1998. "Isolation of Peirce's [sic] Disease Bacteria from Grapevines in Europe." *European Journal of Plant Pathology* 104 (5): 427–33. https://doi.org/10.1023/A:1008655621235.

Bernardes, Giovana, Gustavo Fernando Ferreira Gonçalves, Ruan Lima Santos, Marcelo Mattos Cavallari, Federico de Pina Matta, and Marcos Rafael Gusmao. 2022. "Antibiose de *Paspalum* spp. (L.) às cigarrinhas-das-pastagens *Deois* spp. (Stall [sic] & Fabricius) (Hemiptera: Cercopidae)." *Anais da 14*a *Jornada Científica, Embrapa São Carlos, 14 de Setembro de 2022*. 25. https://www.alice.cnptia.embrapa.br/alice/bitstream/doc/1150043/1/AntibiosePaspalumSpp.pdf.

Bernardo, Eduardo Roberto de Almeida, Vander de Freitas Rocha, Osni Puga, and Ricardo Adaime da Silva. 2003. "Espécies de cigarrinhas-das-pastagens (Hemiptera: Cercopidae) no Meio-norte do Mato Grosso." *Ciência Rural* 33 (2): 369–71. https://doi.org/10.1590/S0103-84782003000200030.

Berry, A. J., and P. G. Willmer. 1986. "Temperature and the Colour Polymorphism of *Philaenus spumarius* (Homoptera: Aphrophoridae)." *Ecological Entomology* 11 (3): 251–59. https://doi.org/10.1111/j.1365-2311.1986.tb00301.x.

Biedermann, Robert. 2002. "Mating Success in the Spittlebug *Cercopis sanguinolenta* (Scopoli, 1763) (Homoptera, Cercopidae): The Role of Body Size and Mobility." *Journal of Ethology* 20 (1): 13–18. https://doi.org/10.1007/s10164-002-0048-9.

Biedermann, Robert. 2003. "Aggregation and Survival of *Neophilaenus albipennis* (Hemiptera: Cercopidae) Spittlebug Nymphs." *European Journal of Entomology* 100 (4): 493–500.

Bini, Andressa Peres, Marina Carnaz Duarte Serra, Izadora Farina Pastore et al. 2023. "Transmission of *Xanthomonas albilineans* by the Spittlebug, *Mahanarva fimbriolata* (Hemiptera: Cercopidae), in Brazil: First Report of an Insect Vector for the Causal Agent of Sugarcane Leaf Scald." *Journal of Insect Science* 23 (6): 28. https://doi.org/10.1093/jisesa/iead116.

Birkinshaw, Christopher R. 1999. "Use of Millipedes by Black Lemurs to Anoint Their Bodies." *Folia Primatologica* 70 (3): 170–71. https://doi.org/10.1159/000021691.

Blackburn, F. H. B. 1954. "Further Developments in Froghopper Control." In *Proceedings of the 1954 meeting of the British West Indies Sugar Technologists*, 137–40. Imperial College of Tropical Agriculture. https://www.cabdirect.org/cabdirect/abstract/19560304230.

Blades, H., P. Compton, P. Smith, C. Persad, and A. Khan. 1997. *Report from the Independent Committee to Investigate All Aspects of the Froghopper Integrated Pest Management Program at Caroni (1975) Ltd.* Trinidad: Unpublished material, 42-page printed document.

Boas, Artemisia M. Vilas, and Rejane M. Andrade. 1990. "Observações preliminares sobre a biologia de *Acmopolynema hervali* gomes (Hymenoptera: Mymaridae) parasitóide da cigarrinha *Mahanarva posticata* Stal (Homoptera: Cercopidae)." *Anais da Sociedade Entomológica do Brasil* 19 (2): 307–13. https://doi.org/10.37486/0301-8059.v19i2.660.

Bodino, Nicola, Riccardo Barbera, Natalia González-Mas, Stefano Demichelis, Domenico Bosco, and Paola Dolci. 2024. "Activity of Natural Occurring Entomopathogenic Fungi on Nymphal and Adult Stages of *Philaenus spumarius*." *Journal of Invertebrate Pathology* 204:108078. https://doi.org/10.1016/j.jip.2024.108078.

Bodino, Nicola, Vincenzo Cavalieri, Crescenza Dongiovanni, Elisa Plazio et al. 2019. "Phenology, Seasonal Abundance and Stage-Structure of Spittlebug (Hemiptera: Aphrophoridae) Populations in Olive Groves in Italy." *Scientific Reports* 9 (1): 17725. https://doi.org/10.1038/s41598-019-54279-8.

Bodino, Nicola, Vincenzo Cavalieri, Crescenza Dongiovanni, Matteo Alessandro Saladini et al. (2020) "Spittlebugs of Mediterranean Olive Groves: Host-Plant Exploitation throughout the Year." *Insects* 11 (2): 130. https://doi.org/10.3390/insects11020130.

Bodino, Nicola, Vincenzo Cavalieri, Crescenza Dongiovanni, Maria Saponari, and Domenico Bosco. 2023. "Bioecological Traits of Spittlebugs and Their Implications for the Epidemiology and Control of the *Xylella fastidiosa* Epidemic in Apulia (Southern Italy)." *Phytopathology* 113 (9): 1647–60. https://doi.org/10.1094/PHYTO-12-22-0460-IA.

Bodino, Nicola, Vincenzo Cavalieri, Maria Saponari, Crescenza Dongiovanni, Giuseppe Altamura, and Domenico Bosco. 2022. "Transmission of *Xylella fastidiosa* subsp. *pauca* St53 by the Sharpshooter *Cicadella viridis* from Different Source Plants and Artificial Diets." *Journal of Economic Entomology* 115 (6): 1852–58. https://doi.org/10.1093/jee/toac172.

Bohart, R. M. 1946. "A New Species of *Halictophagus* Parasitic on Cercopidae." *Proceedings of the Entomological Society of Washington* 48 (8): 200.

Boodram, Kim. 1999. "Caroni Helps Overseas Firms Fight Crop Pests." *Internet Express, Trinidad Express Business*, 6 November 1999.

Borges, Ricardo Aquino, Alexander Machado Auad, Marcy Das Graças Fonseca et al. 2020. "Coloration Patterns of the Tegmina of *Mahanarva spectabilis* (Hemiptera: Cercopidae): Biological, Morphological and Genetic Bases." *Florida Entomologist* 103 (3): 376–83. https://doi.org/10.1653/024.103.0310.

Boscia, D., V. Cavalieri, C. Dongiovanni et al. 2023. "The Current Status of *Xylella fastidiosa* in Salento (Italy): Are We Approaching a New Phase of the Epidemic?" Paper presented at the 4th European Conference on *Xylella fastidiosa*: Researching Sustainable Solutions, Lyon, France, August 19–20. https://www.efsa.europa.eu/sites/default/files/2023-09/2023-09-boscia.pdf.

Bosso, Luciano, Mirko Di Febbraro, Gennaro Cristinzio, Astolfo Zoina, and Danilo Russo. 2016. "Shedding Light on the Effects of Climate Change on the Potential Distribution of *Xylella fastidiosa* in the Mediterranean Basin." *Biological Invasions* 18 (6): 1759–68. https://doi.org/10.1007/s10530-016-1118-1.

Boukhris-Bouhachem, Sonia, Rebha Souissi, Raied Abou Kubaa, Maroun El Moujabber, and Vladimir Gnezdilov. 2023. "Aphrophoridae as Potential Vectors of *Xylella fastidiosa* in Tunisia." *Insects* 14 (2): 119. https://doi.org/10.3390/insects14020119.

Boulaouad, Belkacem Aimene, Oussama Ailam, S. Daoudi-Hacini, and S. Doumandji. 2018. "Trophic Ecology of Short-Toed Treecreeper's *Certhia brachydactyla* (Brehm, 1820) Nestling in the Suburban Area of Elharrach (Algeria)." *Journal of Animal & Plant Sciences* 28 (1): 321–26.

Boulard, M. 1969. "Hémiptéroïdes nuisibles ou associés aux cacaoyers en République Centrafricaine. Deuxième partie. Homoptères Auchénorhynches." *Café, Cacao, Thé (Paris)* 13 (4): 310–24.

Boulard, M. 1991. "L'urine des homoptères, un matériau utilisé ou recyclé de façons étonnantes. 1ère partie." *Insectes* 81: 7–8.

Boulard, M., and J. Boulard. 1979. "Un organe énigmatique chez les femelles de *Locris* (Homoptera Cercopidae)." *Les Annales de La Société Entomologique de France (N.S.)* 15 (3): 513–23.

Box, H. E. 1929. "Una relación de nuestros conocimientos de la familia Cercopidae en la Argentina." *Revista Industrial y Agrícola de Tucumán* 20 (1–2): 1–18.

Box, H. E. 1950. "The More Important Insect Pests of Sugar Cane in Northen Venezuela." *Proceedings of the Hawaiian Entomological Society* 14 (1): 41–49.

Box, H. E. 1953. "The History and Changing Status of Some Neotropical Insect Pests of Sugar Cane." In *Transactions of the IXth International Congress of Entomology*, Volume 2 254–59. Amsterdam.

Braman, Kristine S., and Rew F. Pendley. 1993. "Relative and Seasonal Abundance of Beneficial Arthropods in Centipede Grass as Influenced by Management Practices." *Journal of Economic Entomology* 86 (2): 494–504. https://doi.org/10.1093/jee/86.2.494.

Braman, Kristine S., and John M. Ruter. 1997. "Preference of Twolined Spittlebug for *Ilex* Species, Hybrids and Cultivars." *Journal of Environmental Horticulture* 15 (4): 211–14. https://doi.org/10.24266/0738-2898-15.4.211.

Brindley, M. D. 1934. "Sparrows Feeding Their Young on the 'Cuckoo-Spit' (Homoptera, Cercopidae)." *Proceedings of the Royal Entomological Society of London* 9: 69–70.

Brodbeck, Brent V., Peter C. Andersen, and Russell F. Mizell. 1995. "Differential Utilization of Nutrients during Development by the Xylophagous Leafhopper, *Homalodisca coagulata*." *Entomologia Experimentalis et Applicata* 75 (3): 279–89. https://doi.org/10.1111/j.1570-7458.1995.tb01938.x.

Brodbeck, Brent V., Russell F. Mizell, and Peter C. Andersen. 1993. "Physiological and Behavioral Adaptations of Three Species of Leafhoppers in Response to the Dilute Nutrient Content of Xylem Fluid." *Journal of Insect Physiology* 39 (1): 73–81. https://doi.org/10.1016/0022-1910(93)90020-R.

Brodbeck, Brent V., Russell F. Mizell, William J. French, Peter C. Andersen, and James H. Aldrich. 1990. "Amino Acids as Determinants of Host Preference for the Xylem Feeding Leafhopper, *Homalodisca coagulata* (Homoptera: Cicadellidae)." *Oecologia* 83 (3): 338–45. https://doi.org/10.1007/BF00317557.

Brooks, G. L., and J. B. Whittaker. 1999. "Responses of Three Generations of a Xylem-Feeding Insect, *Neophilaenus lineatus* (Homoptera), to Elevated CO2." *Global Change Biology* 5 (4): 395–401. https://doi.org/10.1046/j.1365-2486.1999.00239.x.

Brown, A. D. 1974. "Spittle Bug Plague Threatens South Texas Range." *The Cattleman* 60: 46–48.

Brown, Harvey R. 2013. "The Theory of the Rise of Sap in Trees: Some Historical and Conceptual Remarks." *Physics in Perspective* 15: 320–58.

Brunck, F. 1958. *Compte-rendu d'un déplacement effectué dans l'Ouest Cameroun en Décembre 1958. Die Back de l'*Eucalyptus saligna; *Attaques de cicadelles du genre* Ptyelus *sur* Eucalyptus smithii. CTFT. https://agritrop.cirad.fr/363760/.

Buckton, George B. 1890. *Monograph of the British Cicadœ or Tettigidœ*. Macmillan & Company.

Bueno, V. H. P. 1986. "Aspectos etológicos de *Porasilus barbiellinii* Curran, 1934 (Diptera, Asilidae) em pastagem de *Brachiaria decumbens* Stapf." *Revista Brasileira de Entomologia* 30 (1): 17–26.

Bullón F., O. A. 1956. "*Tomaspis calicata*, Jácobi un insecto nuevo parásito de la maleza kikuyo." *Agronomía (Lima, Peru)* 22: 55–60.

Burrola-Morales, Jesús R., Gerardo García-Nevárez, and Leslie Carnero-Avilés. 2022. "Salivazo *Clastoptera achatina* Germar (Hemiptera: Clastopteridae) en nogal pecanero: fluctuación poblacional, distribución y eficacia de insecticidas." *Ecosistemas y Recursos Agropecuarios* 9 (2): e3335. https://doi.org/10.19136/era.a9n2.3335.

Burrows, Malcolm. 2003. "Froghopper Insects Leap to New Heights." *Nature* 424 (6948): 509. https://doi.org/10.1038/424509a.

Burrows, Malcolm. 2006a. "Jumping Performance of Froghopper Insects." *Journal of Experimental Biology* 209 (23): 4607–21.

Burrows, Malcolm. 2006b. "Morphology and Action of the Hind Leg Joints Controlling Jumping in Froghopper Insects." *Journal of Experimental Biology* 209 (23): 4622–37. https://doi.org/10.1242/jeb.02554.

Burrows, Malcolm, Stephen R. Shaw, and Gregory P. Sutton. 2008. "Resilin and Chitinous Cuticle Form a Composite Structure for Energy Storage in Jumping by Froghopper Insects." *BMC Biology* 6 (1): 41. https://doi.org/10.1186/1741-7007-6-41.

Butler, Samuel. 1877. "Letter to Francis Darwin. Darwin Correspondence Project, Letter No. 11254." 25 November 1877. https://www.darwinproject.ac.uk/letter/?docId=letters/DCP-LETT-11254.xml.

Butler, Samuel. 1878. *Life and Habit*. London: Trübner & Co.

Buxton, Katrina N., Michael J. Clearwater, K. Giles-Hansen, E. W. Hewett, and Ian B. Ferguson. 2007. "Comparison of Xylem Sap Mineral Concentrations between Kiwifruit Shoot Types Using Spittlebugs for Non-Destructive Sampling of Sap." *Functional Plant Biology* 34 (11): 1029–37. https://doi.org/10.1071/FP07091.

Byers, R. A., and H. D. Wells. 1966. "Phytotoxemia of Coastal Bermudagrass Caused by the Two-Lined Spittlebug, *Prosapia bicincta* (Homoptera: Cercopidae)." *Annals of the Entomological Society of America* 59 (6): 1067–71. https://doi.org/10.1093/aesa/59.6.1067.

Caimi, Jean-Marc, and Valentina Piccinni. 2022. *Fastidiosa*. London: Overlapse.

Callan, E. M. 1980. "Nesting Behavior and Prey of *Argogorytes* Ashmead (Hymenoptera: Sphecidae)." *Journal of the Washington Academy of Sciences* 70 (4): 160–65.

Callan, E. M. 1991. "Nesting Behaviour and Prey of *Pseudoplisus* Ashmead (Hymenoptera: Sphecidae)." *Journal of Natural History* 25 (6): 1611–15. https://doi.org/10.1080/00222939100771001.

Campagnani, Michelle O., Alexander M. Auad, Rogério M. Maurício, et al. 2024. "Endophytic Capacity of Entomopathogenic Fungi in a Pasture Grass and Their

Potential to Control the Spittlebug *Mahanarva spectabilis* (Hemiptera: Cercopidae)." *Agronomy* 14 (5): 943.

Campo, M. L. del, J. T. King, and M. R. Gronquist. 2011. "Defensive and Chemical Characterization of the Froth Produced by the Cercopid *Aphrophora cribrata*." *Chemoecology* 21: 1–8.

Canny, M. J. 1998. "Transporting Water in Plants." *American Scientist* 86 (2): 152–59.

Cao, Kun-Fang, Shi-Jian Yang, Yong-Jiang Zhang, and Tim J. Brodribb. 2012. "The Maximum Height of Grasses Is Determined by Roots." *Ecology Letters* 15 (7): 666–72. https://doi.org/10.1111/j.1461-0248.2012.01783.x.

Cao, Yanghui, and Christopher H. Dietrich. 2022. "Phylogenomics of Flavobacterial Insect Nutritional Endosymbionts with Implications for Auchenorrhyncha Phylogeny." *Cladistics* 38 (1): 38–58. https://doi.org/10.1111/cla.12474.

Cardona, Cesar, Paola Fory, Guillermo Sotelo, Alejandro Pabon, Giovanna Diaz, and John W. Miles. 2004. "Antibiosis and Tolerance to Five Species of Spittlebug (Homoptera: Cercopidae) in *Brachiaria* spp.: Implications for Breeding for Resistance." *Journal of Economic Entomology* 97 (2): 635–45. https://doi.org/10.1093/jee/97.2.635.

Cariddi, C., M. Saponari, D. Boscia, et al. 2014. "Isolation of a *Xylella fastidiosa* Strain Infecting Olive and Oleander in Apulia, Italy." *Journal of Plant Pathology* 96 (3): 1–5.

Carle, P., and G. Moutous. 1965. "Observations sur le mode de nutrition sur vigne de quatre espèces de cicadelles." *Annales des Epiphyties* 16 (4): 333–54.

Carson, Walter P., and Richard B. Root. 1999. "Top-down Effects of Insect Herbivores during Early Succession: Influence on Biomass and Plant Dominance." *Oecologia* 121 (2): 260–72. https://doi.org/10.1007/s004420050928.

Carvalho, Albiane D., Gabriela C. Alves, Bruno J. R. Alves, Leandro A. Santos, and Veronica M. Reis. 2023. "Root Development and Nitrogen Acquisition of Maize Inoculated with Two Species of Diazotrophs." *Bragantia* 82:e20230060. https://doi.org/10.1590/1678-4499.20230060.

Carvalho, Gervásio S., and Michael D. Webb. 2005. *Cercopid Spittle Bugs of the New World: Hemiptera, Auchenorrhyncha, Cercopidae*. Sofía: Coronet Books.

Carvalho, Gervásio S., and Juliana S. G. Teixeira. 2011. "Two New Species of *Deois* Fennah, 1949 (Hemiptera: Cercopidae) from Brazil." *Zootaxa* 2808 (1): 64–68.

Carvalho, Thais L. G., E. Balsemão-Pires, R. M. Saraiva, P. C. G. Ferreira, and A. S. Hemerly. 2014. "Nitrogen Signalling in Plant Interactions with Associative and Endophytic Diazotrophic Bacteria." *Journal of Experimental Botany* 65 (19): 5631–42. https://doi.org/10.1093/jxb/eru319.

Carvalho, Thais Louise G., Aline C. Rosman, Clícia Grativol, Eduardo de M. Nogueira, José I. Baldani, and Adriana S. Hemerly. 2022. "Sugarcane Genotypes with Contrasting Biological Nitrogen Fixation Efficiencies Differentially Modulate Nitrogen Metabolism, Auxin Signaling, and Microorganism Perception Pathways." *Plants* 11 (15): 1971. https://doi.org/10.3390/plants11151971.

Carvalho-Filho, Fernando da Silva, Gabriela Pirani, and Thiago G. Kloss. 2018. "A New Species and Notes on Unusual Natural History of *Cladochaeta* Coquillett, 1900 (Diptera: Drosophilidae)." *Zootaxa* 4410 (3): 483–96.

Casarin, Noemi, Séverine Hasbroucq, Gabriel Carestia, Audrey Glibert, Claude Bragard, and Jean-Claude Grégoire. 2023. "Investigating Dispersal Abilities of Aphrophoridae in European Temperate Regions to Assess the Threat of Potential *Xylella fastidiosa*-Based Pathosystems." *Journal of Pest Science* 96 (2): 471–88.

Castillo, Andreina I., Carlos Chacón-Díaz, Neysa Rodríguez-Murillo, Helvecio D. Coletta-Filho, and Rodrigo P. P. Almeida. 2020. "Impacts of Local Population

History and Ecology on the Evolution of a Globally Dispersed Pathogen." *BMC Genomics* 21 (1): 369. https://doi.org/10.1186/s12864-020-06778-6.

Castro, R. 1981. "Study and Control of the Sugarcane Froghopper in Mexico." In *Proceedings, Second Inter-American Sugar Cane Seminar: Insect and Rodent Pests, Miami, Florida, 6–7 October 1981,13–18.* Miami, FL: Inter-American Sugar Cane Seminar.

Castro, Sérgio Gustavo Quassi de, Leila L. Dinardo-Miranda, Juliano V. Fracasso, et al. 2019. "Changes in Soil Pest Populations Caused by Sugarcane Straw Removal in Brazil." *BioEnergy Research* 12 (4): 878–87. https://doi.org/10.1007/s12155-019-10019-4.

Castro, Ulises, L. A. Gómez, Y. Gutiérrez, L. P. Andrade, A. Villegas, and N. Bernal. 2009. "Distribución y especies de salivazo (Hemiptera: Cercopidae) asociados con la caña de azúcar (*Saccharum officinarum* L.) en el Valle Del Cauca y Colombia." In *Memorias Tecnicaña VIII, Congreso, Cali, Colombia*, 144–51.

Castro, Ulises, Anuar Morales, and Daniel C. Peck. 2005. "Dinámica poblacional y fenología del salivazo de los pastos *Zulia carbonaria* (Lallemand) (Homoptera: Cercopidae) en el valle geográfico del Río Cauca, Colombia." *Neotropical Entomology* 34: 459–70. https://doi.org/10.1590/S1519-566X2005000300015.

Castro-Valderrama, Ulises, Daniel C. Peck, Gervasio S. Carvalho, Jorge M. Valdez-Carrasco, and Jesús Romero-Nápoles. 2018. "A New Species of the Genus *Mahanarva* Distant, 1909 (Hemiptera: Cercopoidea: Cercopidae), with a Key to the Species from Central America and Mexico." *Journal of Insect Biodiversity.* 6 (2): 1–7.

Castro-Valderrama, Ulises, Daniel C. Peck, Gervasio S. Carvalho, Jorge M. Valdez-Carrasco, and Jesús Romero-Nápoles. 2020. "Description of Two New Species of *Prosapia* (Hemiptera: Cercopidae) from the Nearctic and Neotropics, with a Key to Species of the *P. inferens* Species Group." *Canadian Entomologist* 152 (3): 288–97. https://doi.org/10.4039/tce.2020.20.

Castro-Valderrama, Ulises, Jesús Romero-Nápoles, Daniel C. Peck, et al. 2017. "First Report of Spittlebug Species (Hemiptera: Cercopidae) Associated with *Pinus* Species (Pinaceae) in Mexico." *Florida Entomologist* 100 (1): 206–8. https://doi.org/10.1653/024.100.0136.

Cavalieri, Vincenzo, Giuseppe Altamura, Giulio Fumarola, et al. 2019. "Transmission of *Xylella fastidiosa* subspecies *pauca* Sequence Type 53 by Different Insect Species." *Insects* 10 (10): 324. https://doi.org/10.3390/insects10100324.

Cawich, A. 1981. "Bionomics and Forecasting of the Sugarcane Froghopper, *Aeneolamia postica jugata* (Fowler) [Belize]." In *Proceedings, Second Inter-American Sugar Cane Seminar: Insect and Rodent Pests, Miami, Florida, 6–7 October 1981,5–12.* Miami, FL: Inter-American Sugar Cane Seminar.

Cecil, R. "A Biological and Morphological Study of a Cercopid, *Philaenus leucophthalmus* Linnaeus." PhD diss., Ohio State University, 1930.

Chalk, Phillip M. 2016. "The Strategic Role of 15N in Quantifying the Contribution of Endophytic N2 Fixation to the N Nutrition of Non-Legumes." *Symbiosis* 69 (2): 63–80. https://doi.org/10.1007/s13199-016-0397-8.

Challita, Elio J., Prateek Sehgal, Rodrigo Krugner, and M. Saad Bhamla. 2023. "Droplet Superpropulsion in an Energetically Constrained Insect." *Nature Communications* 14 (1): 860. https://doi.org/10.1038/s41467-023-36376-5.

Chang, Shu-Chen, Hsien-Tzung Shih, and Kuang-Hui Lu. 2019. "Antifungal Effect and Chitinase Activities of the Froth of Spittlebug *Poophilus Costalis* (Walker) (Hemiptera: Cercopoidea: Aphrophoridae)." *Journal of Asia-Pacific Entomology* 22 (1): 269–73. https://doi.org/10.1016/j.aspen.2019.01.006.

Chapin, Edward A. 1925. *Food Habits of the Vireos: A Family of Insectivorous Birds*. Washington, DC: US Department of Agriculture.

Chartois, Marguerite, Xavier Mesmin, Ileana Quiquerez, et al. 2023. "Environmental Factors Driving the Abundance of *Philaenus Spumarius* in Mesomediterranean Habitats of Corsica (France)." *Scientific Reports* 13 (1): 1901. https://doi.org/10.1038/s41598-023-28601-4.

Chatterjee, N. C., and M. Bose. 1933. "Entomological Investigations on the Spike Disease of Sandal (13). Membracidae and Cercopidae (Homopt.). Supplementary Data." *Indian Forest Records (Entomological Series)* 19 (2): 1–10.

Chaves S., M. "Semblanza Ing. Agr. Carlos Sáenz Acosta." Paper presented at XIV Exposición Nacional de la Caña de Azúcar (EXPO-CAÑA 2004), San Carlos, Alajuela, Costa Rica, February 1–3, 2004.

Chaves, Vinicius de Vicente, Guilherme V. Pimentel, Alejandro H. Pabón Valverde, Lidiane A. Silva, Márcio H. P. Barbosa, and Luiz A. Peternelli. 2014. "Biology and Preferred Oviposition Site of the *Mahanarva indentata* Froghopper (Hemiptera: Cercopidae) on Sugarcane." *Florida Entomologist* 97 (1): 73–79. https://doi.org/10.1653/024.097.0109.

Chen, Hong, Gexia Qiao, and Aiping Liang. 2022. "Chromosome-Level Genome Assembly of *Callitettix versicolor* (Rice Spittlebug)." *Genome Biology and Evolution* 14 (9): evac130. https://doi.org/10.1093/gbe/evac130.

Chen, J., Hui Jiang, Yuling Li, et al. 2023. "Overview of Sinoalidae in Cenomanian Kachin Amber from Northern Myanmar (Cicadomorpha, Cercopoidea)." *Cretaceous Research* 142: 105399. https://doi.org/10.1016/j.cretres.2022.105399.

Chen, J., J. Szwedo, B. Wang, et al. 2018. "The First Mesozoic Froghopper in Amber from Northern Myanmar (Hemiptera, Cercopoidea, Sinoalidae)." *Cretaceous Research* 85: 243–49. https://doi.org/10.1016/j.cretres.2017.10.029.

Chen, J., B. Wang, H. Zhang, et al. 2019. "A Remarkable New Sinoalid Froghopper with Probable Disruptive Colouration in Mid-Cretaceous Burmese Amber (Hemiptera, Cicadomorpha)." *Cretaceous Research* 98: 9–17. https://doi.org/10.1016/j.cretres.2019.02.004.

Chen, J., B. Wang, Y. Zheng, et al. 2020. "Taxonomic Review and Phylogenetic Inference Elucidate the Evolutionary History of Mesozoic Procercopidae, with New Data from the Cretaceous Jehol Biota of NE China (Hemiptera, Cicadomorpha)." *Journal of Zoological Systematics and Evolutionary Research* 58 (1): 174–93. https://doi.org/10.1111/jzs.12349.

Chen, J., B. Wang, Y. Zheng, et al. 2019. "New Fossil Data and Phylogenetic Inferences Shed Light on the Morphological Disparity of Mesozoic Sinoalidae (Hemiptera, Cicadomorpha)." *Organisms Diversity & Evolution* 19: 287–302.

Chen, Xu, and Ai-Ping Liang. 2012. "Laboratory Rearing of *Callitettix versicolor* (Hemiptera: Cicadomorpha: Cercopidae), with Descriptions of the Immature Stages." *Annqls of the Entomological Society of America* 105 (5): 664–70. https://doi.org/10.1603/AN12052.

Chen, Xu, and Ai-Ping Liang. 2015. "Identification of a Self-Regulatory Pheromone System That Controls Nymph Aggregation Behavior of Rice Spittlebug *Callitettix versicolor*." *Frontiers in Zoology* 12 (1): 10. https://doi.org/10.1186/s12983-015-0102-4.

Chen, Xu, V. Benno Meyer-Rochow, Alberto Fereres, Marina Morente, and Ai-Ping Liang. 2018. "The Role of Biofoam in Shielding Spittlebug Nymphs (Insecta, Hemiptera, Cercopidae) against Bright Light." *Ecological Entomology* 43 (3): 273–81. https://doi.org/10.1111/een.12496.

Chengula, Fasco, and Briggite Nyambo. 2016. "The Significance of Indigenous Weather Forecast Knowledge and Practices under Weather Variability and Climate Change:

A Case Study of Smallholder Farmers on the Slopes of Mount Kilimanjaro." *International Journal of Agricultural Education and Extension* 2 (2): 31–43.

Cheung, W. W. K., and A. T. Marshall. 1973. "Water and Ion Regulation in Cicadas in Relation to Xylem Feeding." *Journal of Insect Physiology* 19 (9): 1801–16. https://doi.org/10.1016/0022-1910(73)90049-8.

Chiaradia, L. A., M. Miranda, and V. J. Fedatto. 2013. "Flutuação populacional de cigarrinhas-das-pastagens na grama-missioneira-gigante." *Agropecuária Catarinense* 26 (1): 70–75.

China, William E. 1927. "Some Strange Relatives of the Froghopper or Cuckoo-Spit Bugs." *Natural History Magazine* 1 (3): 71–81.

China, William E. 1942. "Dr. J. G. Myers." *Nature* 149 (3780): 406. https://doi.org/10.1038/149406a0.

Choueiri, Elia, Raied A. Kubaa, Franco Valentini, et al. 2023. "First Report of *Xylella fastidiosa* on Almond (*Prunus dulcis*) in Lebanon." *Journal of Plant Pathology* 105 (3): 1157–1157. https://doi.org/10.1007/s42161-023-01361-w.

Christenhusz, Maarten J. M., and James W. Byng. 2016. "The Number of Known Plants Species in the World and Its Annual Increase." *Phytotaxa* 261 (3): 201–17. https://doi.org/10.11646/phytotaxa.261.3.1.

Chuche, Julien, Elaine A. Backus, Denis Thiéry, and Nicolas Sauvion. 2017. "First Finding of a Dual-Meaning X Wave for Phloem and Xylem Fluid Ingestion: Characterization of *Scaphoideus titanus* (Hemiptera: Cicadellidae) EPG Waveforms." *Journal of Insect Physiology* 102: 50–61. https://doi.org/10.1016/j.jinsphys.2017.01.013.

Cid-Muñoz, Raquel, Montserrat Cervantes-Espinoza, Ulises Castro-Valderrama, Gerardo Cuellar-Rodríguez, David Cibrián-Tovar, and Francisco Armendáriz-Toledano. 2022. "Pasado y presente de los salivazos del complejo *Ocoaxo* de los pinos (Hemiptera: Cercopidae): Distribución espacial y temporal." *Revista Mexicana de Biodiversidad* 93:e934130. https://doi.org/10.22201/ib.20078706e.2022.93.4130.

Cid-Muñoz, Raquel, David Cibrián-Tovar, Ernestina Valadez-Moctezuma, Emma Estrada-Martínez, and Francisco Armendáriz-Toledano. 2020. "Biology and Life Stages of Pine Spittle Bug *Ocoaxo assimilis* Walker (Hemiptera: Cercopidae)." *Insects* 11 (2): 96.

Clark, Elizabeth G., Daniele Cornara, Craig R. Brodersen, Andrew J. McElrone, Dilworth Y. Parkinson, and Rodrigo P. P. Almeida. 2023. "Anatomy of an Agricultural Antagonist: Feeding Complex Structure and Function of Three Xylem Sap-Feeding Insects Illuminated with Synchrotron-Based 3D Imaging." *Journal of Morphology* 284 (10): e21639. https://doi.org/10.1002/jmor.21639.

Clark, W. E., G. E. Ibarra-Diaz, and H. W. Van Cleave. 1976. "Taxonomy and Biology of Spittlebugs of the Genera *Aeneolamia* Fennah and *Prosapia* Fennah (Cercopidae) in Northeastern Mexico." *Folia Entomologica Mexicana* 34: 13–24.

Clausen, Curtis P. 1931. *Insects Injurious to Agriculture in Japan*. Washington, DC, US Department of Agriculture.

Clavijo-Coppens, Fernando, Nicolas Ginet, Sophie Cesbron, Martial Briand, Marie-Agnès Jacques, and Mireille Ansaldi. 2021. "Novel Virulent Bacteriophages Infecting Mediterranean Isolates of the Plant Pest *Xylella fastidiosa* and *Xanthomonas albilineans*." *Viruses* 13 (5): 725. https://doi.org/10.3390/v13050725.

Clem, P. D., and J. O. Whitaker Jr. 1995. "Distribution of the Mosquitofish, *Gambusia affinis* (Baird & Girard), in Indiana, with Comments on Resource Competition." *Proceedings of the Indiana Academy of Science* 104 (3–4): 249–58.

Clusella Trullas, Susana, Johannes H. van Wyk, and James R. Spotila. 2007. "Thermal Melanism in Ectotherms." *Journal of Thermal Biology* 32 (5): 235–45. https://doi.org/10.1016/j.jtherbio.2007.01.013.

Cock, M. J. W. 1985. *A Review of Biological Control of Pests in the Commonwealth Caribbean and Bermuda up to 1982*. Slough, UK: Commonwealth Agricultural Bureaux.

Cock, M.J.W., and Fred D. Bennett. 2011. "John Golding Myers (1897–1942), an Extraordinary Exploratory Entomologist." *CABI Reviews* 2011 (May): 1–18. https://doi.org/10.1079/PAVSNNR20116008.

Coe, R. L. 1966. *Diptera: Pipunculidae*. First Edition. London: Royal Entomological Society of London.

Collatz, G. James, Joseph A. Berry, and James S. Clark. 1998. "Effects of Climate and Atmospheric CO2 Partial Pressure on the Global Distribution of C4 Grasses: Present, Past, and Future." *Oecologia* 114 (4): 441–54. https://doi.org/10.1007/s004420050468.

Collier, Alexander, Joe B. Keiper, and Lowell P. Orr. 1998. "The Invertebrate Prey of the Northern Leopard Frog, *Rana pipiens*, in a Northeastern Ohio Population." *Ohio Journal of Science* 98 (3): 39–41.

Comitê Organizador. 1976. "Um amigo que se val, uma saudade que fica." [Obituary for Pietro Guagliumi]. In *Resumos do Congresso, III Congresso Brasileiro de Entomologia*. Congresso Brasileiro de Entomologia.

Congio, Guilhermo F. S., Pedro C. de Almeida, Tadeu R. Barreto, et al. 2020. "Spittlebug Damage on Tropical Grass and Its Impact in Pasture-Based Beef Production Systems." *Scientific Reports* 10 (1): 10758.

Cooke, Rob. 2003. "Insect World's High-Jump King." *Newsday*, July 31, 2003. www.newsday.com/news/health/ny-hsflea31q3395751jul31,0,526524.story.

Cooper, Alan, and Malcolm W. Kennedy. 2010. "Biofoams and Natural Protein Surfactants." *Biophysical Chemistry* 151 (3): 96–104. https://doi.org/10.1016/j.bpc.2010.06.006.

Cooper, Alan, Steven J. Vance, Brian O. Smith, and Malcolm W. Kennedy. 2017. "Frog Foams and Natural Protein Surfactants." In *Colloids and Surfaces A: Physicochemical and Engineering Aspects*, A *Collection of Papers Presented at the 11th Eufoam Conference, Dublin, Ireland, 3–6 July, 2016*, 534: 120–29. https://doi.org/10.1016/j.colsurfa.2017.01.049.

Cooper, Robert J. "Dietary Relationships among Insectivorous Birds of an Eastern Deciduous Forest." PhD diss., West Virginia University, 1988.

Córdoba, Freddy, Stephen L. Lapointe, and Miguel S. Serrano. 1990. "Tasa de excrecion y sitio de alimentacioncba de *Aeneolamia reducta* (Lallemand) (Homoptera: Cercopidae) en seis sspecies de gramineas." *Revista Colombiana de Entomología* 16 (2): 15–20.

Cornara, Daniele, Donato Boscia, Giusy D'Attoma, et al. 2024. "An Integrated Strategy for Pathogen Surveillance Unveiled *Xylella fastidiosa* ST1 Outbreak in Hidden Agricultural Compartments in the Apulia Region (Southern Italy)." *European Journal of Plant Pathology*. https://doi.org/10.1007/s10658-024-02945-7.

Cornara, Daniele, Domenico Bosco, and Alberto Fereres. 2018. "*Philaenus spumarius*: When an Old Acquaintance Becomes a New Threat to European Agriculture." *Journal of Pest Science* 91 (3): 957–72. https://doi.org/10.1007/s10340-018-0966-0.

Cornara, Daniele, V. Cavalieri, Crescenza Dongiovanni, et al. 2017. "Transmission of *Xylella fastidiosa* by Naturally Infected *Philaenus spumarius* (Hemiptera, Aphrophoridae) to Different Host Plants." *Journal of Applied Entomology* 141 (1–2): 80–87. https://doi.org/10.1111/jen.12365.

Cornara, Daniele, Elisa Garzo, Marina Morente, Aranzazu Moreno, Javier Alba-Tercedor, and Alberto Fereres. 2018. "EPG Combined with Micro-CT and Video Recording Reveals New Insights on the Feeding Behavior of *Philaenus spumarius*." *PLOS ONE* 13 (7): e0199154. https://doi.org/10.1371/journal.pone.0199154.

Cornara, Daniele, Monica Marra, Biagio Tedone, et al. 2020. "No Evidence for Cicadas' Implication in *Xylella fastidiosa* Epidemiology." *Entomologia Generalis* 40 (2): 125–32.

Cornara, Daniele, Marina Morente, Anna Markheiser, et al. 2019. "An Overview on the Worldwide Vectors of *Xylella fastidiosa*." *Entomologia Generalis*, 39 (3–4): 157–81. https://doi.org/10.1127/entomologia/2019/0811.

Cornara, Daniele, Onofrio Panzarino, Giacomo Santoiemma, et al. 2021. "Natural Areas as Reservoir of Candidate Vectors of *Xylella Fastidiosa*." *Bulletin of Insectology* 74: 173–80.

Cornara, Daniele, Anne Sicard, Adam R. Zeilinger, Francesco Porcelli, Alexander H. Purcell, and Rodrigo P. P. Almeida. 2016. "Transmission of *Xylella fastidiosa* to Grapevine by the Meadow Spittlebug." *Phytopathology* 106 (11): 1285–90. https://doi.org/10.1094/PHYTO-05-16-0202-R.

Cornara, Daniele, Valentina Zaffaroni-Caorsi, Zeinab Hamouche, et al. 2024. "Host Finding and Probing Behavior by *Philaenus spumarius* on Olive Varieties with a Different Degree of Susceptibility to *Xylella fastidiosa*." *Journal of Pest Science* 97 (4): 2101–13. https://doi.org/10.1007/s10340-024-01743-8.

Coronado P., Ricardo. 1978. "Memoria de la campaña contra la mosca pinta." Secretaria de Agricultura y Recursos Hidraulicos.

Costa Lima, A. da. 1942. *Homópteros. Insetos do Brasil 3*. Escola Nacional de Agronomia.

Couturier, Guy, Herminio I. Sanchez, and Elva T. Flores. 1996. "Insectos plaga del arroz en Loreto, Amazonía Peruana." *Revista Peruana de Entomología* 39: 131–34.

Crews, Laura J., Margaret E. McCully, Martin J. Canny, Cheng X. Huang, and Lewis E. C. Ling. 1998. "Xylem Feeding by Spittlebug Nymphs: Some Observations by Optical and Cryo-Scanning Electron Microscopy." *American Journal of Botany* 85 (4): 449–60.

Crispolon Jr., Elorde S., Adeline Soulier-Perkins, and Eric Guilbert. 2023. "Molecular Phylogeny of Cercopidae (Hemiptera, Cercopoidea)." *Zoologica Scripta* 52 (5): 494–516. https://doi.org/10.1111/zsc.12597.

Cruz, Ivan, M. Figueredo, Miguel M. G. Neto, and Rafaek B. da Silva. 2010. *Danos da cigarrinha-das-pastagens,* Deois favopicta *Stal (Homoptera: Cercopidae) em milho consorciado com braquiárias. Circula Technica 144*. Embrapa. https://www.infoteca.cnptia.embrapa.br/infoteca/handle/doc/880841.

Cruz, Wilfrido de la, and Wilson Cajilema. 2012. "Control biológico del salivazo (*Mahanarva andigena*) en caña de azúcar (*Saccharum officinarum*) con *Metarhizium* sp. (Fungi: Ascomycota: Clavicipitaceae)." *Revista Amazónica Ciencia y Tecnología* 1 (1): 18–26.

Cruz-Zapata, Gabriela de la, Estonia García-López, Saúl Sánchez-Soto, Néstor Bautista-Martínez, Juan J. Ortiz-Díaz, and Rodolfo Osorio-Osorio. 2016. "Identidad de mosca pinta (Hemiptera: Cercopidae) y sus hospederas en cañaverales en Cárdenas, Tabasco, México." *Southwestern Entomologist* 41 (1): 145–51. https://doi.org/10.3958/059.041.0116.

Cryan, Jason R. 2005. "Molecular Phylogeny of Cicadomorpha (Insecta: Hemiptera: Cicadoidea, Cercopoidea and Membracoidea): Adding Evidence to the Controversy." *Systematic Entomology* 30 (4): 563–74. https://doi.org/10.1111/j.1365-3113.2004.00285.x.

Cryan, Jason R., and Gavin J. Svenson. 2010. "Family-Level Relationships of the Spittlebugs and Froghoppers (Hemiptera: Cicadomorpha: Cercopoidea)." *Systematic Entomology* 35 (3): 393–415. https://doi.org/10.1111/j.1365-3113.2009.00520.x.

Cuthrell, D. L. 1999. "Special Animal Abstract for *Prosapia ignipectus* (Red-Legged Spittlebug)." Lansing, Michigan: Michigan Natural Features Inventory. https://mnfi.anr.msu.edu/abstracts/zoology/Prosapia_ignipectus.pdf.

Dammerman, Karol W. 1948. *The Fauna of Krakatau*. Amsterdam: N.V. Noord-Hollandsche Uitgevers Maatschappij. https://digitalcommons.usu.edu/bee_lab_da/435.

Daramola, Ajibola M. 1984. "The Insect Pest Complex of the Edible Roselle *Hibiscus sabdariffa* var. *sabdariffa* L and the Damage They Cause in Southwestern Nigeria." *Nigerian Journal of Entomology* 5: 62–69.

Darling, H. S. 1946. *Annual Report of the Agricultural Entomologist*. Department of Agriculture, Uganda.

Darwin, Francis. 1896. "Report of a Discussion on the Ascent of Water in Trees." *Annals of Botany Old Series* 10 (4): 630–61. https://doi.org/10.1093/oxfordjournals.aob.a088636.

David, H., and V. Nandagopal. 1986. "Pests of Sugarcane–Distribution, Symptomatology of Attack and Identification." In *Sugarcane Entomology in India*, edited by H. David, S. Easwaramoorthy, and R Jayanthi. Coimbatore, India: Sugarcane Breeding Institute.

Davis, C. J., and A. L. Mitchell. 1946. "Host Records of *Philaenus spumarius* (Linn.) at Kilauea, Hawaii National Park (Homoptera: Cercopidae)." *Proceedings of the Hawaiian Entomological Society* 12 (3): 515–516.

Davis, Michael J., Alex H. Purcell, and Sherman V. Thomson. 1978. "Pierce's Disease of Grapevines: Isolation of the Causal Bacterium." *Science* 199 (4324): 75–77. https://doi.org/10.1126/science.199.4324.75.

Davranoglou, Leonidas-Romanos, Beth Mortimer, Graham K. Taylor, and Igor Malenovský. 2020. "On the Morphology and Evolution of Cicadomorphan Tymbal Organs." *Arthropod Structure & Development* 55: 100918. https://doi.org/10.1016/j.asd.2020.100918.

De Geer, Charles. 1773. *Memoires pour servir a l'historie des Insectes*. Vol. 3. Stockholm: Pierre Hesselberg.

De La Fuente, Leonardo, Juan A. Navas-Cortés, and Blanca B. Landa. 2024. "Ten Challenges to Understanding and Managing the Insect-Transmitted, Xylem-Limited Bacterial Pathogen *Xylella fastidiosa*." *Phytopathology* 114 (5): 869–84. https://doi.org/10.1094/PHYTO-12-23-0476-KC.

DeAngelis, Paul L. 2002. "Evolution of Glycosaminoglycans and Their Glycosyltransferases: Implications for the Extracellular Matrices of Animals and the Capsules of Pathogenic Bacteria." *The Anatomical Record* 268 (3): 317–26. https://doi.org/10.1002/ar.10163.

Decary, R. 1928. "Contribution à la biologie du *Ptyelus goudoti* Benn. [Hem.]." *Bulletin de la Société Entomologique de France* 33 (16): 259–63.

Delclòs, Xavier, Enrique Peñalver, Eduardo Barrón, et al. 2023. "Amber and the Cretaceous Resinous Interval." *Earth-Science Reviews* 243: 104486. https://doi.org/10.1016/j.earscirev.2023.104486.

Delgado, C., G. Couturier, L. Balcazar, and A. Chichipe. 2023. "Insect Pests of *Theobroma cacao* (Malvaceae) in the Peruvian Amazon." *Tropical Agriculture* 100 (2): 110–14.

DeLong, Dwight Moore, and Henry Herman Paul Severin. 1950. "Spittle-insect Vectors of Pierce's Disease Virus. I. Characters, Distribution, and Food Plants." *Hilgardia* 19(11): 339–56

Deloria-Sheffield, Christie M., Kelly F. Millenbah, Carol I. Bocetti, Paul W. Sykes Jr., and Cameron B. Kepler. 2001. "Kirtland's Warbler Diet as Determined through Fecal Analysis." *The Wilson Bulletin* 113 (4): 384–87. https://doi.org/10.1676/0043-5643(2001)113[0384:KSWDAD]2.0.CO;2.

Dennis, D. S., Robert J. Lavigne, and Jeanne G. Dennis. 2010. "Hemiptera (Heteroptera/Homoptera) as Prey of Robber Flies (Diptera: Asilidae) with Unpublished Records." *Journal of the Entomological Research Society* 12 (1): 27–47.

Despréz-Loustau, Marie-Laure, Yilmaz Balci, Daniele Cornara, Paolo Gonthier, Cécile Robin, and Marie-Agnès Jacques. 2021. "Is *Xylella fastidiosa* a Serious Threat to European Forests?" *Forestry* 94 (1): 1–17. https://doi.org/10.1093/forestry/cpaa029.
Detilleux, L., R. N. Poligui, L. Iannello, T. Dogot, F. Francis, and R. C. Megido. 2022. "Entomophagy in Gabon across the African Context." *Journal of Insects as Food and Feed* 8 (7): 711–20. https://doi.org/10.3920/JIFF2021.0064.
Dhileepan, K. 2014. "Prospects for the Classical Biological Control of *Calotropis procera* (Apocynaceae) Using Coevolved Insects." *Biocontrol Science and Technology* 24 (9): 977–98. https://doi.org/10.1080/09583157.2014.912611.
Dietrich, Christopher H. 2002. "Evolution of the Cidcaomorpha (Insecta, Hemiptera)." *Denisia* 4 (76): 155–70.
Dietrich, Christopher H., Julie M. Allen, Alan R. Lemmon, et al. 2017. "Anchored Hybrid Enrichment-Based Phylogenomics of Leafhoppers and Treehoppers (Hemiptera: Cicadomorpha: Membracoidea)." *Insect Systematics and Diversity* 1 (1): 57–72.
Dietrich, Christopher H., Serguei A. Simutnik, and Evgeny E. Perkovsky. 2023. "Typhlocybinae Leafhoppers (Hemiptera, Cicadellidae) from Eocene Rovno Amber Reveal a Transition in Wing Venation and a Defensive Adaptation." *Journal of Paleontology* 97 (2): 366–79. https://doi.org/10.1017/jpa.2023.3.
Dike, M. C., O. Ajayi, and T. O. Ajiboye. 2002. "Incidence of Millet Stem Borer, *Coniesta ignefusalis* Hampson (Lepidoptera: Pyralidae) in Farmers' Fields in Northern Nigeria." *Nigerian Journal of Entomology* 19: 51–60. https://doi.org/36108/NJE/2002/91.0160.
Dinardo-Miranda, Leila L., and Juliano V. Fracasso. 2013. "Sugarcane Straw and the Populations of Pests and Nematodes." *Scientia Agricola* 70 (5): 305–10. https://doi.org/10.1590/S0103-90162013000500012.
Dinardo-Miranda, Leila L., and Juliano V. Fracasso, Dilermando Perecin, Isabella D. Miranda, Nádia M. Réchia, and Driele F. dos Santos Buzzatto. 2018. "Resistance of Sugarcane Cultivars to *Mahanarva fimbriolata*." *Bragantia* 77 (2): 314–25. https://doi.org/10.1590/1678-4499.2017162.
Distant, W. L. 1908. *The Fauna of British India Including Ceylon and Burma, Rhynchota. Vol. IV.* London: Taylor and Francis. http://archive.org/details/dli.csl.8636.
Distant, W. L. 1909. "XLII.—Rhynchotal Notes.—XLVII." *Annals and Magazine of Natural History* 8: 187–213. https://doi.org/10.1080/00222930908692586.
Distant, W. L. 1916. *The Fauna of British India Including Ceylon and Burma, Rhynchota, Homoptera: Appendix*. Vol. 6. London: Taylor and Francis. https://doi.org/10.5962/bhl.title.9193.
Dixon, Henry H., and John Joly. 1894. "On the Ascent of Sap." *Proceedings of the Royal Society of London* 57: 3–5.
Dmitriev, D. A., R. Angelova, G. A. Anufriev, et al. 2024. "World Auchenorrhyncha Database." On-line database. TaxonWorks. Accessed November 16, 2024. https://hoppers.speciesfile.org/.
Doering, K. C. 1928. "The Genus *Clastoptera* in America North of Mexico." *University of Kansas Science Bulletin* 18 (1): 5–153.
Doering, K. C. 1931. "Some Biological Notes on the Cercopidae North of Mexico (Homoptera)." *Journal of the Kansas Entomological Society* 4 (2): 48–51.
Doering, K. C. 1941. "A Revision of Two Genera of North American Cercopidae (Homoptera) (Continued)." *Journal of the Kansas Entomological Society* 14 (4): 109–34.
Doering, K. C. 1942. "Host Plant Records of Cercopidae in North America, North of Mexico (Homoptera)." *Journal of the Kansas Entomological Society* 15 (2): 65–72.

Dong, Meng, Zongtao Yang, Guangyuan Cheng, Lei Peng, Qian Xu, and Jingsheng Xu. 2018. "Diversity of the Bacterial Microbiome in the Roots of Four *Saccharum* Species: *S. spontaneum*, *S. robustum*, *S. barberi*, and *S. officinarum*." *Frontiers in Microbiology* 9:Article 267. https://doi.org/10.3389/fmicb.2018.00267.

Dongiovanni, Crescenza, Vincenzo Cavalieri, Nicola Bodino, et al. 2019. "Plant Selection and Population Trend of Spittlebug Immatures (Hemiptera: Aphrophoridae) in Olive Groves of the Apulia Region of Italy." *Journal of Economic Entomology* 112 (1): 67–74. https://doi.org/10.1093/jee/toy289.

Dongiovanni, Crescenza, Michele Di Carolo, Giulio Fumarola, et al. 2023. "Comparing Different Sticky Traps to Monitor the Occurrence of *Philaenus spumarius* and *Neophilaenus campestris*, Vectors of *Xylella fastidiosa*, in Different Crops." *Insects* 14 (9): 777. https://doi.org/10.3390/insects14090777.

Douglas, A. E. 2006. "Phloem-Sap Feeding by Animals: Problems and Solutions." *Journal of Experimental Botany* 57 (4): 747–54. https://doi.org/10.1093/jxb/erj067.

Drosopoulos, Sakis. 2003. "New Data on the Nature and Origin of Colour Polymorphism in the Spittlebug Genus *Philaenus* (Hemiptera: Aphrophoridae)." *Annales de La Société Entomologique de France (N.S.)* 39 (1): 31–42. https://doi.org/10.1080/00379271.2003.10697360.

Drosopoulos, Sakis, Anna Maryańska-Nadachowska, and Valentina G. Kuznetsova. 2010. "The Mediterranean: Area of Origin of Polymorphism and Speciation in the Spittlebug *Philaenus* (Hemiptera, Aphrophoridae)." *Zoosystematics and Evolution* 86 (1): 125–28. https://doi.org/10.1002/zoos.200900017.

Drosopoulos, Sakis, and J. A. Quartau. 2002. "The Spittle Bug *Philaenus tesselatus* Melichar, 1899 (Hemiptera, Auchenorrhyncha, Cercopidae) Is a Distinct Species." *Zootaxa* 68 (1): 1–8. https://doi.org/10.11646/zootaxa.68.1.1.

Duhaime-Ross, Arielle. 2014. "Manslaughter Conviction Overturned for Italian Geologists, but Other Scientists Are Still Fearful." The Verge. 11 November 2014. https://www.theverge.com/2014/11/11/7193391/italy-judges-clear-geologists-manslaughter-laquila-earthquake-fear.

Dunn, James P., C. J. Summerfield, and M. Johnson. 2002. "Distribution, Seasonal Cycle, Host Plant Records, and Habitat Evaluation of a Michigan Threatened Insect: The Great Plains Spittlebug, *Lepyronia gibbosa* (Homoptera: Cercopidae)." *Great Lakes Entomologist* 35 (2): 121–30.

Ebert, Timothy A., Elaine A. Backus, Holly J. Shugart, and Michael E. Rogers. 2018. "Behavioral Plasticity in Probing by *Diaphorina citri* (Hemiptera, Liviidae): Ingestion from Phloem Versus Xylem Is Influenced by Leaf Age and Surface." *Journal of Insect Behavior* 31 (2): 119–37. https://doi.org/10.1007/s10905-018-9666-0.

Eden-Green, S. J., R. Balfas, T. Sutarjo, and Jamalius. 1992. "Characteristics of the Transmission of Sumatra Disease of Cloves by Tube-building Cercopoids, *Hindola* spp." *Plant Pathology* 41 (6): 702–12. https://doi.org/10.1111/j.1365-3059.1992.tb02553.x.

EFSA (European Food Safety Authority), Vincenzo Cavalieri, Elisa Fasanelli, et al. 2024. "Update of the *Xylella* spp. Host Plant Database–Systematic Literature Search up to 31 December 2023." *EFSA Journal* 22 (7): e8898. https://doi.org/10.2903/j.efsa.2024.8898.

EFSA Panel on Plant Health (PLH). 2016. "Four Statements Questioning the EU Control Strategy against *Xylella fastidiosa*." *EFSA Journal* 14 (3): 4450. https://doi.org/10.2903/j.efsa.2016.4450.

Egerton, Frank N. 2008. "A History of the Ecological Sciences, Part 30: Invertebrate Zoology and Parasitology during the 1700s." *Bulletin of the Ecological Society of America* 89 (4): 407–33.

Egwuatu, R. I., and C. B. Ita. 1982. "Some Effects of Single and Split Applications of Carbofuran on the Incidence of and Damage by *Locris maculata*, *Busseola fusca* and *Sesamia calamistis* on Maize." *Tropical Pest Management* 28 (3): 277–83. https://doi.org/10.1080/09670878209370721.

Ehlers, Sarah, Daniel Baum, Roland Mühlethaler, Hannelore Hoch, and Peter Bräunig. 2022. "Large Abdominal Mechanoreceptive Sense Organs in Small Plant-Dwelling Insects." *Biology Letters* 18 (4): 20220078. https://doi.org/10.1098/rsbl.2022.0078.

Eisner, Thomas, Randy C. Morgan, Athula B. Attygalle, Scott R. Smedley, Kithsiri B. Herath, and Jerrold Meinwald. 1997. "Defensive Production of Quinoline by a Phasmid Insect (*Oreophoetes peruana*)." *Journal of Experimental Biology* 200 (19): 2493–2500. https://doi.org/10.1242/jeb.200.19.2493.

El-Khoury, Yara, Giuseppe Bari, Daniele Cornara, Michele Poliseno, and Eustachio Tarasco. 2024. "Susceptibility of *Philaenus spumarius* (Rhynchota, Aphrophoridae) to Entomopathogenic Nematodes and Fungi." *Redia* 107: 97–98. https://doi.org/10.19263/REDIA-107.24.14.

Emosairu, S. O., and J. O. Shiyam. 2000. "A Compendium of Insect Pest and Natural Enemies Associated with Lowland Rice in Southern Nigeria." *Global Journal of Pure and Applied Sciences* 6 (3): 385–88.

Endrestøl, Anders. 2013. "Contribution to the Icelandic Fauna of Auchenorrhyncha (Hemiptera, Fulgoromorpha & Cicadomorpha)." *Norwegian Journal of Entomology* 60 (1): 95–107.

Enkerlin, D., and J. A. Morales. 1980. "The Grass Spittlebug Complex *Aeneolamia albofasciata* and *Prosapia simulans* in Northeastern Mexico and Its Possible Control by Resistant Buffelgrass [*Cenchrus ciliaris*] Hybrids." In *Biology and Breeding for Resistance to Arthropods and Pathogens in Agricultural Plans*, edited by M. K. Harris. Texas Agricultural Experiment Station.

Erwin, Terry L. 1983. "Tropical Forest Canopies: The Last Biotic Frontier." *Bulletin of the Entomological Society of America* 29 (1): 14–20. https://doi.org/10.1093/besa/29.1.14.

Espitia Buitrago, Paula A., Maria R. Manzano, and Luis M. Hernández. 2022. "Spittlebugs (Hemiptera: Cercopidae): Integrated Pest Management on Gramineous Crops in the Neotropical Ecozone." *Frontiers in Sustainable Food Systems* 6:Article 891417.

European Commission. 2014. "Commission Implementing Decision of 23 July 2014 as Regards Measures to Prevent the Introduction into and the Spread within the Union of *Xylella fastidiosa* (Well[s] and Raju)." Official Journal of the European Union. https://eur-lex.europa.eu/legal-content/en/TXT/?uri=CELEX:32014D0497.

European Commission. 2015. "Commission Implementing Decision (EU) 2015/789 of 18 May 2015 as Regards Measures to Prevent the Introduction into and the Spread within the Union of *Xylella fastidiosa* (Wells et al.) (Notified under Document C(2015) 3415)." Official Journal of the European Union. https://eur-lex.europa.eu/legal-content/en/TXT/?uri=CELEX:32015D0789.

European Commission. 2020. "Commission Implementing Regulation (EU) 2020/1201 of 14 August 2020 as Regards Measures to Prevent the Introduction into and the Spread within the Union of *Xylella fastidiosa* (Wells et Al.)." Official Journal of the European Union. https://eur-lex.europa.eu/legal-content/en/TXT/?uri=CELEX:32020R1201.

European Commission. 2024. "*Xylella fastidiosa*." 2024. https://food.ec.europa.eu/plants/plant-health-and-biosecurity/legislation/control-measures/xylella-fastidiosa_en.

Evans, D. E., and D. A. Buxo. 1972. "Insecticidal Control of the Sugar-Cane Froghopper." In *Proceedings of the 14th Congress Internatioanl Society of Sugar Cane Technologists*, 507–15. Carapichaima, Trinidad.

Evans, D. L., N. Castoriades, and H. Muhtasib. 1985. "Is the Six-Spotted Burnet Moth, *Zygaena filipendulae*, a Part of *Cercopis intermedia* Mullerian Mimicry Complex?" *Lebanese Science Bulletin* 1 (1): 65–71.

Evans, Howard E., and Kevin M. O'Neill. 2007. *The Sand Wasps: Natural History and Behavior.* Harvard University Press. https://doi.org/10.4159/9780674036611.

Evans, J. W. 1940. "The Morphology of *Tettigarcta tornentosa* White,(Homoptera, Cicadidae)." In *Papers and Proceedings of the Royal Society of Tasmania*, 35–50. http://eprints.utas.edu.au/13273/1/1940-evans-morphology.pdf.

Evans, J. W. 1966. "The Leafhoppers and Froghoppers of Australia and New Zealand (Homoptera: Cicadelloidea and Cercopoidea)." *The Australian Museum Memoir* 12: 1–347. https://doi.org/10.3853/j.0067-1967.12.1966.425.

Evans, J. W. 1972. "Some Remarks on the Family Jascopidae (Homoptera, Auchenorrhyncha)." *Psyche* 79: 120–21.

Everly, R. T. 1958. "Evaluation of Population Estimates and Rate of Loss of Forage for the Meadow Spittlebug, *Philaenus Leucophthalmus* (L.)." *Proceedings of the Indiana Academy of Science* 68: 171–85.

Fagan, E. Brad, and L. C. Kuitert. 1969. "Biology of the Two-Lined Spittlebug, *Prosapia bicincta*, on Florida Pastures (Homoptera: Cercopidae)." *Florida Entomologist* 52 (3): 199–206. https://doi.org/10.2307/3493854.

Fagan, E. Brad, and Ovidio V. Picado. 1971. "The Influence of Adult *Prosapia distanti* Feeding on the Forage Quality of Kikuyugrass in Costa Rica." *Turrialba (Costa Rica)* 21 (2): 181–83.

Faliarivola, Manoa L., M. J. Raherilalao, A. Andrianarimisa, and S. M. Goodman. 2020. "The Diet of Malagasy Dry Forest Understory Birds Based on Faecal Samples." *Ostrich* 91 (1): 35–44.

Farigoule, Pauline, Marguerite Chartois, Xavier Mesmin, et al. 2022. "Vectors as Sentinels: Rising Temperatures Increase the Risk of *Xylella fastidiosa* Outbreaks." *Biology* 11 (9): 1299. https://doi.org/10.3390/biology11091299.

Fasano, Stefano. 2020. "An Olive Tree Plague in Italy Can Teach Us a Lot about the Pandemic." *Wired*, December 12, 2020. https://www.wired.com/story/italy-olive-trees-epidemic/.

Fazolin, M., and C. R. Costa. 2009. "Levantamento de insetos-praga associados aos capins Tanner-grass, Tangola e Estrela-Africana no Acre." *Revista Amazônia: Ciência & Desenvolvimento* 4: 161–73.

Felt, Ephraim P., and Kenyon F. Chamberlain. 1935. *The Occurrence of Insects at Some Height in the Air, Especially on the Roofs of High Buildings. Circular 17.* New York State Museum.

Fennah, R. G. 1948. "XLIII.—New Genera and Species of Neotropical Cercopoidea (Homoptera)." *Annals and Magazine of Natural History Series 12* 1 (9): 605–20. https://doi.org/10.1080/00222934808653935.

Fernández, M., and A. Ramos. 1986. "Biología de *Prosapia* (Monecphora) *bicincta fraterna* (Homoptera: Cercopidae) en caña de azúcar y pasto." *Revista de Protección Vegetal* 1 (1): 43–50.

Fernandéz T., M., and C. Benjamin. 2010. *Socioeconomic Baseline of the Sugar Related Areas in the Framework of the National Sugar Adaptation Strategy Trinidad and Tobago. Final Report. European Union Delegation to Trinidad and Tobago.* Ambiente Risorse Sviluppo Progetti S. P. A.

Ferraz, Edmir C. de A. 1985. "Chemical Control of *Clastoptera* sp. (Homoptera, Cercopidae), the Spittlebug of the Flower Cushion in Cacao." In *Proceedings 9th International Cocoa Research Conference, Lomé, Togo, 12–18 February 1984*, 573–78. Lagos, Nigeria: Cocoa Producers' Alliance.

Ferreira, Evane, José A. F. Barrigossi, and Emílio da M. de Castro. 2003. *Homópteros associados ao arroz. Documento 152*. Empresa Brasileira de Pesquisa Agropecuária Centro Nacional de Pesquisa de Arroz e Feijão. https://www.infoteca.cnptia.embrapa.br/bitstream/doc/212490/1/doc152.pdf.

Ferreira, Rebecca Caroline U., Letícia A. de Castro Lara, Lucimara Chiari, et al. 2019. "Genetic Mapping with Allele Dosage Information in Tetraploid *Urochloa decumbens* (Stapf) R. D. Webster Reveals Insights into Spittlebug (*Notozulia entreriana* Berg) Resistance." *Frontiers in Plant Science* 10:Article 92.

Ferrer, Francisco R., and Magdalia H. Torres. 1984. "Sinopsis histórica sobre el control de la candelilla (*Aeneolamia* spp.) en Venezuela." *Barquisimeto* Noviembre 1984: 105–40.

Ferrer Wurs, Francisco. 2021. "Control biológico de plagas agrícolas en Venezuela: Los logros históricos de la Empresa Servicio Biológico (SERVBIO)." *Revista de Ciencias Ambientales* 55 (1): 327–44.

Fewkes, D. W. 1969a. "The Biology of Sugar Cane Froghoppers." In *Pests of Sugar Cane*, edited by J. R. Williams, J.R. Metcalf. R. W. Montgomery, and R. Mathes. Elsevier Science Publishers.

Fewkes, D. W. 1969b. "The Control of Froghoppers in Sugar Cane Plantations." In *Pests of Sugar Cane*, edited by J. R. Williams, J. R. Metcalf. R. W. Montgomery, and R. Mathes, Elsevier Science Publishers.

Fewkes, D. W, and D. A. Buxo. 1964. "Biological Control Investigations." In *Tate & Lyle Central Agricultural Research Station Annual Report 1964*, Carapichaima, Trinidad and Tobago.

Fewkes, D. W., E. A. C. Hagley, and D. A. Buxo. 1965. "Yield Losses in Sugar Cane Due to Froghopper Infestations." In *Tate & Lyle Central Agricultural Research Station Annual Report 1965*, Carapichaima, Trinidad and Tobago.

Fidelis, Elisangela G., Gervásio S. Carvalho, Regina Oliveira, Rosinei A. Santos, and José R. Valério. 2021. "*Aeneolamia* Fennah, 1949 (Hemiptera: Cercopidae) Attacks Pastures in Roraima, with First Report in Brazil." *Entomological Communications* 3:ec03055. https://doi.org/10.37486/2675-1305.ec03055.

Fielding, C. A., J. B. Whittaker, J. E. L. Butterfield, and J. C. Coulson. 1999. "Predicting Responses to Climate Change: The Effect of Altitude and Latitude on the Phenology of the Spittlebug *Neophilaenus lineatus*." *Functional Ecology* 13 (s1): 65–73. https://doi.org/10.1046/j.1365-2435.1999.00009.x.

Figueredo, Luis, Adriana Villa-Murillo, Yelitza Colmenarez, and Carlos Vásquez. 2021. "A Hybrid Artificial Intelligence Model for *Aeneolamia varia* (Hemiptera: Cercopidae) Populations in Sugarcane Crops." *Journal of Insect Science* 21 (2): 1–6. https://doi.org/10.1093/jisesa/ieab017.

Filho, Adrião C. 1944. *A cigarrinha dos canavais:* Tomaspis liturata *(Lepeletier et Serville, 1825) var.* ruforiavulata *Stal, 1854*. Instituto do Açúcar e do Álcool.

Fitch, Asa. 1851. "Catalogue with References and Descriptions of the Insects Collected and Arranged for the State Cabinet of Natural History." *Fourth Annual Report, New York State Cabinet of Natural History*, 443–69. https://www.biodiversitylibrary.org/item/109934.

Fletcher, Thomas B. 1914. *Some South Indian Insects and Other Animals of Importance Considered Especially from an Economic Point of View*. Madras, India: Superintendent, Government Press.

Fletcher, Thomas B. 1920. "Annotated List of Indian Crop-Pests." In *Proceedings of the Third Entomological Meeting, Held at Pusa 13–15 February 1919*, 3: 33–314. Calcutta, India: Superintendent Government Printing.

Flores Velastegui, Tonny Marlon. 2005. "Diagnóstico fitosanitario en bosques implantados de *Tectona grandis* (Teca) en la Zona de Balzar, Provincia del Guayas." Thesis de Grado, Universidad Técnica Estatal de Quevedo, 2005.

Flores F., J. D., and H. Velasco P. 1974. "Danos causados por adultos de mosca pinta, *Aeneolamia postica* (Wlk.), a diversas especies de zacates forrajeros." *Folia Entomológica Mexicana* 1974 (28): 71–75.

Fogarty, Michael J., and Willa M. Hetrick. 1973. "Summer Foods of Cattle Egrets in North Central Florida." *The Auk* 90 (2): 268–80.

Foieri, Álvaro. "Taxonomía y biología de los *Cercopidae* (Hemiptera) asociados a pasturas nativas y cultivadas del centro y norte de la Argentina, e identificación de sus enemigos naturales." Doctoral diss., Universidad Nacional de La Plata, 2017. https://doi.org/10.35537/10915/59231.

Foieri, Álvaro, Eduardo G. Virla, Arnaldo Maciá, and Ana Maria M. de Remes Lenicov. 2020. "Effect of Host Plant on the Fitness of the Spittlebug *Notozulia entreriana*: Alternative Method for Rearing." *Entomologia Experimentalis et Applicata* 168 (8): 618–25. https://doi.org/10.1111/eea.12950.

Fonseca, M. G., A. M. Auad, T. T. Resende, M. C. Hott, and Cristiano A. V. Borges. 2016. "How Will *Mahanarva spectabilis* (Hemiptera: Cercopidae) Respond to Global Warming?" *Journal of Insect Science* 16 (1): 32. https://doi.org/10.1093/jisesa/iew005.

Fonseca, M. G., A. M. Auad, T. T. Resende, B. A. Veríssimo, and C. M. Oliveira. 2023. "Exposure of Insects and Host Plants to Different Concentrations of CO2 Affects the Performance of *Mahanarva spectabilis* (Hemiptera: Cercopidae) in Successive Insect Generations." *Brazilian Journal of Biology* 83:e273470.

Fontes, Eliana G., Carmen S. S. Pires, and Edison R. Sujii. 1995. "Mixed Risk-Spreading Strategies and the Population Dynamics of a Brazilian Pasture Pest, *Deois flavopicta* (Homoptera: Cercopidae)." *Journal of Economic Entomology* 88 (5): 1256–62. https://doi.org/10.1093/jee/88.5.1256.

Formisano, Giorgio, Luigi Iodice, Pasquale Cascone, et al. 2022. "*Wolbachia* Infection and Genetic Diversity of Italian Populations of *Philaenus spumarius*, the Main Vector of *Xylella fastidiosa* in Europe." *PLOS ONE* 17 (8): e0272028. https://doi.org/10.1371/journal.pone.0272028.

Fountain, H. 2003. "By Leaps and Bounds." *The New York Times*. August 5, 2003.

Fowler, W. W. 1897. "Insecta. Rhynchota. Hemiptera-Homoptera." *Biologia Centrali-Americana*. Volume 1, Part 1: 174–206.

French, Bruce R. 2006. *Insects Pests of Food Plants of Papua New Guinea: A Compendium*. Food Plants International Inc.

Froza, Joyce A., Pedro H. A. Moura, Luiz F. O. Silva, and João R. S. Lopes. 2023. "Ecological Aspects of Spittlebugs of the Genus *Clastoptera* (Hemiptera: Cercopoidea: Clastopteridade), Possible Vectors of *Xylella fastidiosa* in Olives in Southeastern Brazil." Paper presented at the 4th European Conference on *Xylella fastidiosa*: Researching sustainable solutions, Lyon, France, August 19–20. https://www.efsa.europa.eu/sites/default/files/2023-09/2023-09-lopes.pdf.

Froza, Joyce A., Pedro H. A. Moura, Luiz F. O. Silva, Gabriel Mejdalani, and João R. S. Lopes. 2024. "Species Composition and Prevalence of Sharpshooters and Spittlebugs Potential Vectors of *Xylella fastidiosa* in Olive Orchards of Southeastern Brazil." *Revista Brasileira de Entomologia* 68:e20240017. https://doi.org/10.1590/1806-9665-RBENT-2024-0017.

Fu, Yanzhe, Chenyang Cai, and Diying Huang. 2020. "First Mesozoic Procercopids in Mid-Cretaceous Amber from Northern Myanmar (Hemiptera: Cercopoidea)." *Geological Magazine* 157 (3): 506–12. https://doi.org/10.1017/S0016756819001018.

Fu, Yanzhe, and Diying Huang. 2018. "New Fossil Genus and Species of Sinoalidae (Hemiptera: Cercopoidea) from the Middle to Upper Jurassic Deposits in Northeastern China." *European Journal of Entomology* 115: 127–133. doi: 10.14411/eje.2018.011.

Fu, Yanzhe, Diying Huang, and Michael S. Engel. 2018. "A New Species of the Extinct Family Procercopidae (Hemiptera: Cercopoidea) from the Jurassic of Northeastern China." *Palaeoentomology* 1 (1): 51–57. https://doi.org/10.11646/palaeoentomology.1.1.7.

Galey, Miranda. "Predator to Prey to Poop: Bats as Microbial Hosts and Insectivorous Hunters." MS diss., University of Minnesota, 2020.

Gallego, Diego, Sandra C. Sabah, José L. Lencina, and Antonio F. Carrillo. 2023. "Bioclimatic and Landscape Factors Drive the Potential Distribution of *Philaenus spumarius*, *Neophilaenus campestris* and *N. lineatus* (Hemiptera, Aphrophoridae) in Southeastern Iberian Peninsula." *Insects* 14 (7): 592. https://doi.org/10.3390/insects14070592.

Gao, Yunhu, and André Cabrera Serrenho. 2023. "Greenhouse gas emissions from nitrogen fertilizers could be reduced by up to one-fifth of current levels by 2050 with combined interventions." *Nature Food* 4: 170–178. https://doi.org/10.1038/s43016-023-00698-w.

García J., J. R. 1982. "Interacción entre *Carabunia myersi* Waterston (Hymenoptera: Encyrtidae) y *Clastoptera globosa* Fowler (Homoptera:Cercopidae) en La Chontalpa, Tabasco, México." *Folia Entomológica Mexicana* 54: 86–87.

Garcia, José F., Eliane Grisoto, Paulo S. M. Botelho, José R. P. Parra, and Beatriz Appezzato-da-Glória. 2007. "Feeding Site of the Spittlebug *Mahanarva fimbriolata* (Stål) (Hemiptera: Cercopidae) on Sugarcane." *Scientia Agricola* 64: 555–57.

García-González, Juan C., Jose López-Collado, Carlos G. García-García, Juan A. Villanueva-Jiménez, and Martha E. Nava-Tablada. 2017. "Biotic, Abiotic and Agronomic Factors That Affect Adult Populations of Spittlebug (Hemiptera: Cercopidae) in Sugar Cane Crops in Veracruz, México." *Acta Zoológica Mexicana* 33 (3): 508–17.

Garman, Philip. 1923. "Notes on the Life History of *Clastoptera obtusa* and *Lepyronia quadrangularis* (Order Hemiptera, Family Cercopidae)." *Annals of the Entomological Society of America* 16 (2): 153–63. https://doi.org/10.1093/aesa/16.2.153.

Gawel, Lauren, Erin C. Powell, Michelle Brock, and Lisa A. Taylor. 2023. "Conspicuous Stripes on Prey Capture Attention and Reduce Attacks by Foraging Jumping Spiders." *Royal Society Open Science* 10 (11): 230907. https://doi.org/10.1098/rsos.230907.

Genung, W.G., and V.E. Green Jr. 1974. "Food Habits of the Meadowlark in the Everglades in Relation to Agriculture." *Environmental Entomology* 3 (1): 39–42. https://doi.org/10.1093/ee/3.1.39.

Genung, William G., M. J. Janes, and Victor E. Green. 1976. "Insects and Other Dietary Items of Maynard's Red-Wing Blackbird in Relation to Agriculture." *Florida Entomologist* 59 (3): 309–16. https://doi.org/10.2307/3494269.

Germinara, Giancinto S., Sonia Ganassi, Marco O. Pistillo, Carmela Di Domenico, Antonio De Cristofaro, and Antonella M. Di Palma. 2017. "Antennal Olfactory Responses of Adult Meadow Spittlebug, *Philaenus spumarius*, to Volatile Organic Compounds (VOCs)." *PLOS ONE* 12 (12): e0190454. https://doi.org/10.1371/journal.pone.0190454.

Gerónimo L., Verónica, Saúl Sánchez Soto, Néstor Bautista M., Manuel Pérez de la Cruz, and José H. R. Mendoza H. 2013. "Fluctuación poblacional de *Clastoptera laenata* (Hemiptera: Clastopteridae) en el cultivo del cacao en Tabasco, México." *Fitosanidad* 17 (3): 131–37.

Ghosh, Lakshmi K. 2008. *Handbook on Hemipteran Pests in India*. Kolkata: Zoological Survey of India. http://archive.org/details/dli.zoological.hpg.038.

Gibson, Dianne O. 1974. "Batesian Mimicry without Distastefulness?" *Nature* 250 (5461): 77–79. https://doi.org/10.1038/250077a0.

Gibson, Dianne O. 1980. "The Role of Escape in Mimicry and Polymorphism: I. The Response of Captive Birds to Artificial Prey." *Biological Journal of the Linnean Society* 14 (2): 201–14. https://doi.org/10.1111/j.1095-8312.1980.tb00105.x.

Giménez-Romero, Alex, Javier Galván, Marina Montesinos, et al. 2022. "Global Predictions for the Risk of Establishment of Pierce's Disease of Grapevines." *Communications Biology* 5 (1): 1–13. https://doi.org/10.1038/s42003-022-04358-w.

Glenn, D. M., and F. Takeda. 1989. "Guttation as a Technique to Evaluate the Water Status of Strawberry." *HortScience* 24 (4): 599–601. https://doi.org/10.21273/HORTSCI.24.4.599.

Glick, P. A. 1939. *The Distribution of Insects, Spiders, and Mites in the Air*. Technical Bulletin 673. Washinton DC: US Department of Agriculture. https://ageconsearch.umn.edu/record/168268/files/tb673.pdf.

Global Desertification Research Network. 2022. "Pharmacological Control Technology of *Aphrophora flavipes* Uhler of *Pinus sylvestris* var. Mongolica in Horqin Sandy Land." June 8, 2022. http://www.eo4ldn.org.cn/cn/article/id/9e852903-f522-4dbd-93bb-f1267fdfb525?map=1.

Godefroid, Martin, M. Morente, T. Schartel, et al. 2021. "Climate Tolerances of *Philaenus spumarius* Should Be Considered in Risk Assessment of Disease Outbreaks Related to *Xylella fastidiosa*." *Journal of Pest Science* 95 (2): 855–68. https://doi.org/10.1007/s10340-021-01413-z.

Godefroid, Martin, Astrid Cruaud, Jean-Claude Streito, Jean-Yves Rasplus, and Jean-Pierre Rossi. 2022. "Forecasting Future Range Shifts of *Xylella fastidiosa* under Climate Change." *Plant Pathology* 71 (9): 1839–48. https://doi.org/10.1111/ppa.13637.

Goetzke, Hanns H., Jonathan G. Pattrick, and Walter Federle. 2019. "Froghoppers Jump from Smooth Plant Surfaces by Piercing Them with Sharp Spines." *Proceedings of the National Academy of Sciences* 116 (8): 3012–17. https://doi.org/10.1073/pnas.1814183116.

Golfari, L. 1963. "Observaciones sobre *Cephisus siccifolius* en eucaliptos de Misiones." *Idia* (189): 9–14.

Gómez, Yannery. 2005. "Efecto de la dosis y formulación de *Metarhizium anisopliae* (Mertsch) en el combate de Cercópidos en el pasto brachiaria (*Brachiaria ruzizensis*)." *Alcances Tecnológicos* 3 (1): 14–24.

Goodchild, A. J. P. 1966. "Evolution of the Alimentary Canal in the Hemiptera." *Biological Reviews* 41 (1): 97–139. https://doi.org/10.1111/j.1469-185X.1966.tb01540.x.

Gough, L. H. 1910. "Results of Experiments with the 'Froghopper Fungus.'" *Proceedings of the Agricultural Society of Trinidad and Tobago* 10: 463–65.

Graesser, Jordan, T. M. Aide, H. R. Grau, and Navin Ramankutty. 2015. "Cropland/Pastureland Dynamics and the Slowdown of Deforestation in Latin America." *Environmental Research Letters* 10 (3): 034017. https://doi.org/10.1088/1748-9326/10/3/034017.

Gravely, F. H. 1915. "Notes on the Habits of Indian Insects, Myriapods and Arachnids." *Records of the Indian Museum* 11: 483–539.

Greathead, D. J. 1970. "A Study of the Host Relations of *Halictophagus pontifex* Fox (Strepsiptera), a Parasite of Cercopidae (Hem., Aphrophorinae), in Uganda." *Bulletin of Entomological Research* 60 (1): 33–42. https://doi.org/10.1017/S000748530003412X.

Greco, Davide, Erika Sabella, Angelo G. D. Donne, Luigi De Bellis, Andrea Luvisi, and Giambattista Carluccio. 2023. "*Xylella fastidiosa*, Possible New Threat to Chestnut (*Castanea sativa* Mill.) in Italy." *Horticulturae* 2023: 1315. https://doi.org/10.3390/horticulturae9121315.

Greene, Albert. 1983. "Comparative Early Growth and Foraging of Two Naturally Established Vespine Wasp Colonies." In *The Biology of Social Insects*, edited by M. D. Breed, C. D. Michener, and H. E. Evans. CRC Press.

Grifaldo-Alcántara, P. F., R. Alatorre-Rosas, J. A. Villanueva-Jiménez, F. Hernández-Rosas, S. P. Stock, and G. Ramírez-Valverde. 2019. "Evaluación de dos cepas de nematodos entomopatógenos (Steinernematidae, Heterorhabditidae) para el control del salivazo (Hemiptera: Cercopidae) en caña de azúcar." *Nematropica* 49 (1): 83–90.

Griffith, Katherine, Daniel C. Peck, and J. Stuckey. 2000. "Agriculture in Monteverde: Moving toward Sustainability." In *Monteverde: Ecology and Conservation of a Tropical Cloud Forest*, edited by N. M. Nadkarni and N. T. Wheelwright. Oxford University Press. https://www.cabdirect.org/cabdirect/abstract/20001812766.

Griffith, Lizabeth A., and J. E. Gates. 1985. "Food Habits of Cave-Dwelling Bats in the Central Appalachians." *Journal of Mammalogy* 66 (3): 451–60. https://doi.org/10.2307/1380919.

Griffiths, Andy, Stephen Moran, Liam M. Crowley, et al. 2024. "The Genome Sequence of the Alder Spittlebug, *Aphrophora alni* (Fallén, 1805) [Version 1; Peer Review: 2 Approved]." *Wellcome Open Research* 9: 632. https://doi.org/10.12688/wellcomeopenres.23248.1.

Grimaldi, David A., and Tam Nguyen. 1999. "Monograph on the Spittlebug Flies, Genus *Cladochaeta* (Diptera, Drosophilidae, Cladochaetini)." *Bulletin of the American Museum of Natural History* No. 241: 1–326.

Grimaldi, David A., and Michael S. Engel. 2005. *Evolution of the Insects*. First Edition. Cambridge University Press.

Grimaldi, David A., and Courtney Richenbacher. 2023. "Exceptional Species Diversity of Drosophilidae (Diptera) in a Neotropical Forest." *American Museum Novitates* (3997): 1–28.

Grisoto, Eliane, José D. Vendramim, André L. Lourenção, José A. Usberti, and Ricardo A. de Olinda. 2018. "Evaluation of Forage Grass Resistance to *Mahanarva fimbriolata* (Stål)." *Bragantia* 77 (1): 107–15. https://doi.org/10.1590/1678-4499.2016280.

Gruner, Max. "Biologische Untersuchungen an Schaumcikaden: Gatt. *Aphrophora* Germ. und *Philaenus* Stal." PhD diss., Friedrich-Wilhelms-Universität, 1901.

Guagliumi, P. 1954. "Contribuciones al studio de la candelilla (*Aeneolamia* spp. y *Delassor* spp. Homoptera Cercopidae) en Venezuela. I.–Historia del insecto en el Pais." *Agronomia Tropical* 4 (3): 151–61.

Guagliumi, P. 1955a. "Contribuciones al studio de la candelilla de las gramineas en Venezuela. II. Los cercopidos causantes de la candelilla. A) *Aeneolamia (=Tomaspis) varia* (F.) y sus subspecies." *Agronomia Tropical* 5 (3): 315–194.

Guagliumi, P. 1955b. "Perspectivas para el combate biológico de la candelilla. *AGRO: Revista Sociedad Estación Ingeniero Agrónomo (Venezuela)* 10 (37–40): 23–39.

Guagliumi, P. 1955c. Suplemento a la lista de los insectos de la caña de azúcar en Venezuela (1953–1955). *Agronomia Tropical* 5 (2): 99–104.

Guagliumi, P. 1956a. "Contribuciones al studio de la candelilla de las gramineas en Venezuela. II. Los cercopidos causantes de la candelilla. B) *Aeneolamia flavilatera* (Urich) y sus subspecies." *Agronomia Tropical* 6 (2): 51–73.

Guagliumi, P. 1956b. "Contribuciones al studio de la candelilla de las gramineas en Venezuela. II. Los cercopidos causantes de la candelilla. C) *Ae. reducta montana* Fennah. D) *Ae. lepidior* (Fowl.)." *Agronomia Tropical* 6 (3): 123–33.

Guagliumi, P. 1962. *Las plagas de la caña de azúcar en Venezuela*. 2 vols. Maracay, Venezuela: Ministerio de Agricultura y Cría.

Guagliumi, P. 1968. "As cigarrinhas dos carnavais no Brasil." *Brasil Açucareiro* 72 (3): 34–43.

Guagliumi, P. 1970. "Cigarrinha das pastagens ataca a cana-de-açúcar no Nordeste." *Brasil Açucareiro* 76 (4): 89–91.

Guerra, Abby, Silvia Lopez, Iroel Rodriguez, et al. 2023. "Evaluation of the Efficiency of Color Sticky Traps in Sampling Time for the Capture of *Aeneolamia* spp. (Hemiptera: Cercopidae) and Arthropods Present in Sugar Cane in Panama." *Tropical and Subtropical Agroecosystems* 26 (3): 21–27. https://doi.org/10.56369/tsaes.4848.

Guilbeau, Braxton H. 1908. "The Origin and Formation of the Froth in Spittle-Insects." *The American Naturalist* 42 (504): 783–98. https://doi.org/10.1086/279010.

Güldür, M., B. Çağlar, M. Castellano, et al. 2005. "First Report of Almond Leaf Scorch in Turkey." *Journal of Plant Pathology* 87 (3): 246.

Guo, Kaiyan, Jun Yang, Nan Yu, Li Luo, and Ertao Wang. 2023. "Biological Nitrogen Fixation in Cereal Crops: Progress, Strategies, and Perspectives." *Plant Communications* 4 (2). https://doi.org/10.1016/j.xplc.2022.100499.

Guppy, P. L. 1913. "Life-History of Syrphid Fly Predaceous on Froghopper Nymphs." *Bulletin Department Agriculture Trinidad and Tobogo* 12: 159–61.

Gusmão, M. R., J. R. Valério, F. P. Matta, F. H. D. Souza, B. B. Z. Vigna, A. P. Fávero, W. Barioni Jr., and G. R. Inácio. 2016. "Warm-Season (C4) Turfgrass Genotypes Resistant to Spittlebugs (Hemiptera: Cercopidae)." *Journal of Economic Entomology* 109 (4): 1914–21.

Haddad, Najat, Imane Mrabti, Ahmed Douaik, et al. 2024. "*Philaenus tesselatus* (Hemiptera: Aphrophoridae), the Main Potential Vector of *Xylella fastidiosa* in Morocco: Seasonal Abundance, Phenology and Host-Plant Colonization by Nymphs." *Annales de La Société Entomologique de France (N.S.)* 60 (5): 461–76. https://doi.org/10.1080/00379271.2024.2364963.

Hagar, Joan C., Katie M. Dugger, and Edward E. Starkey. 2007. "Arthropod Prey of Wilson's Warblers in the Understory of Douglas-Fir Forests." *The Wilson Journal of Ornithology* 119 (4): 533–46.

Hagerty, A. M., and J. E. McPherson. 2000. "Life History and Laboratory Rearing of *Arilus cristatus* (Heteroptera: Reduviidae) in Southern Illinois." *Florida Entomologist* 83 (1): 58–63. https://doi.org/10.2307/3496229.

Hagley, Elmer A. C. 1966. "Studies of the Feeding and Nutrition of the Sugarcane Froghopper." Tate and Lyle, Central Agricultural Research Station Annual Report. Carapichaima, Trinidad and Tobago.

Hagley, Elmer A. C., and J. A. Blackman. 1966. "Site of Feeding of the Sugarcane Froghopper, *Aeneolamia varia saccharina* (Homoptera: Cercopidae)." *Annals of the Entomological Society of America* 59 (6): 1289–91. https://doi.org/10.1093/aesa/59.6.1289.

Haines, F. H. 1942. "The Enemies of Cercopidae (Hem.)." *Entomologist's Monthly Magazine* 78: 205.

Halkka, Antti, Liisa Halkka, Olli Halkka, Kaisa Roukka, and Jussi Pokki. 2006. "Lagged Effects of North Atlantic Oscillation on Spittlebug *Philaenus spumarius* (Homoptera) Abundance and Survival." *Global Change Biology* 12 (12): 2250–62. https://doi.org/10.1111/j.1365-2486.2006.01266.x.

Halkka, Olli, and Liisa Halkka. 1990. "Population Genetics of the Polymorphic Meadow Spittlebug, *Philaenus spumarius* (L.)." *Evolutionary Biology.* 24: 149–91.

Halkka, Olli, Liisa Halkka, Mikko Raatikainen, and R. Hovinen. 1973. "The Genetic Basis of Balanced Polymorphism in *Philaenus* (Homoptera)." *Hereditas* 74 (1): 69–80. https://doi.org/10.1111/j.1601-5223.1973.tb01105.x.

Halkka, Olli, and Tarja Kohila. 1976. "Persistence of Visual Polymorphism, Despite a Low Rate of Predation, in *Philaenus spumarius* (L.) (Homoptera, Aphrophoridae)." *Annales Zoologici Fennici* 13 (4): 185–88.

Halkka, Olli, and Riitta Lallukka. 1969. "The Origin of Balanced Polymorphism in the Spittlebugs (*Philaenus*, Homoptera)." *Annales Zoologici Fennici* 6: 431–34.

Halkka, Olli, Mikko Raatikainen, Liisa Halkka, and Riitta Lallukka. 1970. "The Founder Principle, Genetic Drift and Selection in Isolated Populations of *Philaenus spumarius* (L.) (Homoptera)." *Annales Zoologici Fennici* 7 (2): 221–38.

Halkka, Olli, Mikko Raatikainen, Liisa Halkka, and J. Lokki. 1971. "Factors Determining the Size and Composition of Island Populations of *Philaenus spumarius* (L.) (Homoptera)." *Acta Entomologica Fennica* 28: 83–100.

Halkka, Olli, Mikko Raatikainen, Liisa Halkka, and T. Raatikainen. 1977. "Coexistence of Four Species of Spittle-Producing Homoptera." *Annales Zoologici Fennici* 14 (4): 228–31.

Halkka, Olli, Mikko Raatikainen, Arja Vasarainen, and L. Heinonen. 1967. "Ecology and Ecological Genetics of *Philaenus spumarius* (L.) (Homoptera)." *Annales Zoologici Fennici* 4 (1): 1–18.

Hamedi, Azadeh, Shirin Farjadian, and Mohammad R. Karami. 2015. "Immunomodulatory Properties of Taranjebin (Camel's Thorn) Manna and Its Isolated Carbohydrate Macromolecules." *Journal of Evidence-Based Complementary & Alternative Medicine* 20 (4): 269–74. https://doi.org/10.1177/2156587215580490.

Hamilton, K. G. A. 1971. "A Remarkable Fossil Homopteran from Canadian Cretaceous Amber Representing a New Family." *Canadian Entomologist* 103 (7): 943–46. https://doi.org/10.4039/Ent103943-7.

Hamilton, K. G. A. 1977. "Review of the World Species of *Prosapi*a Fennah (Rhynchota: Homoptera: Cercopidae)." *Canadian Entomologist* 109 (4): 621–30. https://doi.org/10.4039/Ent109621-4.

Hamilton, K. G. A. 1980. "Aphrophorinae of Polynesia (Rhynchota: Homoptera: Cercopidae)." *Pacific Insects* 22 (3–4): 347–60.

Hamilton, K. G. A. 1981a. "Aphrophorinae of New Caledonia and the Loyalty Islands (Rhynchota: Homoptera: Cercopidae)." *Pacific Insects* 23 (3–4): 451–64.

Hamilton, K. G. A. 1981b. "Aphrophorinae of the Fiji, New Hebrides and Banks Islands (Rhynchota: Homoptera: Cercopidae)." *Pacific Insects* 23 (3–4): 465–77.

Hamilton, K. G. A. 1982. *The Spittlebugs of Canada. Homoptera: Cercopidae. The Insects and Arachnids of Canada. Part 10. Publication 1740.* Ottawa: Agriculture Canada.

Hamilton, K. G. A. 1990. "Homoptera." In *Insects from the Santana Formation, Lower Cretaceous, of Brazil*, edited by D. A. Grimaldi. *Bulletin of the American Museum of Natural History* No. 195: 82–122.

Hamilton, K. G. A. 2001. "A New Family of Froghoppers from the American Tropics (Hemiptera: Cercopoidea: Epipygidae)." *Biodiversity* 2 (3): 15–21. https://doi.org/10.1080/14888386.2001.9712551.

Hamilton, K. G. A. 2014. "The Old-World Zygonini Tr. Nov. (Hemiptera, Cercopoidea, Clastopteridae), with New Taxa from the Related Machaerotinae." *Zootaxa* 3768 (4): 437–59. https://doi.org/10.11646/zootaxa.3768.4.3.

Hamilton, K. G. A. 2015. "A New Tribe and Species of Clastopterinae (Hemiptera: Cercopoidea: Clastopteridae) from Africa, Asia and North America." *Zootaxa* 3946 (2): 151–89. https://doi.org/10.11646/zootaxa.3946.2.1.

Hamilton, K. G. A. 2016. "Neotropical Spittlebugs Related to Neaenini (Hemiptera, Cercopidae) and the Origins of Subfamily Cercopinae." *Zootaxa* 4169 (2): 201–50. https://doi.org/10.11646/zootaxa.4169.2.1.

Hamilton, K. G. A., and Vinton Thompson. 2007. "Evidence for Spittlebug Warning Colouration and Mimicry between Aphrophorine and Cercopine Species in a Peruvian Valley." *Biodiversity* 8 (2): 3–6. https://doi.org/10.1080/14888386.2007.9712864.

Hansson, Christer. 2009. "Eulophidae of Costa Rica Hymenoptera:Chalcidoidea), 3 The Genus *Horsimenus*." *Memoirs of the American Entomological Institute* 82: 916.

Hardy, James. 1862. "Notes on Cuckoo-Flowers and the Cuckoo Spit." *The Edinburgh New Philosophical Journal* 16 (N.S.): 70–76.

Harper, G., and J. B. Whittaker. 1976. "The Role of Natural Enemies in the Colour Polymorphism of *Philaenus spumarius* (L.)." *Journal of Animal Ecology* 45 (1): 91–104. https://doi.org/10.2307/3769.

Hayashi, Masakazu. 2023. "*Lebidia octoguttata* Morawitz (Coleoptera, Carabidae): Predation on Nymphs of Spittlebugs (Hemiptera: Aphrophoridae)." *Japanese Journal of Entomology (N S.)* 26 (3): 175–78.

Heidemann, Otto. 1901. "Remarks on the Spittle Insect, *Clastoptera xanthocephala* Germ." *Proceedings of the Entomological Society of Washington* 4 (4): 399–402.

Heinrichs, E. A., and Alberto T. Barrion. 2004. *Rice-Feeding Insects and Selected Natural Enemies in West Africa: Biology, Ecology, Identification*. International Rice Research Institute and the African Rice Center.

Helmberger, M. 2021. "Science One Sweep at a Time." *The Timberjay*, 11August 2021. http://www.timberjay.com/stories/science-one-sweep-at-a-time,17990.

Helmberger, M. 2023. "Spittlebug Coloration Signals a Changing Climate." *The Timberjay*, 22 March 2023. http://www.timberjay.com/stories/spittlebug-coloration-signals-a-changing-climate,20056.

Henderson, Gregg, George D. Hoffman, and Robert L. Jeanne. 1990. "Predation on Cercopids and Material Use of the Spittle in Aphid-Tent Construction by Prairie Ants." *Psyche* 97: 43–53. https://doi.org/10.1155/1990/28269.

Hernández Alonso, R., E. Martín Bernal, and V. Pérez Fortea. 1992. "*Haematoloma dorsatum* (Ahrens) en Aragon." *Ecología*, 6: 335–40.

Hernández O., J. V., and S. Flores C. 1956. "The Biology and Control of *Aeneolamia postica* (Wlk.)." In *Proceedings of the 9th Congress of the International Society of Sugarcane Technologists, New Delhi, India*, 821–36. Amsterdam: Elsevier.

Hernández, Luis M., Paula Espitia, David Florian, Valheria Castiblanco, Juan A. Cardoso, and María I. Gómez-Jiménez. 2021. "Geographic Distribution of Colombian Spittlebugs (Hemiptera: Cercopidae) via Ecological Niche Modeling: A Prediction for the Main Tropical Forages' Pest in the Neotropics." *Frontiers in Sustainable Food Systems* 5: 725774. https://doi.org/10.3389/fsufs.2021.725774.

Hernández, L. M., X. Bonilla, and P. A. Espitia. 2021. "Primer registro de *Aeneolamia reducta* (Lallemand, 1924) (Hemiptera: Cercopidae) en el Valle del Cauca (Colombia)." *Boletin del Museo de Entomologia de la Universidad del Valle* 20 (1): 1–6.

Hewitt, George B. 1986. "Environmental Factors Affecting Spittlebug Egg Survival during the Dry Season in Central Brazil." *Pesquisa Agropecuaria Brasileira* 21 (12): 1237–43.

Hewitt, George B., and Shashank S. Nilakhe. 1986. "Environmental Factors Affecting the Survival of Eggs and Early Instar Nymphs of Spittlebugs *Zulia entreriana* (Berg) and *Deois flavopicta* Stal during the Rainy Season in Central Brazil." *Anais da Sociedade Entomológica do Brasil* 15 (1): 61–76.

Hickel, Eduardo R. 2023. *Pragas na lavoura catarinense de arroz irrigado: Cigarrinhas-das-pastagens*. Empresa de Pesquisa Agropecuária e Extensão Rural.

Hinde, S. L. 1906. "XIII. On the Habits of a Species of *Ptyelus* in British East Africa. Communicated, with Notes by Professor E.B. Poulton, F.R.S." *Transactions of the Royal Entomological Society of London* 54 (2): 225–29. https://doi.org/10.1111/j.1365-2311.1906.tb02473.x.

Hoch, Hannelore, Martin Pingel, Dagmar Voigt, Urs Wyss, and Stanislav Gorb. 2024. "Adhesive Properties of Aphrophoridae Spittlebug Foam." *Journal of The Royal Society Interface* 21 (210): 20230521. https://doi.org/10.1098/rsif.2023.0521.

Hodkinson, I. D., and D. Casson. 1991. "A Lesser Predilection for Bugs: Hemiptera (Insecta) Diversity in Tropical Rain Forests." *Biological Journal of the Linnean Society* 43 (2): 101–9.

Hoebeke, E. R., and K. G. A. Hamilton. 1983. "*Lepyronia coleoptrata* (L.), a European Spittlebug in Eastern North America: New Locality Records and New Key to the North American Species of *Lepyronia* Amyot and Serville (Homoptera: Cercopidae)." *Proceeding of the Entomological Society of Washington* 85 (2): 263–71.

Hoffman, George D., and Peter B. McEvoy. 1985a. "The Mechanism of Trichome Resistance in *Anaphalis margaritacea* to the Meadow Spittlebug *Philaenus spumarius*." *Entomologia Experimentalis et Applicata* 39 (2): 123–29. https://doi.org/10.1111/j.1570-7458.1985.tb03552.x.

Hoffman, George D., and Peter B. McEvoy. 1985b. "Mechanical Limitations on Feeding by Meadow Spittlebugs *Philaenus spumarius* (Homoptera: Cercopidae) on Wild and Cultivated Host Plants." *Ecological Entomology* 10 (4): 415–26. https://doi.org/10.1111/j.1365-2311.1985.tb00739.x.

Holmann, Federico, and Daniel C. Peck. 2002. "Economic Damage Caused by Spittlebugs (Homoptera: Cercopidae) in Colombia: A First Approximation of Impact on Animal Production in *Brachiaria decumbens* Pastures." *Neotropical Entomology* 31: 275–84. https://doi.org/10.1590/S1519-566X2002000200016.

Horsfield, D. 1977. "Relationships between Feeding of *Philaenus spumarius* (L.) and the Amino Acid Concentration in the Xylem Sap." *Ecological Entomology* 2 (4): 259–66. https://doi.org/10.1111/j.1365-2311.1977.tb00889.x.

Horsfield, D. 1978. "Evidence for Xylem Feeding by *Philaenus spumarius* (l.) (Homoptera: Cercopidae)." *Entomologia Experimentalis et Applicata* 24 (1): 95–99. https://doi.org/10.1111/j.1570-7458.1978.tb02759.x.

Howell, James O., and Robert L. Pienkowski. 1971. "Spider Populations in Alfalfa, with Notes on Spider Prey and Effect of Harvest." *Journal of Economic Entomology* 64 (1): 163–68. https://doi.org/10.1093/jee/64.1.163.

Huang, Yi, Ze Huang, Junpeng Liu, Fei Wu, Qingde Li, and S. C. Yan. 2021. "Optimal Instar and Method of Meta-Tolyl-N-Methyl-Carbamate Application for Killing *Aphrophora costalis* Matsumura." *BioResources* 16 (4): 8082–97.

Hubert, J. F., D. Thomas, A. Cavalier, and J. Gouranton. 1989. "Structural and Biochemical Observations on Specialized Membranes of the 'Filter Chamber', a Water-Shunting Complex in Sap-Sucking Homopteran Insects." *Biology of the Cell* 66 (1): 155–63.

Hudson, W., and J. Dutcher. 2008. "Pecan Insects." In *Summary of Losses from Insect Damage and Cost of Control in Georgia 2006*, 22–23. Miscellaneous Publication, 106. University of Georgia. https://esploro.libs.uga.edu/esploro/outputs/report/Summary-of-losses-from-insect-damage/9949316473402959.

Humphreys, Rosalind K., and Graeme D. Ruxton. 2018. "What Is Known and What Is Not yet Known about Deflection of the Point of a Predator's Attack." *Biological Journal of the Linnean Society* 123 (3): 483–95. https://doi.org/10.1093/biolinnean/blx164.

Huq, Fazlul. 2006. "Molecular Modelling Analysis of the Metabolism of Caffeine." *Asian Journal of Biochemistry* 1 (4): 276–86.

Hutacharern, Chaweewan, and N. Tubtim. 1995. *Checklist of Forest Insects in Thailand.* Bangkok: Office of Environmental Policy and Planning.

Hutchinson, G. E. 1963. "A Note on the Polymorphism of *Philaenus spumarius* (L.) (Homopt., Cercopidae) in Britain." *Entomologist's Monthly Magazine* 99: 175–78.

Instituto Brasileiro de Geografia e Estatística (IBGE). 2022. *Monitoring the Coverage and Land Use of Brazil: 2018/2020.* Instituto Brasileiro de Geografia e Estatística,

Ministério da Economia. https://biblioteca.ibge.gov.br/index.php/biblioteca-catalogo?view=detalhes&id=2101966.

ICAR-National Research Centre for Banana. n.d. "Cercopid, *Phymatostetha deschampsi.*" Accessed September 5, 2024. https://nrcb.icar.gov.in/album/Cercopid%2C%20Phymatostetha%20deschampsi/index.html.

Imbler, Sabrina. 2021. "This Insect Drinks Your Milkshake." *The New York Times.* July 13, 2021. https://www.nytimes.com/2021/07/13/science/froghopper-xylem-straw.html.

Institute of Medicine, Committee on Military Nutrition Research. 2001. *Caffeine for the Sustainment of Mental Task Performance: Formulations for Military Operations.* National Academies Press. http://www.ncbi.nlm.nih.gov/books/NBK223802/.

Jana, Gautam, K. K. Chaki, and K. K. Misra. 2012. "Quantitative Estimation of Insect Diversity Inhabiting *Calotropis procera* in Industrial and Nonindustrial Areas of West Bengal, India." *Ecological Research* 27 (1): 153–62. https://doi.org/10.1007/s11284-011-0883-7.

Jank, L., C. B. Valle, and R. M. S. Resende. 2011. "Breeding Tropical Forages." *Crop Breeding and Applied Biotechnology* S1: 27–34. https://doi.org/10.1590/S1984-70332011000500005.

Jank, Liana, Sanzio C. Barrios, Cacilda B. do Valle, Rosangela M. Simeão, and Geovani F. Alves. 2014. "The Value of Improved Pastures to Brazilian Beef Production." *Crop and Pasture Science* 65 (11): 1132–37. https://doi.org/10.1071/CP13319.

Jank, Liana, Mateus F. Santos, and G. J. Braga. 2022. *O capim-BRS Zuri (*Panicum maximum *Jacq.) na diversificação e intensificação das pastagens.* Embrapa.

Jeal, Tim. 2001. *Livingstone.* Yale University Press.

Jenkinson, F. 1903. "*Verrallia aucta* and Its Host." *Entomologist's Monthly Magazine* 39: 222–23.

Jensen, K. H., K. Berg-Sørensen, H. Bruus, et al. 2016. "Sap Flow and Sugar Transport in Plants." *Reviews of Modern Physics* 88 (3): 1–70. https://doi.org/10.1103/RevModPhys.88.035007.

Jenser, G., A. M. Hegab, and M. Dezséry. 1981. "*Philaenus spumarius* Linne as a Vector of the Causative Pathogen of Rubus Stunt Disease." *Acta Phytopathologica Academiae Scientiarum Hungaricae* 16 (1–2): 233–37.

Jiménez-G., Jaime A. 1978. "Estudios tendientes a establecer el control integrado de las salivitas de los pastos." *Revista Colombiana de Entomología* 4 (1–2): 19–33. https://doi.org/10.25100/socolen.v4i1-2.10372.

Johnson, Kevin P., Christopher H. Dietrich, Frank Friedrich, et al. 2018. "Phylogenomics and the Evolution of Hemipteroid Insects." *Proceedings of the National Academy of Sciences* 115 (50): 12775–80. https://doi.org/10.1073/pnas.1815820115.

Jones, J. S., B. H. Leith, and P. Rawlings. 1977. "Polymorphism in *Cepaea*: A Problem with Too Many Solutions?" *Annual Review of Ecology and Systematics* 8: 109–43.

Kahn, Alexandra K., and Rodrigo P. P. Almeida. 2022. "Phylogenetics of Historical Host Switches in a Bacterial Plant Pathogen." *Applied and Environmental Microbiology* 88 (7): e02356-21. https://doi.org/10.1128/aem.02356-21.

Kajak, Anna. 1965. "An Analysis of Food Relations between the Spiders *Araneus cornutus* Clerck and *Araneus quadratus* Clerck and Their Prey in Meadows." *Ekologia Polska. Seria A* 13 (32): 717–64.

Karban, Richard. 1989. "Community Organization of *Erigeron glaucus* Folivores: Effects of Competition, Predation, and Host Plant." *Ecology* 70 (4): 1028–39. https://doi.org/10.2307/1941371.

Karban, Richard, and Mikaela Huntzinger. 2018. "Decline of Meadow Spittlebugs, a Previously Abundant Insect, along the California Coast." *Ecology* 99 (11): 2614–16.

Karban, Richard, and Mikaela Huntzinger. 2021. "Spatial and Temporal Refugia for an Insect Population Declining Due to Climate Change" *Ecosphere* 12.11 (2021): e03820.

Karban, Richard, and Sharon Y. Strauss. 2004. "Physiological Tolerance, Climate Change, and a Northward Range Shift in the Spittlebug, *Philaenus spumarius*." *Ecological Entomology* 29 (2): 251–54. https://doi.org/10.1111/j.1365-2311.2004.00576.x.

Kareem, A. A., R. M. Nachiappan, T. R. Subramaniam, and V. Govindarajan. 1971. "Severe Occurrence of a Spittle Bug, *Cosmoscarta relata*, Dist. (Cercopidae: Homoptera) on Jack at Kolli Hills in Tamil Nadu and Its Control." *South Indian Horticulture* 19: 87–89.

Kassab, Samir O., Elisângela De S. Loureiro, Camila Rossoni, et al. 2014. "Combinations of *Metarhizium anisopliae* with Chemical Insecticides and Their Effectiveness in *Mahanarva fimbriolata* (Hemiptera: Cercopidae) Control on Sugarcane." *Florida Entomologist* 97 (1): 146–54. https://www.jstor.org/stable/24362448.

Kehr, Julia. 2006. "Phloem Sap Proteins: Their Identities and Potential Roles in the Interaction between Plants and Phloem-Feeding Insects." *Journal of Experimental Botany* 57 (4): 767–74. https://doi.org/10.1093/jxb/erj087.

Keller-Grein, Gerhard, Brigitte L. Maass, and Jean Hanson. 1996. "Natural Variation in *Brachiaria* and Existing Germplasm Collections." In Brachiaria: *Biology, Agronomy, and Improvement, CIAT Publication No. 259*, 16–35. Centro International de Agricultura Tropical.

Kelson, W. E. 1964. "The Biology of *Aphrophora permutata* and Some Observations on *Aphrophora canadensis* Attacking Monterey Pine in California (Homoptera: Cercopidae)." *Pan-Pacific Entomologist* 40 (3): 135–46.

Kennedy, Patrick C., and Louis F. Wilson. 1971. "Understory Vegetation Associated with Saratoga Spittlebug Damage in Michigan Red Pine Plantations." *Canadian Entomologist* 103 (10): 1421–26. https://doi.org/10.4039/Ent1031421-10.

Kershaw, J. C. 1913. "Recommendations for Dealing with the Froghopper Pest, from Work Thereon during 1913." *Department of Agriculture Trinidad and Tobago: Bulletin* 12: 197–203.

Kershaw, J. C. 1914. "The Alimentary Canal of a Cercopid." *Psyche* 21: 65–72.

Kerstetter, Rex E., Richard G. Zepp, and L. H. Carreira. 1998. "Peroxidases in Grass Dew Derived from Guttation: Possible Role in Polymerization of Soil Organic Matter." *Biogeochemistry* 42: 311–23.

Khan, A., and W. A. P. Isaac. 2020. "Biological Control in Trinidad and Tobago." In *Biological Control in Latin America and the Caribbean*, edited by J. C. van Lenteren, V. H. P. Bueno, M. G. Luna, and Y. C. Colmenarez. CAB International.

Khan, A. M. 1974. "Observations on the Mycetomes of *Ptyelus sexvittatus* (Walker) (Hemiptera: Cercopidae)." *Mushi* 48 (4): 21–28.

Khan, E. J. A., and W. H. Mark. 1983. "Some Initial Results of Field Trials Conducted in Establishing a Pasture-Seed Industry in Cojedes State, Venezuela." In *Proceedings of the XIV International Grassland Congress, June 15–24, 1981, Lexington, Kentucky*. Westview Press.

Kim, Wonjung. 2013. "Mechanics of Xylem Sap Drinking." *Biomedical Engineering Letters* 3: 144–48.

King, R. D. 1969. "Spring and Summer Foods of Ruffed Grouse on Vancouver Island." *The Journal of Wildlife Management*, 33 (2): 440–42.

Kingsley, J. S. 1926. "Edward Sylvester Morse (1838–1925)." *Proceedings of the American Academy of Arts and Sciences* 61 (12): 549–55.

Kittelberger, Kyle D., Solomon V. Hendrix, and Çağan H. Şekercioğlu. 2021. "The Value of Citizen Science in Increasing Our Knowledge of Under-Sampled Biodiversity:

An Overview of Public Documentation of Auchenorrhyncha and the Hoppers of North Carolina." *Frontiers in Environmental Science* 9: Article 710396. https://doi.org/10.3389/fenvs.2021.710396.

Kleintjes, P. K., and D. L. Dahlsten. 1994. "Foraging Behavior and Nestling Diet of Chestnut-Backed Chickadees in Monterey Pine." *The Condor* 96 (3): 647–53. https://doi.org/10.2307/1369468.

Klimaszewski, S. M. 1993. "The Changes in the Frequency of Occurrence of the Colour Forms of *Philaenus spumarius* [L.] [Homoptera, Aphrophoridae] within the Areas of Elevated Radiation after Chernobyl Disaster." *Acta Biologica Silesiana* 22: 69–77.

Kment, Petr, and Torsten van der Heyden. 2022. "*Zelus renardii* (Hemiptera: Heteroptera: Reduviidae): First Records from Croatia, Montenegro, and an Accidental Introduction to the Czech Republic." *Heteroptera Poloniae–Acta Faunistica* 16: 7–14.

Knowlton, Geroge F. 1942. "Range Lizards as Insect Predators." *Journal of Economic Entomology* 35 (4): 602.

Knowlton, George F., and W. P. Nye. 1948. "Insect Food of the Vesper Sparrow." *Journal of Economic Entomology* 41 (5): 821–821.

Knutson, L. V. 1971. "Puparia of *Salpingogaste conopida* and *S. texana* with Notes on Prey (Diptera: Syrphidae)." *Entomological News* 82: 29–38.

Koga, Ryuichi, Gordon M. Bennett, Jason R. Cryan, and Nancy A. Moran. 2013. "Evolutionary Replacement of Obligate Symbionts in an Ancient and Diverse Insect Lineage." *Environmental Microbiology* 15 (7): 2073–81. https://doi.org/10.1111/1462-2920.12121.

Koga, Ryuichi, and Nancy A. Moran. 2014. "Swapping Symbionts in Spittlebugs: Evolutionary Replacement of a Reduced Genome Symbiont." *The ISME Journal* 8 (6): 1237–46. https://doi.org/10.1038/ismej.2013.235.

Kokiçi, Hysen, Ejup Çota, Michel Frem, et al. 2024. "Knowledge of Xylem Feeders (Hemiptera: Auchenorrhyncha) in Albanian Olive Orchards in Readiness for a Potential Outbreak with *Xylella fastidiosa* Invasion." *Annales de La Société Entomologique de France (N.S.)* 60 (1): 1–14. https://doi.org/10.1080/00379271.2023.2299469.

Körner, Monika. "Zur Rolle der Hämolymph-Inhaltsstoffe bei der Feindabwehr von Zikaden (Cicadomorpha et Fulgoromorpha) unter besonderer Berücksichtigung der Blutzikade *Cercopis vulnerata* Rossi." Doctoral diss., Bayreuth University, 2006. https://epub.uni-bayreuth.de/id/eprint/756/.

Korytkowski G., Cheslavo, and Manuel Torres B. 1966. "Insectos que atacan al cultivo del frijol de palo (*Cajanus cajan*) en el Perú." *Revista Peruana de Entomología*, 9 (1): 3–5.

Kottelenberg, David, Lia Hemerik, Maria Saponari, and Wopke van der Werf. 2021. "Shape and Rate of Movement of the Invasion Front of *Xylella fastidiosa* spp. *pauca* in Puglia." *Scientific Reports* 11 (1): 1061. https://doi.org/10.1038/s41598-020-79279-x.

Koudamiloro, Augustin, Francis E. Nwilene, Abou Togola, and Martin Akogbeto. 2015. "Insect Vectors of Rice Yellow Mottle Virus." *Journal of Insects* 2015:Article 721751. https://doi.org/10.1155/2015/721751.

Kramer, Sol. 1950. "The Morphology and Phylogeny of Auchenorhynchous Homoptera (Insecta)." *Illinois Biological Monographs* 20 (4): 1–111.

Krištín, Anton. 1995. "The Diet and Foraging Ecology of the Penduline Tit (*Remiz pendulinus*)." *Folia Zoologica (Prague)* 44 (1): 23–29.

Kuenzi, F. M., and H. C. Coppel. 1985a. "The Biology of *Clastoptera arborina* Ball (Homoptera: Cercopidae) in Wisconsin." *Transactions of the Wisconsin Academy of Sciences, Arts, and Letters* 73: 144–53.

Kuenzi, F. M., and H. C. Coppel. 1985b. "Feeding Site and Spittle of *Clastoptera arborina* Ball (Homoptera: Cercopidae)." *Transactions of the Wisconsin Academy of Sciences, Arts, and Letters* 73: 154–59.

Kumar, Sushil, Sachin M. Chavan, and K. H. Kabade. 2014. "A Report of Spittlebug, *Clovia* sp. (Aphrophoridae: Homoptera) Infesting Teak in Gujarat." *Insect Environment* 20 (2): 32–34.

La Notte, Pierfederico, Maria Saponari, Soraya Mousavi, et al. 2024. "A Survey in Natural Olive Resources Exposed to High Inoculum Pressure Indicates the Presence of Traits of Resistance to *Xylella fastidiosa* in Leccino Offspring." *Frontiers in Plant Science* 15: 1457831. https://doi.org/10.3389/fpls.2024.1457831.

Labadessa, Rocco, Maria Adamo, Cristina Tarantino, and Saverio Vicario. 2024. "The Side Effects of the Cure: Large-Scale Risks of a Phytosanitary Action Plan on Protected Habitats and Species." *Journal of Environmental Management* 371: 123285.

Labandeira, Conrad C. 1997. "Insect Mouthparts: Ascertaining the Paleobiology of Insect Feeding Strategies." *Annual Review of Ecology and Systematics* 28 (1): 153–93. https://doi.org/10.1146/annurev.ecolsys.28.1.153.

Labandeira, Conrad C. 2019. "The Fossil Record of Insect Mouthparts: Innovation, Functional Convergence, and Associations with Other Organisms." In *Insect Mouthparts: Form, Function, Development and Performance*, edited by H. W. Krenn. Springer International Publishing. https://doi.org/10.1007/978-3-030-29654-4_17.

Labroussaa, Fabien, Michael Ionescu, Adam R. Zeilinger, Steven E. Lindow, and Rodrigo P. P. Almeida. 2017. "A Chitinase Is Required for *Xylella fastidiosa* Colonization of Its Insect and Plant Hosts." *Microbiology* 163 (4): 502–9. https://doi.org/10.1099/mic.0.000438.

Lacki, Michael J., Joseph S. Johnson, Luke E. Dodd, and Michael D. Baker. 2007. "Prey Consumption of Insectivorous Bats in Coniferous Forests of North-Central Idaho." *Northwest Science* 81 (3): 199–205. https://doi.org/10.3955/0029-344X-81.3.199.

Lago, Clara, Elisa Garzo, Aránzazu Moreno, et al. 2021. "Flight Performance and the Factors Affecting the Flight Behaviour of *Philaenus spumarius* the Main Vector of *Xylella fastidiosa* in Europe." *Scientific Reports* 11 (1): 17608. https://doi.org/10.1038/s41598-021-96904-5.

Lago, Clara, Marina Morente, Daniel De las Heras-Bravo, et al. 2021. "Dispersal of *Neophilaenus campestris*, a Vector of *Xylella fastidiosa*, from Olive Groves to Over-summering Hosts." *Journal of Applied Entomology* 145 (7): 648–59. https://doi.org/10.1111/jen.12888.

Lahbib, Nada, Ugo Picciotti, Valdete Sefa, Sonia Boukhris-Bouhachem, Francesco Porcelli, and Francesca Garganese. 2022. "*Zelus renardii* Roaming in Southern Italy." *Insects* 13 (2): 158.

Lallemand, Victor. 1912. *Homoptera Fam. Cercopidae. Genera Insectorum*. Bruxelles: P. Wytsman.

Lallemand, Victor. 1933. "Entomological Investigations on the Spike Disease of Sandal (4). Cercopidae (Homopt.)." *Indian Forest Records* 18 (1): 1–4.

Lallemand, Victor. 1949. "Révision des cercopinae (Hemiptera Homoptera). Première partie." *Mémoire Institut Royal des Sciences Naturelles de Belgique* 32: 1–193.

Lambert, Jonathan. 2021. "Froghoppers Are the Super-Suckers of the Animal World." Science News. 13 July 2021. https://www.sciencenews.org/article/froghoppers-suction-insects-biophysics.

Landon, Bengt-Olof. (undated) "Geer, Charles De | Encyclopedia.Com." Accessed 1 May 2023. https://www.encyclopedia.com/science/dictionaries-thesauruses-pictures-and-press-releases/geer-charles-de.

Lara E., F., and R. D. Shenefelt. 1961. "A Preliminary Survey of Insect Relationships with Cushion Gall of Cacao in Costa Rica." *Cacao* 6 (1): 1–14.

Latha, Thomas, and Karen Bautista. 2020. "Preliminary Investigation on Insect Pests of Sugarcane in the Northern Sugar Belt Region of Belize." *Journal of Entomology and Zoology Studies* 8 (2): 1334–40.

Latini, Arianna, Cipriano Foxi, Flavio Borfecchia, et al. 2019. "Tacking the Vector of *Xylella fastidiosa*: Geo-Statistical Analysis of Long-Term Field Observations on Host Plants Influencing the Distribution of *Phylaenus* [sic] *spumarius* Nymphs." *Environmental Science and Pollution Research* 26 (7): 6503–16. https://doi.org/10.1007/s11356-018-3870-5.

Laurent, J. 1964. "Étude de l'entomofaune des légumineuses cultívées de Madagascar." *Bulletin de l'École Nationale Supérieure Agronomique de Nancy* 6 (2): 109–27.

Le Caherec, Françoise, Marie-Thérèse Guillam, Fabienne Beuron, et al. 1997. "Aquaporin-Related Proteins in the Filter Chamber of Homopteran Insects." *Cell and Tissue Research* 290 (1): 143–51. https://doi.org/10.1007/s004410050916.

Le Cesne, Maxime, Elorde Crispolon, and Adeline Soulier-Perkins. 2021. "Male Terminalia of Cercopidae (Hemiptera, Cicadomorpha): Towards a Consensus Terminology." *Scientific Reports* 11 (1): 10412. https://doi.org/10.1038/s41598-021-89759-3.

Ledesma Antón, L. 1979. "Insectos más importantes que atacan al mimbre en la region central de España y su Tratamiento." *Boletín de Sanidad Vegetal–Plagas* 5 (1): 101–24.

Lee, Heejo, Eon K. Dong, Mi-Jeong Kim, and Young-Gyu Ban. 2018. "The Status of Alien Insect Species in the Southern Region of Korean Peninsula." *Korean Society of Applied Entomology Conference Proceedings*, April, 107. https://db.koreascholar.com/Article/Detail/347096.

Legaspi, J. C., and J. Irvine. 1996. "Occurrence of the Spittlebug, *Prosapia simulans*, a Potential Pest in Texas Sugarcane." *Sugar Journal* 59 (1): 20–21.

Lei, T. S., J. G. Lu, D. F. Zhou, and X. C. Liu. 1992. "Studies on the Behavior of *Callitettix versicolor* (Fabr.) and Its Control." *Entomological Knowledge* 29 (6): 334–36.

Leite, Luís G., Laerte A. Machado, Roberto M. Goulart, Fernando M. Tavares, and Antonio B. Filho. 2005. "Screening of Entomopathogenic Nematodes (Nemata: Rhabditida) and the Efficiency of *Heterorhabditis* sp. against the Sugarcane Root Spittlebug *Mahanarva fimbriolata* (Fabr.) (Hemiptera: Cercopidae)." *Neotropical Entomology* 34 (5): 785–90. https://doi.org/10.1590/S1519-566X2005000500010.

Lepage, H. S., and O. Monte. 1942. "As cigarrhinas do capim 'Kikuio.'" *Biologico* 8 (10): 255–59.

Lepesme, P. 1947. "Damage by *Ptyelus flavescens* to *Eucalyptus* Species in Oubangi-Chari." *Revue Internationale de Botanique Appliquée et d'Agriculture Tropicale*, 27 (293–294): 162–63.

Li, J.-Q., Z. M. Zhao, S.-Y. Wu, K. Ming, and L.-N. Hou. 2001. "Biology and Ecology of Rice Spittle Bug (*Callitettix versicolor*)." *Journal of Southwest Agricultural University* 23 (2): 156–59.

Li, Qinglong, Haiying Zhong, Yalin Zhang, and Cong Wei. 2015. "Comparative Morphology of the Distal Segments of Malpighian Tubules in Cicadas and Spittlebugs, with Reference to Their Functions and Evolutionary Indications to Cicadomorpha

(Hemiptera: Auchenorrhyncha)." *Zoologischer Anzeiger* 258: 54–68. https://doi.org/10.1016/j.jcz.2015.07.002.

Li, Shu, Chungkun Shih, Chen Wang, Hong Pang, and Dong Ren. 2013. "Forever Love: The Hitherto Earliest Record of Copulating Insects from the Middle Jurassic of China." *PLOS ONE* 8 (11): e78188. https://doi.org/10.1371/journal.pone.0078188.

Li, Shu, Y. Wang, Ren Dong, Jacek Szwedo, and Hong Pang. 2010. "Froghoppers, Leafhoppers, Planthoppers and Their Allies from the Mesozoic of Northeastern China (Hemiptera: Cicadomorpha and Fulgoromorpha)." *Earth Science Frontiers* 17: 250–51.

Li, Zengzhi, Sérgio B. Alves, Donald W. Roberts, et al. 2010. "Biological Control of Insects in Brazil and China: History, Current Programs and Reasons for Their Successes Using Entomopathogenic Fungi." *Biocontrol Science and Technology* 20 (2): 117–36. https://doi.org/10.1080/09583150903431665.

Liang, Ai-Ping. 2003. "A New Tibial Gland in Male Spittlebugs with Descriptions of Two New Species of *Augustohahnia* Schmidt (Hemiptera: Cercopidae) from New Guinea." *Journal of Zoology* 261 (2): 173–80. https://doi.org/10.1017/S0952836903004072.

Liang, Ai-Ping. 2020. "A New Structure on the Frons of Male Adults of the Asian Rice Spittlebug *Callitettix versicolor* (Hemiptera: Auchenorrhyncha: Cercopidae)." *Zootaxa* 4801 (3): 591–99. https://doi.org/10.11646/zootaxa.4801.3.12.

Liang, Ai-Ping, and Murray J. Fletcher. 2002. "Morphology of the Antennal Sensilla in Four Australian Spittlebug Species (Hemiptera: Cercopidae) with Implications for Phylogeny." *Australian Journal of Entomology* 41 (1): 39–44. https://doi.org/10.1046/j.1440-6055.2002.00266.x.

Licent, P. É. 1912. "Recherches d'anatomie et de physiologie comparées sur le tube digestif des Homoptères supérieurs." *La Cellule* 28 (1): 1–161.

Licht, Lawrence E. "The Ecology of Coexistence in Two Closely Related Species of Frogs (Rana)." PhD diss., University of British Columbia, 1971. https://doi.org/10.14288/1.0101309.

Lim, G. T., and Y. C. Pan. 1980. "Entomofauna of Sugarcane in Malaysia." In *Proceedings of the XVII Congress of the International Society of Sugar Cane Technologists*, 1658–79. Manila, Philippines. https://www.cabdirect.org/cabdirect/abstract/20057026038.

Lin, Y.-S., and Y.-L. Chang. 2012. "The Insect Vectors of Pierce's Disease on Grapevines in Taiwan." *Formosan Entomologist* 32: 155–67.

Linares F., Blas A., and Gerardo Pérez. 1985. "Gramíneas hospederas de *Aeneolamia* spp (Homoptera: Cercopidae) en la Región Centro Occidental de Venezuela." *Caña de Azúcar* 3 (1): 34–42.

Linck, Holger, and Annette Reineke. 2019. "Rubus Stunt: A Review of an Important Phytoplasma Disease in *Rubus* spp." *Journal of Plant Diseases and Protection* 126 (5): 393–99.

Lindsay, K. L., and A. T. Marshall. 1980. "Ultrastructure of the Filter Chamber Complex in the Alimentary Canal of *Eurymela distincta* Signoret (Homoptera, Eurymelidae)." *International Journal of Insect Morphology and Embryology* 9 (3): 179–98. https://doi.org/10.1016/0020-7322(80)90014-8.

Linnavuori, Rauno. 1969. "A New Brachypterous Cercopid Genus from Ethiopia." *Annales Zoologici Fennici* 33 (4): 232–33.

Linné, Carl von. 1913. *Iter Lapponicum*. 2nd Edition, with Appendices and Notes by Theodor Magnus Fries. Uppsala: Almqvist & Wiksells Boktrycker-A.-B. https://doi.org/10.5962/bhl.title.150751.

Liu, Jie, Cuiping Bu, Benjamin Wipfler, and Aiping Liang. 2014. "Comparative Analysis of the Mitochondrial Genomes of Callitettixini Spittlebugs (Hemiptera: Cercopidae) Confirms the Overall High Evolutionary Speed of the AT-Rich Region but

Reveals the Presence of Short Conservative Elements at the Tribal Level." *PLOS ONE* 9 (10): e109140. https://doi.org/10.1371/journal.pone.0109140.

Livingstone, David. 1857. *Missionary Travels and Researches in South Africa*. London: John Murray.

Lomer, C. J., A. B. Stride, and R. Balfas. 1993. "The Biology of *Hindola* spp. (Homoptera: Machaerotidae) Vectors of Sumatra Disease of Clove and Some Related Species in Indonesia." *Bulletin of Entomological Research* 83 (2): 213–19. https://doi.org/10.1017/S0007485300034696.

López Machado, F. E., D. C. Peck, and Lerma J. Montoya. 2001. "Importancia de la comunicación vibracional en el comportamiento reproductivo del salivazo de los pastos (Homoptera: Cercopidae)." *Revista Colombiana de Entomología* 27 (1–2): 9–15. https://hdl.handle.net/10568/44342.

López-Mercadal, Júlia, Jacinto Benhadi-Marín, Isabel Rodrigues, Miguel Á. Miranda, and José A. Pereira. 2024. "Does the Superfluous Killing Exhibited by Spiders Mask the Functional Response Parameters? A Case Study from the Perspective of Biological Control." *Journal of Applied Entomology* 148 (4): 364–70. https://doi.org/10.1111/jen.13233.

Loureiro, Talita, Luís Serra, Ângela Martins, Isabel Cortez, and Patrícia Poeta. 2024. "*Xylella fastidiosa* Dispersion on Vegetal Hosts in Demarcated Zones in the North Region of Portugal." *Microbiology Research* 15 (3): 1050–72. https://doi.org/10.3390/microbiolres15030069.

Lü, Ruoqing, and Tiansen Xu. 1992. "Studies on the Bionomics of *Aphrophora horizontalis* Kato and Its Control." *Forest Research* 5 (6): 687–92.

Maa, Tsing-Chao T. 1949. "On Some Machaerotinae from South-Eastern Asia (Homoptera)." *Treubia* 20: 11–24.

Maa, Tsing-Chao T. 1963. "A Review of the Machaerotidae (Hemiptera: Cercopoidea)." *Pacific Insects Monograph* 5: 1–166.

Macklin, Paul R. 1958. "Spittle Insects as Food of the Red-Winged Blackbird." *Auk* 75 (2): 225.

Madaleno, Leonardo L., Gisele C. Ravaneli, Leandro E. Presotti, Miguel A. Mutton, Odair A. Fernandes, and Márcia J. R. Mutton. 2008. "Influence of *Mahanarva fimbriolata* (Stål) (Hemiptera: Cercopidae) Injury on the Quality of Cane Juice." *Neotropical Entomology* 37 (1): 68–73. https://doi.org/10.1590/S1519-566X2008000100010.

Maes, Jean-Michael, and J. Tellez Robleto. 1988. "Catálogo de los insectos y artrópodos terrestres asociados a las principales plantas de importancia económica en Nicaragua." *Revista Nicaragüense de Entomología*, (5): 1–95.

Magalhães, Bonifácio P. "Biologia da cigarrinha-das-pastagens *Deois* (Acanthodeois) *incompleta* (Walker 1851) (Homoptera: Cercopidae) em *Brachiaria* spp." MS diss., Piracicaba: Universidade de São Paulo, 1982. https://doi.org/10.11606/D.11.1983.tde-20231122-100315.

Magalhães, Bonifácio P., José R. P. Parra, and A. B. Silva. 1984. "Biology of the Pasture Froghopper *Deois incompleta* on *Brachiaria* spp." In *Resumos IX Congresso Brasileiro de Entomologia, Londrina, Paraná, 22–27 July 1984*. https://www.cabidigitallibrary.org/doi/full/10.5555/19850525687.

Mallet, James. 2010. "Shift Happens! Shifting Balance and the Evolution of Diversity in Warning Colour and Mimicry." *Ecological Entomology* 35 (s1): 90–104. https://doi.org/10.1111/j.1365-2311.2009.01137.x.

Mallott, Elizabeth K., Paul A. Garber, and Ripan S. Malhi. 2017. "Integrating Feeding Behavior, Ecological Data, and DNA Barcoding to Identify Developmental Differences in Invertebrate Foraging Strategies in Wild White-Faced Capuchins

(*Cebus capucinus*)." *American Journal of Physical Anthropology* 162 (2): 241–54. https://doi.org/10.1002/ajpa.23113.

Malone, Michael, R. Watson, and J. Pritchard. 1999. "The Spittlebug *Philaenus spumarius* Feeds from Mature Xylem at the Full Hydraulic Tension of the Transpiration Stream." *The New Phytologist* 143 (2): 261–71.

Malone, Michael, Michelle Herron, and M.-Angeles Morales. 2002. "Continuous Measurement of Macronutrient Ions in the Transpiration Stream of Intact Plants Using the Meadow Spittlebug Coupled with Ion Chromatography." *Plant Physiology* 130 (3): 1436–42. https://doi.org/10.1104/pp.007740.

Mangan, R. L., and A. Wutz. 1983. "Aggregation Patterns of Meadow Spittlebugs, *Philaenus spumarius* L. (Homoptera: Cercopidae), on Old-Field Alfalfa Plants" *Environmental Entomology* 12 (1): 151–57. https://doi.org/10.1093/ee/12.1.151.

Manickavasagam, Sagadai, and A. Rameshkumar. 2010. "New Records of Genera and Species of Encyrtidae (Hymenoptera: Chalcidoidea) from Tamil Nadu, India." *Madras Agricultural Journal* 97 (10–12): 402–7.

Mann, G. S. 1979. "Record of *Machaerota spangbergii* Signoret as a Pest of Jujube *Zizyphus mauritiana* (Lamk) Cultivars at Ludhiana (Punjab)." *Indian Journal of Horticulture* 36 (4): 427–29.

Marco-Noales, E., S. Barbé, A. Monterde, et al. 2021. "Evidence That *Xylella fastidiosa* Is the Causal Agent of Almond Leaf Scorch Disease in Alicante, Mainland Spain (Iberian Peninsula)." *Plant Disease* 105 (11): 3349–52. https://doi.org/10.1094/PDIS-03-21-0625-SC.

Markey, Roger F. J. "The Structure and Permeability of Protective Coverings Secreted by Certain Insect Species." PhD diss., London University, 1965.

Markheiser, Anna, Giacomo Santoiemma, Alberto Fereres, Sanela Kugler, Michael Maixner, and Daniele Cornara. 2023. "DC-EPG Assisted Comparison of European Spittlebugs and Sharpshooters Feeding Behaviour on Grapevine." *Journal of Applied Entomology* 147 (3): 192–204. https://doi.org/10.1111/jen.13098.

Marshall, Alan T. 1964a. "Spittle-Production and Tube-Building by Cercopoid Nymphs (Homoptera). I. The Cytology of the Malpighian Tubules of Spittle-Bug Nymphs." *Journal of Cell Science* 3 (70): 257–62.

Marshall, Alan T. 1964b. "Spittle-Production and Tube-Building by Cercopoid Nymphs (Homoptera). 2. The Cytology and Function of the Granule Zone of the Malpighian Tubules of the Tube-building Nymphs." *Quarterly Journal of Microscopical Science* 105 (4): 415–22.

Marshall, Alan T. 1965a. "Batelli Glands of Cercopoid Nymphs (Homoptera)." *Nature* 205 (4974): 925–925.

Marshall, Alan T. 1965b. "Spittle-Production and Tube-Building by Cercopoid Nymphs (Homoptera). 3. The Cytology and Function of the Fibril Zone of the Malpighian Tubules of Tube-Building Nymphs." *Journal of Cell Science* 3 (73): 37–44.

Marshall, Alan T. 1966a. "Histochemical Studies on a Mucocomplex in the Malpighian Tubules of Cercopid Larvae." *Journal of Insect Physiology* 12 (8): 925–32. https://doi.org/10.1016/0022-1910(66)90079-5.

Marshall, Alan T. 1966b. "Spittle-Production and Tube-Building by Cercopid Larvae (Homoptera)—IV. Mucopolysaccharide Associated with Spittle-Production." *Journal of Insect Physiology* 12 (6): 635–44. https://doi.org/10.1016/0022-1910(66)90109-0.

Marshall, Alan T. 1968. "The Chemical Nature of Malpighian Tubule Mucofibrils in Cercopoid Dwelling-Tubes." *Journal of Insect Physiology* 14 (10): 1435–44.

Marshall, Alan T, and W. W. K. Cheung. 1974. "Studies on Water and Ion Transport in Homopteran Insects: Ultrastructure and Cytochemistry of the Cicadoid and

Cercopoid Malpighian Tubules and Filter Chamber." *Tissue and Cell* 6 (1): 153–71. https://doi.org/10.1016/0040-8166(74)90029-9.

Marshall, Alan T, and W. W. K. Cheung. 1975. "Ionic Balance of Homoptera in Relation to Feeding Site and Plant Sap Composition." *Entomologia Experimentalis et Applicata* 18: 117–20.

Martin, R. M., J. R. Cox, D. G. Alston, and F. Ibarra. 1995. "Spittlebug (Homoptera: Cercopidae) Life Cycle on Buffelgrass in Northwestern Mexico." *Annals of the Entomological Society of America* 88 (4): 471–78. https://doi.org/10.1093/aesa/88.4.471.

Martin-R., Martha, J. R. Cox, Fernando Ibarra-F., D. G. Alston, R. E. Banner, and John C. Malecheck. 1999. "Spittlebug and Buffelgrass Responses to Summer Fires in Mexico." *Journal of Range Management* 52 (6): 621–25.

Maryańska-Nadachowska, Anna, Sakis Drosopoulos, Dorota Lachowska, Łukasz Kajtoch, and Valentina G. Kuznetsova. 2010. "Molecular Phylogeny of the Mediterranean Species of *Philaenus* (Hemiptera: Auchenorrhyncha: Aphrophoridae) Using Mitochondrial and Nuclear DNA Sequences." *Systematic Entomology* 35 (2): 318–28.

Mascarin, Gabriel M., Rogerio B. Lopes, Ítalo Delalibera, Éverton K. K. Fernandes, Christian Luz, and Marcos Faria. 2019. "Current Status and Perspectives of Fungal Entomopathogens Used for Microbial Control of Arthropod Pests in Brazil." *Journal of Invertebrate Pathology* 165: 46–53. https://doi.org/10.1016/j.jip.2018.01.001.

Masters, G. J., V. K. Brown, I. P. Clarke, J. B. Whittaker, and J. A. Hollier. 1998. "Direct and Indirect Effects of Climate Change on Insect Herbivores: Auchenorrhyncha (Homoptera)." *Ecological Entomology* 23 (1): 45–52. https://doi.org/10.1046/j.1365-2311.1998.00109.x.

Matsumoto, Kazuma. 1990. "Effects of Density on the Survival and Development of Nymphs in the Pine Spittlebug, *Aphrophora flavipes* (Homoptera: Cercopidae) Experiments in an Outdoor Cage." *Applied Entomology and Zoology* 25 (3): 339–46. https://doi.org/10.1303/aez.25.339.

Mattson, William J. 1980. "Herbivory in Relation to Plant Nitrogen Content." *Annual Review of Ecology and Systematics* 11: 119–61.

Mayné, R., and J. Ghesquière. 1934. *Hémiptères nuisibles aux végétaux du Congo Belge*. Bruxelles: Imprimerie Breuer.

McAtee, W. L. 1932. "Effectiveness in Nature of the So-Called Protective Adaptations in the Animal Kingdom, Chiefly as Illustrated by the Food Habits of Nearctic Birds." *Smithsonian Miscellaneous Collections, Publication 3125* 85 (7): 1–201.

McCabe, Robert A. 1991. *The Little Green Bird, Ecology of the Willow Flycatcher*. Rusty Rock Press.

McCaw, Beth A., Rebecca Cairns, Samuel P. Rogerson, and Katherine Lester. 2024. "Prevalence of the Endoparasitoid *Verrallia aucta* in Central Scotland: A Potential Biocontrol for *Philaenus spumarius* (Hemiptera, Aphrophoridae), an Important Vector of *Xylella fastidiosa*." *Journal of Applied Entomology* 148 (2): 158–70. https://doi.org/10.1111/jen.13217.

McEvoy, Peter B. 1986. "Niche Partitioning in Spittlebugs (Homoptera: Cercopidae) Sharing Shelters on Host Plants." *Ecology* 67 (2): 465–78. https://doi.org/10.2307/1938590.

McKinnon, Jeffrey S., and Michele E. R. Pierotti. 2010. "Colour Polymorphism and Correlated Characters: Genetic Mechanisms and Evolution." *Molecular Ecology* 19 (23): 5101–25. https://doi.org/10.1111/j.1365-294X.2010.04846.x.

Mead, F. W., and Frederick D. Bennett. 1987. "Casuarina Spittlebug, *Clastoptera undulata* Uhler (Homoptera: Cercopidae). Entomology Circular No. 294." Florida Department of Agriculture and Consumer Services.

Medina U., Claudia A. 1995. "Hormigas depredadoras de huevos de salivazo de los pastos *Aeneolamia varia* (Hemiptera: Cercopidae) en pasturas de *Brachiaria*, en los Llanos Orientales de Colombia." *Boletín del Museo de Entomología Universidad del Valle* 3 (1): 1–13.

Medler, J. T., and A. O. Adenuga. 1969. "Observations on Larvae of *Leucophenga proxima* Adams (Diptera: Drosophilidae) Living in Spittle-Masses of *Ptyelus grossus* F. (Homoptera: Cercopidae)." *Bulletin of the Entomological Society of Nigeria* 2: 51–53. https://doi.org/10.5281/zenodo.10757974.

Meena, Sugan C., Kewal K. Sharma, Arumugam Mohanasundaram, Sweet Verma, and Mohammad D. Monobrullah. 2014. "Insect-Pest Complex of *Flemingia semialata* Roxb–a Bushy Host for Lac Cultivation." *The Bioscan* 9 (4): 1375–81.

Mei, Lijuan, Xuewen Wang, Ying Yin, Guirong Tang, and Chengshu Wang. 2021. "Conservative Production of Galactosaminogalactan in *Metarhizium* Is Responsible for Appressorium Mucilage Production and Topical Infection of Insect Hosts." *PLOS Pathogens* 17 (6): e1009656. https://doi.org/10.1371/journal.ppat.1009656.

Mello, Maria Luiza S. 1987. "Effect of Some Enzymes, Chemicals and Insecticides on the Macromolecular Structure of the Froth of the Spittlebug, the Cercopid *Deois* sp." *Entomologia Experimentalis et Applicata* 44 (2): 139–44. https://doi.org/10.1111/j.1570-7458.1987.tb01058.x.

Mello, Maria Luiza S., E. R. Pimentel, A. T. Yamada, and A. Storopoli-Neto. 1987. "Composition and Structure of the Froth of the Spittlebug, *Deois* sp." *Insect Biochemistry* 17 (3): 493–502. https://doi.org/10.1016/0020-1790(87)90009-6.

Mello, M., H. Carvalho, and L. A. N. Sá. 1996. "Topochemistry and Structure of the Froth of Different Spittlebug Species." *Revista Brasileira de Biologia* 56: 177–81.

Melo, C. G., A. C. Tomaz, B. O. Soares, K. N. Kuki, L. A. Peternelli, and M. H. P. Barbosa. 2018. "Anatomical, Morphological, and Physiological Responses of Two Sugarcane Genotypes of Contrasting Susceptibility to *Mahanarva fimbriolata* (Hemiptera: Cercopidae)." *Bulletin of Entomological Research* 108 (4): 556–64. https://doi.org/10.1017/S0007485317001110.

Mendes, A. C. de B., and J. de J. da S. Garcia. 1985. "Ação de inseticidas sobre insetos nocivos e aranhas associados ao cacaueiro na Região de Ouro Preto d'Oeste, Rondônia, Brasil." *Revista Theobroma* 15 (2): 57–63.

Mendoza Mora, Jorge, Kleiner Mejía M., and Darío Gualle A. 2004. "El salivazo de la caña de azúcar, *Mahanarva andigena*." Publicación Tecnica No. 4. Centro de Investigación de la Caña de Azúcar del Ecuador.

Mesmin, Xavier, Marguerite Chartois, Guénaëlle Genson, Jean-Pierre Rossi, Astrid Cruaud, and Jean-Yves Rasplus. 2020. "*Ooctonus vulgatus* (Hymenoptera, Mymaridae), a Potential Biocontrol Agent to Reduce Populations of *Philaenus spumarius* (Hemiptera, Aphrophoridae) the Main Vector of *Xylella fastidiosa* in Europe." *PeerJ* 8:e8591. https://doi.org/10.7717/peerj.8591.

Mesmin, Xavier, Maxime Lambert, Marguerite Chartois, et al. 2023. "No Detection of *Xylella fastidiosa* in Cicadas (Hemiptera, Cicadidae) Sampled in Infected Areas of Corsica (France)." *Journal of Applied Entomology* 147 (7): 559–63. https://doi.org/10.1111/jen.13120.

Messner, B., and Joachim Adis. 1992. "Kutikuläre Wachsausscheidungen als plastronhaltende Strukturen bei Larven von Schaum- und Singzikaden (Auchenorrhyncha: Cercopidae und Cicadidae)." *Revue Suisse de Zoologie* 99 (3): 713–20. https://doi.org/10.5962/bhl.part.79847.

Metcalf, Z. P. 1960–1962. *General Catalogue of the Homoptera. Fascicle VII. Cercopoidea. Parts 1–4*. Waverly Press: Baltimore.

Miles, John W., Brigitte L. Maas, and Cacilda B. do Valle. 1996. Brachiaria: *Biology, Agronomy, and Improvement, CIAT Publication No. 259.* Cali, Colombia: Centro International de Agricultura Tropical.

Mitchell, R. T. 1972. Letter to Vinton Thompson in response to inquiry about bird predators of spittlebugs. 19 April 1972.

Mittler, T. E. 1967. "Water Tensions in Plants—an Entomological Approach." *Annals of the Entomological Society of America* 60 (5): 1074–76. https://doi.org/10.1093/aesa/60.5.1074.

Molinatto, Giulia, Stefano Demichelis, Nicola Bodino, Massimo Giorgini, Nicola Mori, and Domenico Bosco. 2020. "Biology and Prevalence in Northern Italy of *Verrallia aucta* (Diptera, Pipunculidae), a Parasitoid of *Philaenus spumarius* (Hemiptera, Aphrophoridae), the Main Vector of *Xylella fastidiosa* in Europe." *Insects* 11 (9): 607.

Moore, Barry P., W. Vance Brown, and Miriam Rothschild. 1990. "Methylalkylpyrazines in Aposematic Insects, Their Hostplants and Mimics." *Chemoecology* 1 (2): 43–51. https://doi.org/10.1007/BF01325227.

Moore, Thomas E. 1961. "Audiospectrographic Analysis of Sounds of Hemiptera and Homoptera." *Annals of the Entomological Society of America* 54 (2): 273–91.

Moraal, L. G. 1996. "Bionomics of *Haematoloma dorsatum* (Hom., Cercopidae) in Relation to Needle Damage in Pine Forests." *Anzeiger für Schädlingskunde, Pflanzenschutz, Umweltschutz* 69 (5): 114–18. https://doi.org/10.1007/BF01996882.

Moralejo, Eduardo, Margarita Gomila, Marina Montesinos, et al. 2020. "Phylogenetic Inference Enables Reconstruction of a Long-Overlooked Outbreak of Almond Leaf Scorch Disease (*Xylella fastidiosa*) in Europe." *Communications Biology* 3 (1): 1–13. https://doi.org/10.1038/s42003-020-01284-7.

Moreira, Carlos-Canna. 1925. *A cigarrinha vermelha da canna de assucar:* Tomaspis liturata *Lep. et Serv. Boletim No. 4.* Rio de Janeiro: Ministerio da Agricultura, Industria e Commercio.

Morelli, Massimiliano, José M. García-Madero, Ángeles Jos, et al. 2021. "*Xylella fastidiosa* in Olive: A Review of Control Attempts and Current Management." *Microorganisms* 9 (8): 1771. https://doi.org/10.3390/microorganisms9081771.

Moreno S., Carlos, Alex E. Bustillo P., Juan C. López-Nunez, Ulises C. Valderrama, and German D. Ramírez S. 2012. "Virulencia de nematodos entomopatógenos para el control del salivazo *Aeneolamia varia* (Hemiptera: Cercopidae) en caña de azúcar." *Revista Colombiana de Entomología* 38 (2): 260–65.

Morente, Marina, Daniele Cornara, María Plaza, et al. 2018. "Distribution and Relative Abundance of Insect Vectors of *Xylella Fastidiosa* in Olive Groves of the Iberian Peninsula." *Insects* 9 (4): 175. https://doi.org/10.3390/insects9040175.

Morris, Daniel. 1883. *The Colony of British Honduras: Its Resources and Prospects; with Particular Reference to Its Indigenous Plants and Economic Productions.* London: E. Stanford.

Morris, Heather, and D. F. Waterhouse. 2001. *The Distribution and Importance of Arthropod Pests and Weeds of Agriculture in Myanmar.* Canberra, Australia: Australian Center for International Agricultural Research. https://ageconsearch.umn.edu/record/114804/files/67.pdf.

Morse, Edward S. 1876. *First Book of Zoölogy.* New York: D. Appleton and Company.

Morse, Edward S. 1893. "Latrines of the East." *American Architect and Building News* 39 (899): 170–174.

Morse, Edward S. 1900. "A Bubble-Blowing Insect." *Popular Science Monthly* May: 23–29.

Mozaffarian, Fariba, and H. Jafary. 2023. *Potential Role of Auchenorrhyncha in Transmitting* Xylella fastidiosa *Wells, 1987.* Iranian Research Institute of Plant Protection.

Murakami, Yasuaki. 1983. "Factors Determining the Prey Size of the Orb-Web Spider, *Argiope amoena* (L. Koch) (Argiopidae)." *Oecologia* 57 (1): 72–77. https://doi.org/10.1007/BF00379564.

Mus, Florence, Matthew B. Crook, Kevin Garcia, et al. 2016. "Symbiotic Nitrogen Fixation and the Challenges to Its Extension to Nonlegumes." *Applied and Environmental Microbiology* 82 (13): 3698–3710. https://doi.org/10.1128/AEM.01055-16.

Musa, J. L. 1975. "The Immature Stages of *Asarkina ericetorum* (F.) (Dipt., Syrphidae) in Northern Nigeria." *Entomologist's Monthly Magazine* 111 (1331–1333): 109–10.

Myers, J. G. 1930. "*Carabunia myersi*." *Bulletin of Entomological Research* 21 (3): 341–51. https://doi.org/10.1017/S0007485300021878.

Myers, J. G. 1931. *A Preliminary Report on an Investigation into the Biological Control of West Indian Insect Pests.* London: His Majesty's Stationary Office.

Myers, J. G. 1935. "The Ecological Distribution of Some South American Froghoppers of the Genus *Tomaspis* (Hem., Cercopidae)." *Tropical Agriculture* 12 (5): 114–18.

Nachappa, Punya, L. P. Guillebeau, and S. K. Braman. 2006. "Exclusive Use of Hollies as Male-Specific Hosts of Twolined Spittlebug (Hemiptera: Cercopidae)." *Journal of Entomological Science* 41 (3): 261–63.

Nachappa, Punya, L. P. Guillebeau, S. K. Braman, and J. N. All. 2006. "Susceptibility of Twolined Spittlebug (Hemiptera: Cercopidae) Life Stages to Entomophagous Arthropods in Turfgrass." *Journal of Economic Entomology* 99 (5): 1711–16. https://doi.org/10.1603/0022-0493-99.5.1711.

Nair, K. S. S. 2007. *Tropical Forest Insect Pests: Ecology, Impact, and Management.* Cambridge University Press.

Nault, L. R. 1997. "Arthropod Transmission of Plant Viruses: A New Synthesis." *Annals of the Entomological Society of America* 90 (5): 521–41. https://doi.org/10.1093/aesa/90.5.521.

Ndikumana, J., and P. N. de Leeuw. 1996. "Regional Experience with *Brachiaria*: Sub-Sahara Africa." In Brachiaria: *Biology, Agronomy, and Improvement, CIAT Publication No. 259*, edited by J. W. Miles, B. L. Maass, C. B. do Valle, and V. Kumble. Centro International de Agricultura Tropical. https://agris.fao.org/search/en/providers/122432/records/64722b6677fd37171a73650a.

Nencioni, Anita, Elisabetta Gargani, Agostino Strangi, et al. 2024. "Transmission of *Xylella fastidiosa* subspecies *multiplex* from Naturally Infected to Healthy *Rhamnus alaternus* by *Philaenus spumarius* and *Neophilaenus campestris*." *Journal of Pest Science* 97 (3): 1557–67. https://doi.org/10.1007/s10340-024-01775-0.

Netto, A. M. C. de L., A. D. de Araújo, J. F. Warumby, and I. M. R. Marques. 1989. "Flutuação populacional da cigarrinha de pastagens *Deois schach* (Fabricius, 1787) (Homoptera-Cercopidae) em capim pangola (*Digitaria decumbens*) no Estado de Pernambuco." *Anais da Sociedade Entomologica do Brasil* 18 (1): 33–41.

Netto, Pedro M. de Oliveira, Alexander M. Auad, Michelle O. Campagnani de Mendonça, et al. Oliveira. 2024. "Endophytic Potential of Entomopathogenic Fungi Associated with *Urochloa ruziziensis* (Poaceae) for Spittlebug (Hemiptera: Cercopidae) Control." *Florida Entomologist* 107 (1): 20240043. https://doi.org/10.1515/flaent-2024-0043.

Newby, R. C. 1980. "The Use of Insects for Sampling Xylem Sap." *Annals of Botany* 45 (2): 213–15. https://www.jstor.org/stable/42761313.

Newman, Yonna C., Lynn E. Sollenberger, and Carrol G. Chambliss. 2003. "Canopy Characteristics of Continuously Stocked Limpograss Swards Grazed to Different Heights." *Agronomy Journal* 95 (5): 1246–52. https://doi.org/10.2134/agronj2003.1246.

Nguyen, T. C. 2001. "A New Species of *Cladochaeta* Coquillett (Diptera: Drosophilidae) and a New Record for *Cladochaeta sturtevanti* Wheeler and Takada in Arizona, with Notes on Natural History." *Proceedings Entomological Society of Washington* 103 (2): 444–51.

Nichol, H., and M. Locke. 1990. "The Localization of Ferritin in Insects." *Tissue and Cell* 22 (6): 767–77. https://doi.org/10.1016/0040-8166(90)90042-8.

Nie, Jing, and Ai-Ping Liang, 2009. "*Dianmachaerota serratiphalla*, a New Genus and Species of Machaerotini from China (Hemiptera: Cercopoidea: Machaerotidae)." *Annals of the Entomological Society of America* 102 (5): 739–746. https://doi.org/10.1603/008.102.0501.

Nielsen, M., K. Everett, V. Marroni, G. Greer, and S. Bulman. 2019. "Risks to New Zealand's Primary Industries from *Xylella*. A Plant & Food Research Report Prepared for: New Zealand Wine." The New Zealand Institute for Plant and Food Research Limited.

Nilakhe, Shashank S. 1985. "Ecological Observations on Spittlebugs with Emphasis on Their Occurrence in Rice." *Pesquisa Agropecuaria Brasileira* 20 (4): 407–14.

Nilakhe, S. S., and C. M. Buainain. 1988. "Observations on Movement of Spittlebug Adults." *Pesquisa Agropecuária Brasileira* 23 (2): 123–34.

Nolan, Val. 1956. "Spittle Insects as Food of Prairie Warblers." *The Auk* 73 (4): 557. https://doi.org/10.2307/4081955.

Norton, G. A., and D. E. Evans. 1974. "The Economics of Controlling Froghopper (*Aeneolamia varia saccharina* (Dist.) (Hom., Cercopidae)) on Sugar-Cane in Trinidad." *Bulletin of Entomological Research* 63 (4): 619–27. https://doi.org/10.1017/S0007485300047854.

Novotny, Vojtech, and Michael R. Wilson. 1997. "Why Are There No Small Species among Xylem-Sucking Insects?" *Evolutionary Ecology* 11 (4): 419–37. https://doi.org/10.1023/A:1018432807165.

Noyes, John S. 2010. "Encyrtidae of Costa Rica (Hymenoptera Chalcidoidea) 3." *Memoirs of the American Entomological Institute* 84: 1–848.

Noyes, John S. 2019. "Universal Chalcidoidea Database." World Wide Web electronic publication. London: Natural History Museum. https://www.nhm.ac.uk/our-science/data/chalcidoids/database/.

Nwilene, Francis, A. Traore, Alex Asidi, Y. Sere, Onasanya Amos, and M. Abo. 2009. "New Records of Insect Vectors of Rice Yellow Mottle Virus (RYMV) in Cote d'Ivoire, West Africa." *Journal of Entomology* 6: 189–97. https://doi.org/10.3923/je.2009.189.197.

Odhiambo, T. R. 1958. "Drosophilidae (Dipt.) Breeding in Cercopid (Hem.) Spittle Masses." *The Entomologist's Monthly Magazine* 94: 17.

Olmo, Diego, Alicia Nieto, David Borràs, et al. 2021. "Landscape Epidemiology of *Xylella Fastidiosa* in the Balearic Islands." *Agronomy* 11 (3): 473. https://doi.org/10.3390/agronomy11030473.

Oomen, P. A. 1975. "A Population Study of the Spittle Bugs *Aeneolamia occidentalis* (Walk.) and *Prosapia simulans* (Walk.) (Homoptera: Cercopidae) in Mexican Pangola Pastures." *Zeitschrift Für Angewandte Entomologie* 79 (1–4): 225–38. https://doi.org/10.1111/j.1439-0418.1975.tb02337.x.

Orozco-Restrepo, Silvana M, Bruno O. Soares, Vânia M. Xavier, et al. 2017. "Host-Plant Resistance to Spittlebugs: A Life-Table Study with *Mahanarva fimbriolata* (Hemiptera: Cercopidae) in Sugarcane Genotypes." *Journal of Economic Entomology* 110 (6): 2655–61. https://doi.org/10.1093/jee/tox232.

Osborn, Herbert. 1916. *Studies of Life Histories of Froghoppers of Maine. Bulletin 254.* Maine Agricultural Experiment Station.

Otanes, F. O., and F. L. Butac. 1939. "Cotton Pests in the Philippines." *Philippine Journal of Agriculture* 10 (4): 341–71.

Owen, D. F. 1963. "Similar Polymorphisms in an Insect and a Land Snail." *Nature* 198 (4876): 201–3. https://doi.org/10.1038/198201b0.

Owen, D. F., and R. G. Wiegert. 1962. "Balanced Polymorphism in the Meadow Spittlebug, *Philaenus spumarius*." *The American Naturalist* 96 (891): 353–59. https://doi.org/10.1086/282244.

Owen, D. F., and R. G. Wiegert. 1976. "Do Consumers Maximize Plant Fitness?" *Oikos* 27 (3): 488–92. https://doi.org/10.2307/3543467.

Páez, Erika, Janne K. Valkonen, Keith R. Willmott, Pável Matos-Maraví, Marianne Elias, and Johanna Mappes. 2021. "Hard to Catch: Experimental Evidence Supports Evasive Mimicry." *Proceedings of the Royal Society B: Biological Sciences* 288 (1946): 20203052. https://doi.org/10.1098/rspb.2020.3052.

Páez, Pedro J., Germán A. Torres-M., Jaime A. Jiménez-G., And Jesús E. Luque-Z. 1985. "Ciclo biologico y comportamiento del *Salpingogaster nigra* Schiner, predator del "mion" y "salivita" de los pastos." *Revista Colombiana de Entomología* 11 (1): 11–16.

Paladini, Andressa, A. C. Domahovski, D. Krinski, and L. A. Foerster. 2018. "Reports of New Wing Color Polymorphism and Taxonomic Information to Cercopids (Auchenorrhyncha: Cercopidae) from Upland Rice Crop, Pará State, Brazil." *Brazilian Journal of Biology* 78: 728–35. https://doi.org/10.1590/1519-6984.175519.

Paladini, Andressa, Daniela M. Takiya, Rodney R. Cavichioli, and Gervásio S. Carvalho. 2015. "Phylogeny and Biogeography of Neotropical Spittlebugs (Hemiptera: Cercopidae: Ischnorhininae): Revised Tribal Classification Based on Morphological Data." *Systematic Entomology* 40 (1): 82–108. https://doi.org/10.1111/syen.12091.

Paladini, Andressa, Daniela M. Takiya, Julie M. Urban, and Jason R. Cryan. 2018. "New World Spittlebugs (Hemiptera: Cercopidae: Ischnorhininae): Dated Molecular Phylogeny, Classification, and Evolution of Aposematic Coloration." *Molecular Phylogenetics and Evolution* 120: 321–34. https://doi.org/10.1016/j.ympev.2017.12.020.

Paladini, Andressa, Vinton Thompson, Adam J. Bell, and Jason R. Cryan. 2020. "A Remarkable New Species of Spittlebug and a Second Living New World Genus in the Clastopteridae (Hemiptera: Cercopoidea)." *Zootaxa* 4852 (3): 361–71. https://doi.org/10.11646/zootaxa.4852.3.7.

Palmer, Eve, and Norah Pitman. 1961. *Trees of South Africa*. Amsterdam: A.A. Balkema.

Pandit, N. C., and S. Chakravorty. 1986. "Insect Pests of Mesta Plant." *Environment & Ecology*, 4 (2): 398–99.

Pang, Z. H. 2013. "Current Situation and Development in Trend of Eucalypts Pest in China." *Journal of Guangxi Academy of Sciences* 29 (3): 192–206.

Parman, Vince R., and Curtis M. Wilson. 1982. "Alfalfa Crop Responses to Feeding by the Meadow Spittlebug (Homoptera: Cercopidae)." *Journal of Economic Entomology* 75 (3): 481–86. https://doi.org/10.1093/jee/75.3.481.

Parsons, James J. 1972. "Spread of African Pasture Grasses to the American Tropics." *Rangeland Ecology & Management* 25 (1): 12–17.

Partida-Lara, Ruth, Paula L Enríquez, Guillermo Ibarra Nuñez, and Eduardo Chamé Vásquez. 2023. "Consumption of Arthropods by Hummingbirds in the Sierra Madre de Chiapas, Mexico." *Avian Biology Research* 16 (1): 21–31. https://doi.org/10.1177/17581559221144896.

Pascovici, V. D. 1975. "Two New Serious Pests in Osier-Willow Stands in the Iasi Subregion of Rumania: *Phytobia cambii* and *Aphrophora salicina*." *Revista Padurilor–Industria Lemnului, Celuloza si Hirtie, Silvicultura si Exploatarea Padurilo* 90 (1): 40–42.

Pawar, A. D., and O. P. Bhalla. 1977. "Leafhoppers, Planthoppers and Froghoppers as Pests of Rice in Himachal Pradesh, India." *Entomologists' Newsletter* 7: 5.

Peck, Daniel C. 1998a. "Natural History of the Spittlebug *Prosapia* nr. b*icincta* (Homoptera: Cercopidae) in Association with Dairy Pastures of Costa Rica." *Annals of* *the Entomological Society of America* 91 (4): 435–44. https://doi.org/10.1093/aesa/91.4.435.

Peck, Daniel C. 1998b. "Use of Alternative Food Plants Exclusively by Adult Male Froghoppers (Homoptera: Cercopidae)." *Biotropica* 30 (4): 639–44. https://doi.org/10.1111/j.1744-7429.1998.tb00103.x.

Peck, Daniel C. 1999. "Seasonal Fluctuations and Phenology of *Prosapia* Spittlebugs (Homoptera: Cercopidae) in Upland Pastures of Costa Rica." *Environmental Entomology* 28 (3): 372–86. https://doi.org/10.1093/ee/28.3.372.

Peck, Daniel C. 2000a. "Reflex Bleeding in Froghoppers (Homoptera: Cercopidae): Variation in Behavior and Taxonomic Distribution." *Annals of the Entomological Society of America* 93 (5): 1186–94. https://doi.org/10.1603/0013-8746(2000)093[1186:RBIFHC]2.0.CO;2.

Peck, Daniel C. 2000b. "Biology and Diversity of Monteverde's Spittlebug and Froghoppers." In *Monteverde: Ecology and Conservation of a Tropical Cloud Forest*, edited by N. M Nadkarni and N. T. Wheelwright. Oxford University Press.

Peck, Daniel C. 2001. "Diversidad y distribución geográfica del salivazo (Homoptera: Cercopidae) asociado con gramíneas en Colombia y Ecuador." *Revista Colombiana de Entomología* 27 (2): 129–36. https://doi.org/10.25100/socolen.v27i2.9678.

Peck, Daniel C. 2002. "Desafíos y perspectivas para el manejo integrado de la candelilla (Homoptera: Cercopidae) en pastos tropicales." In *VIII Seminario,Manejo y Utilización de Pastos y Forrajes en Sistemas de Producción Animal*, edited by R. Tejos M., W. García, C. Zambrano A., L.E. Mancilla, and N. J. Valbuena. Universidad Nacional Experimental de los Llanos Occidentales.

Peck, Daniel C., A. M. Pérez, and J. W. Medina. 2002. "Biology and Habits of *Aeneolamia reducta* and *A. lepidior* in the Caribbean Coast of Colombia." *Pasturas Tropicales* 24 (1): 16–26.

Peck, Daniel C., Jairo Rodríguez, and Luis A. Gómez. 2004. "Identity and First Record of the Spittlebug *Mahanarva bipars* (Hemiptera: Auchenorrhyncha: Cercopidae) on Sugarcane in Colombia." *Florida Entomologist* 87 (1): 82–84. https://www.jstor.org/stable/3496808.

Peck, Daniel C., Ulises Castro, Francisco Lopez, Anuar Morales, and Jairo Rodriguez. 2001. "First Records of the Sugar Cane and Forage Grass Pest, *Prosapia simulans* (Homoptera: Cercopidae), from South America." *Florida Entomologist* 84 (3): 402–402.

Peck, Daniel C., and V. Thompson. 2008. "Spittlebugs (Hemiptera: Cercopoidea)." In *Encyclopedia of Entomology*, edited by John L. Capinera. Dordrecht: Springer.

Pedreira, B. C., R. M. Pitta, C. M. S. de Andrade, and M. B. D. Filho. 2014. *Degradação de pastagens de braquiarão* (Brachiaria brizantha *cv. Marandu) no estado de Mato Grosso.* Sinop, Mato Grosso: Embrapa Agrossilvipastoril. http://www.infoteca.cnptia.embrapa.br/handle/doc/1013035.

Pelagatti, O., P. F. Roversi, and C. Baccetti. 1995. "Indagini sulla microflora batterica associata alle schiume di *Haematoloma dorsatum* (Ahrens) (Homoptera Cercopidae)." *Redia* 77 (2): 243–51.

Penny, Norman D., and J. R. Arias. 1982. *Insects of an Amazon Forest.* Columbia University Press.

Peralta-Medina, Emiliano, and Howard J. Falcon-Lang. 2012. "Cretaceous Forest Composition and Productivity Inferred from a Global Fossil Wood Database." *Geology* 40 (3): 219–22. https://doi.org/10.1130/G32733.1.

Pereira, George F. "Diversidade e flutuação populacional das cigarrinhas das pastagens: (Hemiptera: Auchenorrhyncha Cercopidae) em *Brachiaria decumbens.*" MS diss., Universidade Estadual do Sudoeste da Bahia, 2013.

Pereira, Marcelo F. A., Alvaro de F. Jr., Alexander M. Auad, and Marilia G. Costa. 2018. "Survival and Injuries of *Deois flavopicta* (Stal., 1854) in Pastures under Seed Treatment with Insecticides and Dry Mass Yield." *Arquivos Do Instituto Biológico* 85:e0722016. https://doi.org/10.1590/1808-1657000722016.

Pérez Milián, José R., Yosel P. Pérez, Jorge F. Álvarez, and Jorge M. R. Rossil. 2018. "Control biológico del salivazo de la caña de azúcar *Aeneolamia* sp. con el nematodo *Heterorhabditis bacteriophora* y los hongos entomopatógenos *Metarhizium anisopliae* y *Beauveria bassiana* como opción económica y sostenible." *Ceiba* 55 (1): 21–27.

Petter, J. J., A. Schilling, and G. Pariente. 1971. "Observations eco éthologiques sur deux lémuriens malgaches nocturnes: *Phaner furcifer* et *Microcebus coquereli*." *Revue d'Ecologie, Terre et Vie* (3): 287–327.

Peuke, Andreas D. 2010. "Correlations in Concentrations, Xylem and Phloem Flows, and Partitioning of Elements and Ions in Intact Plants. A Summary and Statistical Re-Evaluation of Modelling Experiments in *Ricinus communis*." *Journal of Experimental Botany* 61 (3): 635–55.

Pfautsch, S., T. L. Bell, and A. Gessler. 2015. "Uptake, Transport and Redistribution of Amino Nitrogen in Woody Plants." In *Amino Acids in Higher Plants*, edited by J. P. F. D'Mello. CAB International.

Pflucker, Oscar S. 1981. "Pests of Pigeonpea on the Northern Coast of Peru." In *Proceedings of the International Workshop on Pigeonpeas, 5–19 December 1980*, edited by V. Kumble. International Crops Research Institute for the Semi-Arid Tropics Center. https://oar.icrisat.org/1954/1/Proceedings_of_the_international_workshop_on_pigeonpeasV2.pdf#page=364.

Philippe, Geneviève, and Lucy Angenot. 2005. "Recent developments in the field of arrow and dart poisons." *Journal of Ethnopharmacology* 100 (1–2): 85–91. https://doi.org/10.1016/j.jep.2005.05.022.

Picciotti, Ugo, Nada Lahbib, Valdete Sefa, Francesco Porcelli, and Francesca Garganese. 2021. "Aphrophoridae Role in *Xylella fastidiosa* subsp. *pauca* ST53 Invasion in Southern Italy." *Pathogens* 10 (8): 1035. https://doi.org/10.3390/pathogens10081035.

Pickles, Alan. 1930. "LXI.—*Leocomia cinchonæ*, sp. n., a New Cercopid from Jamaica." *Annals and Magazine of Natural History* 6 (35): 586–89. https://doi.org/10.1080/00222933008673255.

Pickles, Alan. 1932. "Notes on the Natural Enemies of the Sugar-Cane Froghopper (*Tomaspis saccharina*, Dist.) in Trinidad, with Descriptions of New Species." *Bulletin of Entomological Research* 23 (2): 203–10. https://doi.org/10.1017/S0007485300004107.

Pickles, Alan. 1933. "Entomological Contributions to the Study of the Sugar-Cane Froghopper." *Tropical Agriculture* 10 (9): 240–45.

Pickles, Alan. 1939. "A New Froghopper Pest of Sugar-Cane." *Tropical Agriculture* 16 (4): 84.

Pickles, Alan. 1937. "The Control of Froghopper Blight." *Tropical Agriculture* 14 (1): 5–9.

Pickles, Alan. 1944. "Observations on Sugar Cane Cultivation in Venezuela." *Tropical Agriculture* 21 (4): 64–66.

Pickles, Alan. 1946. "Field Trials of Insecticides for the Control of the Sugar-Cane Froghopper." *Tropical Agriculture* 23 (1): 9–11.

Pierce, William D. 1918. *A Manual of Dangerous Insects Likely to Be Introduced in the United States Through Importations*. Washington, DC: US Government Printing Office.

Pimentel, André C., Renata O. Dias, Thaís D. Bifano, Fernando A. Genta, Clelia Ferreira, and Walter R. Terra. 2020. "Cathepsins L and B in *Dysdercus peruvianus*,

Rhodnius prolixus, and *Mahanarva fimbriolata*. Looking for Enzyme Adaptations to Digestion." *Insect Biochemistry and Molecular Biology* 127: 103488. https://doi.org/10.1016/j.ibmb.2020.103488.

Pinheiro, C. E. G., A. V. L. Freitas, V. C. Campos, P. J. DeVries, and C. M. Penz. 2016. "Both Palatable and Unpalatable Butterflies Use Bright Colors to Signal Difficulty of Capture to Predators." *Neotropical Entomology* 45: 107–13.

Pirani, Gabriela, and Fernando da S. Carvalho-Filho. 2019. "A New Species of *Cladochaeta* Coquillett, 1900 (Diptera: Drosophilidae) Associated with *Sphodroscarta trivirgata* (Amyot & Serville, 1843) (Auchenorrhyncha: Aphrophoridae) from the Brazilian Amazon Rainforest." *Journal of Natural History* 53 (21–22): 1301–12. https://doi.org/10.1080/00222933.2019.1648701.

Pires, Carmen S. S., Peter W. Price, and Regina C. de Oliveira. 2000. "Distribution of the Spittlebug *Deois flavopicta* Stal (Homoptera: Cercopidae) on Wild and Cultivated Host Species." *Anais da Sociedade Entomológica do Brasil* 29 (3): 401–12. https://doi.org/10.1590/S0301-80592000000300003.

Pires, Carmen S. S., Eliana M. G. Fontes, Edison R. Sujii, Hermínia M. C. Fernandes, and Dagoberto F. Gomes. 1993. "Ocorrência de *Anagrus* sp. (Hymenoptera: Mymaridae) parasitando ovos de *Deois flavopicta* (Stal) (Homoptera: Cercopidae) em pastagens do Brasil central." *Anais da Sociedade Entomológica do Brasil* 22 (2): 411–13.

Plakidas, Antonios G. 1928. *Diseases of Strawberry in Louisiana, Part I.* Circular No. 113. Baton Rouge: Louisiana State University Extension.

Poinar, George, and Alex E. Brown. 2023. "*Araeoanasillus leptosomus* gen. et sp. nov., (Hemiptera, Cercopoidea, Sinoalidae?), A New Froghopper from Mid-Cretaceous Burmese Amber with Evidence of Its Possible Host Plant." *Life* 13 (4): 922. https://doi.org/10.3390/life13040922.

Poinar, George O., K. G. A. Hamilton, and Alex E. Brown. 2014. "*Prisciba*, n. gen., and Two New Species of Fossil Froghoppers (Hemiptera: Cercopoidea: Clastopteridae) in Dominican Amber." *Historical Biology* 26 (1): 1–5. https://doi.org/10.1080/08912963.2012.751102.

Poinar, George O., and Blas Linares. 1985. "*Hexamermis dactylocerus* sp. n. (Mermithidae: Nemotoda), a Parasite of *Aeneolamia varia* (Cercopidae: Homoptera) in Venezuela." *Revue de Nématologie* 8 (2): 109–11.

Pompon, Julien, Dan Quiring, Claudia Goyer, Philippe Giordanengo, and Yvan Pelletier. 2011. "A Phloem-Sap Feeder Mixes Phloem and Xylem Sap to Regulate Osmotic Potential." *Journal of Insect Physiology* 57 (9): 1317–22. https://doi.org/10.1016/j.jinsphys.2011.06.007.

Ponder, K. L., R. J. Watson, M. Malone, and J. Pritchard. 2002. "Mineral Content of Excreta from the Spittlebug *Philaenus spumarius* Closely Matches That of Xylem Sap." *New Phytologist* 153 (2): 237–42. https://doi.org/10.1046/j.0028-646X.2001.00314.x.

Porta, Antonio. 1899. *Ricerche sulla* Aphrophora spumaria *L.* Milan: Tip. Bernardoni di C. Rebeschini e C.

Pott, Christian, and Baoyu Jiang. 2017. "Plant Remains from the Middle–Late Jurassic Daohugou Site of the Yanliao Biota in Inner Mongolia, China." *Acta Palaeobotanica* 57 (2): 185–222. https://doi.org/10.1515/acpa-2017-0012.

Potter, T. E. K., and A. J. Carrington. 1949. "A Further Investigation in the Control of the Trinidad Froghopper (*Tomaspis saccharina* Dist.) by Application to the Nymphs of a 4 percent Benzene Hexachloride Dust." *Tropical Agriculture* 26 (7–12): 113–17.

Powell, Erin C., Collette Cook, Jeffrey Coco, Michelle Brock, Lauren A. Holian, and Lisa A. Taylor. 2019. "Prey Colour Biases in Jumping Spiders (*Habronattus brunneus*) Differ across Populations." *Ethology* 125 (6): 351–61. https://doi.org/10.1111/eth.12859.

Pramanik, M. E. A., and Ai-Ping Liang. 2019. "Description of the Immature and Adult Stages of the Jujube Tube Spittlebug, *Machaerota planitiae* Distant (Hemiptera: Cercopoidea: Machaerotidae) in Bangladesh." *Zootaxa* 4624 (1): 71–86.

Press, Malcolm C., and John B. Whittaker. 1993. "Exploitation of the Xylem Stream by Parasitic Organisms." *Philosophical Transactions of the Royal Society of London. Series B: Biological Sciences* 341 (1295): 101–11. https://doi.org/10.1098/rstb.1993.0095.

Prodhan, M. Z. H., M. S. Noman, M. T. Hasan, and M. A. Maleque. 2016. "Evaluation and Economic Analysis of Some Integrated Pest Management Packages for Suppressing Tube Spittle Bug (*Machaerota planitiae* Distant) Infesting Jujube (*Zizyphus mauritiana* Lamk), in North-Western Bangladesh." *The Agriculturists* 14 (1): 54–62.

Purcell, Alexander H. 2013. "Paradigms: Examples from the Bacterium *Xylella fastidiosa*." *Annual Review of Phytopathology* 51: 339–56. https://doi.org/10.1146/annurev-phyto-082712-102325.

Purcell, Alexander H., 1997. "*Xylella fastidiosa*, a Regional Problem or Global Threat?" *Journal of Plant Pathology* 79 (2): 99–105.

Purcell, Alexander H. 1980. "Environmental Therapy for Pierce's Disease of Grapevines." *Plant Disease* 64 (4): 388–90. https://doi.org/10.1094/PD-64-388.

Purcell, Alexander H., and Allan Finlay. 1979. "Evidence for Noncirculative Transmission of Pierce's Disease Bacterium by Sharpshooter Leafhoppers." *Phytopathology* 69 (4): 393–95. https://doi.org/10.1094/Phyto-69-393.

Purcell, Alexander H., and S. R. Saunders. 1999. "Fate of Pierce's Disease Strains of *Xylella fastidiosa* in Common Riparian Plants in California." *Plant Disease* 83 (9): 825–30. https://doi.org/10.1094/PDIS.1999.83.9.825.

Putman, W. L. 1953. "Notes on the Bionomics of Some Ontario Cercopids (Homoptera)." *Canadian Entomologist* 85 (7): 244–48. https://doi.org/10.4039/Ent85244-7.

Quilici, Serge, Bernard Reynaud, and Jacques Bonfils. 1998. "Sur deux espèces d'Hémiptères Auchénorrhynques, nouvelles pour la Réunion et d'importance économique potentielle." *Bulletin de la Société Entomologique de France*. https://agritrop.cirad.fr/401125/.

Rakitov, Roman A. 2000. "Nymphal Biology and Anointing Behaviors of *Xestocephalus desertorum* (Berg) (Hemiptera: Cicadellidae), a Leafhopper Feeding on Grass Roots." *Journal of the New York Entomological Society* 108 (1): 171–80.

Rakitov, Roman A. 2002. "Structure and Function of the Malpighian Tubules, and Related Behaviors in Juvenile Cicadas: Evidence of Homology with Spittlebugs (Hemiptera: Cicadoidea & Cercopoidea)." *Zoologischer Anzeiger* 241 (2): 117–30. https://doi.org/10.1078/0044-5231-00025.

Ramirez-Gi, Joaquin G., William A. León-Rueda, Mauricio Castro-Franco, and German Vargas. 2023. "Population Dynamics and Estimation of Damage of the Spittlebug *Aeneolamia varia* on Sugarcane in Colombia by Using Remote Sensing and Machine Learning Tools." *Sugar Tech* 25 (5): 1115–33. https://doi.org/10.1007/s12355-023-01247-2.

Ramos, Irene M. 1988. "Curva populacional de ninfas *Deois incompleta* (Walker, 1851) (Homoptera, Cercopidae) na Zona da Mata Úmida de Pernambuco." *Anais da Sociedade Entomológica do Brasil* 17 (1): 41–46.

Ramos, J. A. 1957. "A Review of the Auchenorhynchous Homoptera of Puerto Rico." *The Journal of Agriculture of the University of Puerto Rico* 41 (1): 38–116. https://doi.org/10.46429/jaupr.v41i1.12631.

Ranieri, Emanuele, Sara Ruschioni, Paola Riolo, Nunzio Isidoro, and Roberto Romani. 2016. "Fine Structure of Antennal Sensilla of the Spittlebug *Philaenus spumarius*

L. (Insecta: Hemiptera: Aphrophoridae). I. Chemoreceptors and Thermo-/Hygroreceptors." *Arthropod Structure & Development* 45 (5): 432–39. https://doi.org/10.1016/j.asd.2016.09.005.

Ranieri, Emanuele, Sara Ruschioni, Paola Riolo, Nunzio Isidoro, and Roberto Romani. 2020. "Sensory Receptors Associated with the Labial Tip and Precibarium of *Philaenus spumarius* L. (Hemiptera: Aphrophoridae)." *Microscopy and Microanalysis* 26 (1): 173–81.

Rathore, M. P. S. 1995. *Insect Pests in Agroforestry*. Working Paper No. 70. Nairobi, Kenya: International Centre for Research in Agroforestry.

Ravaneli, Gisele C., Débora B. Garcia, Leonardo L. Madaleno, Miguel Â. Mutton, José P. Stupiello, and Márcia J. R. Mutton. 2011. "Spittlebug Impacts on Sugarcane Quality and Ethanol Production." *Pesquisa Agropecuária Brasileira* 46 (2): 120–29. https://doi.org/10.1590/S0100-204X2011000200002.

Ravaneli, Gisele C., Débora B. Garcia, Leonardo L. Madaleno, Leandro E. Presotti, Miguel Â. Mutton, and Márcia J. R. Mutton. 2006. "Spittlebug Infestation in Sugarcane Affects Ethanolic Fermentation." *Scientia Agricola* 63 (6): 534–39. https://doi.org/10.1590/S0103-90162006000600004.

Raven, J. A. 1983. "Phytophages of Xylem and Phloem: A Comparison of Animal and Plant Sap-Feeders." In *Advances in Ecological Research*, edited by A. MacFadyen and E. D. Ford. Academic Press. https://doi.org/10.1016/S0065-2504(08)60109-9.

Redak, Richard A., Alexander H. Purcell, João R. S. Lopes, Matthew J. Blua, Russell F. Mizell III, and Peter C. Andersen. 2004. "The Biology of Xylem Fluid–Feeding Insect Vectors of *Xylella fastidiosa* and Their Relation to Disease Epidemiology." *Annual Review of Entomology* 49 (1): 243–70. https://doi.org/10.1146/annurev.ento.49.061802.123403.

Reis, P. R., W. Botelho, L. A. S. Melo, and J. Kakida. 1984. "Spittle Bug, *Zulia entreriana* (Berg, 1879) (Homoptera, Cercopidae), in the Northern Region of Minas Gerais State, Brazil. I. Control of Adults by Spraying Insecticides." *Anais Sociedade Entomologica do Brasil* 13 (2): 213–24.

Rémilliet, M., J. F. Silvain, and G. Tavakilian. 1982. "L'Éntomofaune des graminées fourragères en Guyane Française." In *Eighteen Annual Meeting of the Caribbean Food Crops Society, Dover Convention Centre, Barbados, Dover Convention Centre, Barbados. Dover Convention Centre, Barbados. 22–28th August 1982*, Volume XVII 277–82. https://ageconsearch.umn.edu/record/262969/files/18_26.pdf.

Resende, Tiago T., Fausto S. Sobrinho, Michelle O. Campagnani, et al. 2024. "Sixteen Years of Recurrent Selection of Ruzi Grass for Resistance to Spittlebugs (Hemiptera: Cercopidae)." *Agronomy* 14 (7): 1516. https://doi.org/10.3390/agronomy14071516.

Ribeiro, Genésio T., Marcelo da C. Mendonça, João B. de Mesquita, José C. Zanuncio, and Gervasio S. Carvalho. 2005. "Spittlebug *Cephisus siccifolius* Damaging Eucalypt Plants in the State of Bahia, Brazil." *Pesquisa Agropecuária Brasileira* 40: 723–26.

Ribeiro, Leandro do P., Dylan T. T. Amandio, Adriana L. S. Klock, Cristiano N. Nesi, Ângela F. Rech, and Edson L. L. Baldin. 2022. "*Deois flexuosa* (Hemiptera: Cercopidae) in Perennial Forage Species (Poaceae): Quali-Quantitative Impacts and Tolerance Expression." *Neotropical Entomology* 51 (5): 777–94.

Ribeiro, Leandro do P., and Angélica R. Cazarotto. 2019. "Cigarrinhas-das-pastagens em Santa Catarina: Avaliação do complexo de espécies e da incidência natural de fungos entomopatogênicos." *Agropecuária Catarinense* 32 (2): 73–79. https://doi.org/10.22491/RAC.2019.v32n2.11.

Rielli, Daniele. 2023. *Il fuoco invisibile: Storia humana di un disastro naturale.* Milan: Rizzoli.

Robinson, Nicole, Richard Brackin, Kerry Vinall, et al. 2011. "Nitrate Paradigm Does Not Hold up for Sugarcane." *PLOS ONE* 6 (4): e19045. https://doi.org/10.1371/journal.pone.0019045.

Rodman, G. H. 1921. "The Story of the Cuckoo Spit." *Proceedings of the Photomicrographic Society* 10 (2): 31–42.

Rodrigues, Ana S. B., Sara E. Silva, Francisco Pina-Martins, et al. 2016. "Assessing Genotype-Phenotype Associations in Three Dorsal Colour Morphs in the Meadow Spittlebug *Philaenus spumarius* (L.) (Hemiptera: Aphrophoridae) Using Genomic and Transcriptomic Resources." *BMC Genetics* 17 (1): 144. https://doi.org/10.1186/s12863-016-0455-5.

Rodrigues, Isabel, Vítor Ramos, Jacinto Benhadi-Marín, et al. 2022. "A Novel Molecular Diagnostic Method for the Gut Content Analysis of *Philaenus* DNA." *Scientific Reports* 12 (1): 492.

Rodríguez, Alejandro, and Carlos Sáenz. 1999. "Uso biológico del *Metarhizium anisopliae* (Metch.) Sorok. como estrategia para el control del salivazo (Hom: Cercopidae) en la caña de azúcar." In *XI Congreso Nacional Agronómico y de Recursos Naturales, San José, Costa Rica, LAICA-DIECA, July 1999.*

Rodríguez-C., Jairo, Ulises Castro-V., Anuar Morales R., and Daniel C. Peck. 2003. "Biology of the Spittlebug *Prosapia simulans* (Homoptera: Cercopidae), New Pest of Cultivated Graminoids in Colombia." *Revista Colombiana de Entomología* 29 (2): 149–55.

Rodríguez-C., Jairo, and Daniel C. Peck. 2006. "Parámetros poblacionales de *Zulia carbonaria* (Hemiptera: Cercopidae) sobre *Brachiaria ruziziensis. Revista Colombiana de Entomología* 32 (2): 145.

Rodríguez-C., Jairo, and Daniel C. Peck. 2007. "Biología y hábitos de *Mahanarva andigena* (Hemiptera: Cercopidae) en condiciones de casa de malla." *Revista Colombiana de Entomología* 33 (1): 31–35.

Rodríguez-C., Jairo, Daniel C. Peck, and Nelson A. Canal. 2002. "Comparative Biology of Three Species of Grassland Spittlebugs of the Genus *Zulia* (Homoptera: Cercopidae)." *Revista Colombiana de Entomología* 28 (1): 17–25.

Rodríguez-Celma, Jorge, Laura Ceballos-Laita, Michael A. Grusak, Javier Abadía, and Ana-Flor López-Millán. 2016. "Plant Fluid Proteomics: Delving into the Xylem Sap, Phloem Sap and Apoplastic Fluid Proteomes." *Biochimica et Biophysica Acta (BBA)–Proteins and Proteomics* 1864 (8): 991–1002. https://doi.org/10.1016/j.bbapap.2016.03.014.

Rojas, L., C. Godoy, P. Hanson, C. Kleinn, and L. Hilje. 2001. "Hopper (Homoptera: Auchenorrhyncha) Diversity in Shaded Coffee Systems of Turrialba, Costa Rica." *Agroforestry Systems* 53 (2): 171–77. https://doi.org/10.1023/A:1013324520197.

Rorer, J. B. 1913. *The Green Muscardine Fungus and Its Use in Cane Fields.* Department of Agriculture, Trinidad and Tobago.

Rosenstiel, Robert G., and Edward K. Vaughan. 1952. *Insect Pests and Diseases of Strawberries in Oregon.* Bulletin 419. Oregon State College Agricultural Experiment Station.

Rossetto, C. J., J. F. S. Martins, N. C. Schmidt, and L. E. Azzini. 1978. "Damage of Grass Spittlebugs *Deois flavopicta* and *Deois schach* to Rice." *Bragantia* 37 (1): 15–17.

Rothschild, Miriam. 1961. "Defensive Odours and Mullerian Mimicry among Insects." *Transactions of the Royal Entomological Society of London* 113 (5): 101–23. https://doi.org/10.1111/j.1365-2311.1961.tb00803.x.

Roversi, P. F., and C. Baccetti. 1994. "Ecologia ed etologia di *Haematoloma dorsatum* (Ahrens) (Homoptera, Cercopidae)" *Redia* 77 (1): 133–50.

Rowe, Candy, and Christina Halpin. 2013. "Why Are Warning Displays Multimodal?" *Behavioral Ecology and Sociobiology* 67 (9): 1425–39. https://doi.org/10.1007/s00265-013-1515-8.

Roychoudhury, Nilanjan, K. C. Joshi, and Manish Chourasia. 2003. "Insect Pests of Teak in Madhya Pradesh." *Indian Journal of Tropical Biodiversity* 11: 1–7.

Rudel, Thomas K., Birthe Paul, Douglas White, et al. 2015. "LivestockPlus: Forages, Sustainable Intensification, and Food Security in the Tropics." *Ambio* 44: 685–93. https://doi.org/10.1007/s13280-015-0676-2.

Russell, Bertrand. 1935. *In Praise of Idleness and Other Essays.* New York: W.W. Norton.

Ruxton, Graeme D., and Thomas N. Sherratt 2006. "Aggregation, Defence and Warning Signals: the Evolutionary Relationship." Proceedings of the Royal Society B: Biological Sciences 273 (1600): 2417–24.

Ruxton, Graeme D., Thomas N. Sherratt, and Michael Speed. 2004. *Avoiding Attack: The Evolutionary Ecology of Crypsis, Warning Signals, and Mimicry.* 1st Ed. Oxford University Press.

Ruxton, Graeme D., Michael Speed, and Thomas N. Sherratt. 2004. "Evasive Mimicry: When (If Ever) Could Mimicry Based on Difficulty of Capture Evolve?" *Proceedings of the Royal Society of London. Series B: Biological Sciences* 271 (1553): 2135–42. https://doi.org/10.1098/rspb.2004.2816.

Rydin, Catarina, and Else M. Friis. 2010. "A New Early Cretaceous Relative of Gnetales: *Siphonospermum simplexgen* gen. et sp. nov. from the Yixian Formation of Northeast China." *BMC Evolutionary Biology* 10 (1): 183. https://doi.org/10.1186/1471-2148-10-183.

Sá, Luiz A. N. de. "Cigarrinhas das pastagens (Homoptera, Cercopidae): Distribuição geografica e variabilidade genetica." Mastre em Biologia diss., Universidade Estadual de Campinas, 1981. https://repositorio.unicamp.br/acervo/detalhe/51694.

Sabella, Erika, Alessio Aprile, Alessandra Genga, et al. 2019. "Xylem Cavitation Susceptibility and Refilling Mechanisms in Olive Trees Infected by *Xylella fastidiosa.*" *Scientific Reports* 9 (1): 9602. https://doi.org/10.1038/s41598-019-46092-0.

Sáenz, C., D. Alfaro, A. Roríguez, J. D. Salazar, and R. Oviedo. 1999. "Eficiencia en la captura de adultos de salivazo (Hom: Cercopidae) mediante el uso de trampas adhesivas amarillas en caña de azúcar en Costa Rica." In *XI Congreso Nacional Agronómico y de Recursos Naturales, San José, Costa Rica, LAICA-DIECA, July 1999.*

Safni, Irda, Siti Subandiyah, and Mark Fegan. 2018. "Ecology, Epidemiology and Disease Management of *Ralstonia syzygii* in Indonesia." *Frontiers in Microbiology* 9:Article 419. https://doi.org/10.3389/fmicb.2018.00419.

Sage, Rowan F. 2004. "The Evolution of C_4 Photosynthesis." *New Phytologist* 161: 341–370. https://doi.org/10.1111/j.1469-8137.2004.00974.x.

Sage, Rowan F., and David S. Kubien. 2003. "Quo Vadis C4? An Ecophysiological Perspective on Global Change and the Future of C4 Plants." *Photosynthesis Research* 77 (2): 209–25. https://doi.org/10.1023/A:1025882003661.

Sahayaraj, Kitherian, Balakrishnan Saranya, Samy Sayed, Loko Yêyinou L. Estelle, and Koilraj Madasamy. 2021. "Biofoam of Spittlebug, *Poophilus costalis* (Walker): Preferential Sites, Temperature Regulation, Chemical Composition and Antimicrobial Activity." *Insects* 12 (4): 340. https://doi.org/10.3390/insects12040340.

Saifi, Rayane, Hysen Kokiçi, Hadjer Saifi, et al. 2024. "Does Climate Change Heighten the Risk of *Xylella fastidiosa* Infection?" In *Plant Quarantine Challenges under Climate Change Anxiety,* edited by K. A. Abd-Elsalam and S. M. Abdel-Momen. Springer Nature. https://doi.org/10.1007/978-3-031-56011-8_11.

Saikia, Hiranya, Biplab K. Das, and Jatin Kalita. 2015. "Studies on the Insects Associated with *Calotropis gigantea* in Guwahati City of Assam, India." *The Journal of Zoology Studies* 2 (3): 6–13.

Salguero-Navas, V. E., H. Hugo Hidalgo, and J. Ortega. 2000. "Economic Injury Level for Adult Froghoppers, *Aeneolamia* spp. (Hemiptera: Cercopidae), in Sugarcane in Guatemala." In *Sugarcane Pest Management in the New Millennium*, edited by P. G. Allsopp and W. Suasard. International Society of Sugar Cane Technologists. https://www.cabdirect.org/cabdirect/abstract/20003024783.

San José, J., R. Montes, N. Nikonova, J. Grace, and C. Buendía. 2014. "Effect of the Replacement of a Native Savanna by an African *Brachiaria decumbens* Pasture on the CO2 Exchange in the Orinoco Lowlands, Venezuela." *Photosynthetica* 52 (3): 358–70. https://doi.org/10.1007/s11099-014-0039-4.

Sánchez-Soto, Saul. 1995. "Dos plagas del cacao en el Estado de Tabasco, Mexico (Insecta, Lepidoptera: Noctuidae, Hemiptera: Cercopidae)." *Agrotrópica* 7 (3): 71–74.

Sánchez-Soto, Saul, and A. Sol S. 1998. "Plantas hospederas de tres plagas del Cacao en Tabasco, México." *Agrotropica* 10 (2): 119–22.

Sánchez-Bayo, Francisco, and Kris A. G. Wyckhuys. 2019. "Worldwide Decline of the Entomofauna: A Review of Its Drivers." *Biological Conservation* 232: 8–27. https://doi.org/10.1016/j.biocon.2019.01.020.

Sanderlin, R. S., and K. I. Heyderich-Alger. 2000. "Evidence That *Xylella fastidiosa* Can Cause Leaf Scorch Disease of Pecan." *Plant Disease* 84 (12): 1282–86. https://doi.org/10.1094/PDIS.2000.84.12.1282.

Sanderlin, R. S., and R. A. Melanson. 2010. "Insect Transmission of *Xylella fastidiosa* to Pecan." *Plant Disease* 94 (4): 465–70. https://doi.org/10.1094/PDIS-94-4-0465.

Santos, Rodrigo S., and Samara A. da Silva. 2023. "Registro de *Mahanarva spectabilis* (Distant) (Hemiptera: Cercopidae) em *Urochloa brizantha* cv. Piatã no Estado do Acre, Brasil." *Entomology Beginners*, 4:e060. http://www.alice.cnptia.embrapa.br/handle/doc/1156786.

Santos, Rodrigo S., Carlos M. S. de Andrade, and Gervásio S. Carvalho. 2021. "First Record of *Mahanarva spectabilis* (Distant) (Hemiptera: Cercopidae) in the State of Acre, Brazil." *EntomoBrasilis* 14:e962–e962. https://doi.org/10.12741/ebrasilis.v14.e962.

Saponari, Marioa, Dontao Boscia, Franco Nigro, and G. P. Martelli. 2013. "Identification of DNA Sequences Related to *Xylella fastidiosa* in Oleander, Almond and Olive Trees Exhibiting Leaf Scorch Symptoms in Apulia (Southern Italy)." *Journal of Plant Pathology* 95 (3): 668.

Saponari, Marioa, A. Giampetruzzi, G. Loconsole, Donato Boscia, and P. Saldarelli. 2019. "*Xylella fastidiosa* in Olive in Apulia: Where We Stand." *Phytopathology* 109 (2): 175–86. https://doi.org/10.1094/PHYTO-08-18-0319-FI.

Saponari, Maria, Donato Boscia, Giuseppe Altamura, et al. 2016. "Pilot Project on *Xylella fastidiosa* to Reduce Risk Assessment Uncertainties." *EFSA Supporting Publications* 13 (3): 1013E. https://doi.org/10.2903/sp.efsa.2016.EN-1013.

Saponari, Maria, Giuliana Loconsole, Daniele Cornara, et al. 2014. "Infectivity and Transmission of *Xylella fastidiosa* by *Philaenus spumarius* (Hemiptera: Aphrophoridae) in Apulia, Italy." *Journal of Economic Entomology* 107 (4): 1316–19. https://doi.org/10.1603/EC14142.

Saunders, J. L. 1981. "Cacao Pests in Central America." In *Proceedings 7th International Cocoa Research Conference, Douala, Cameroon, 4–12 November 1979*, 429–32.

Sbordoni, V., S. Forestiero, and M. Würmli. 1985. *Weltenzyklopädie der Schmetterlinge: Arten, Verhalten, Lebensräume*. Südwest-Verlag.

Schenk, H. J., Steven Jansen, and Teemu Hölttä. 2021. "Positive Pressure in Xylem and Its Role in Hydraulic Function." *New Phytologist* 230 (1): 27–45. https://doi.org/10.1111/nph.17085.

Schmidt, Harvey. 2014. "Bald-Faced Hornets." *Blue Jay* 72 (3): 149–51.

Schöbel, Christian, and Gervásio S. Carvalho. 2020. "Niche Modeling of Economically Important *Mahanarva* (Hemiptera, Cercopidae) Species in South and Central America: Are Brazilian Spittlebug Sugarcane Pests Potential Invaders of South and Central America?" *Journal of Economic Entomology* 113 (1): 115–25. https://doi.org/10.1093/jee/toz252.

Schöbel, Christian, and Gervásio S. Carvalho. 2021. "The 'State of Art' of *Mahanarva* (Hemiptera: Cercopidae) Research. An Economically Important New World Spittlebug Genus." *Applied Entomology and Zoology* 56: 299–309. https://doi.org/10.1007/s13355-021-00744-8.

Schöbel, Christian, and Gervásio S. Carvalho, Andressa Paladini, and Renato A. Teixeira. 2023. "The Spittlebug *Mahanarva tristis* as a Senior Synonymy of Four Neotropical Species of *Mahanarva* (Hemiptera: Auchenorrhyncha: Cercopidae), and Considerations about *M. tristis* Subspecies." *Neotropical Entomology* 52 (1): 57–66.

Scholten, Rebecca, Laura Martinez S., Alberto Hornero, Juan A. Navas C., Pablo J. Zarco-Tejada, and Pieter S. A. Beck. 2019. "Monitoring the Impact of *Xylella* on Apulia's Olive Orchards Using MODIS Satellite Data Supported by Weather Data." In *Proceedings of the 2nd European Conference on Xylella fastidiosa, Ajaccio, France, 29–30 October 2019.*

Schuch, Sebastian, Karsten Wesche, and Matthias Schaefer. 2012. "Long-Term Decline in the Abundance of Leafhoppers and Planthoppers (Auchenorrhyncha) in Central European Protected Dry Grasslands." *Biological Conservation* 149 (1): 75–83. https://doi.org/10.1016/j.biocon.2012.02.006.

Schumacher, F. 1915. "Der gegenwärtige Stand unserer Kenntnis von der Homopteren-Fauna der Insel Formosa." *Mitteilungen aus dem Zoologischen Museum in Berlin* 8 (1): 73–134.

Scortichini, Marco, and Domenico Ragno. 2024. "Survey on Resilient Olive Groves Previously Severely Damaged by *Xylella fastidiosa* subsp. *pauca* in Salento (Apulia, Italy)." *Agronomy* 14 (9): 2003. https://doi.org/10.3390/agronomy14092003.

Scott, James A. 1986. *The Butterflies of North America: A Natural History and Field Guide.* Stanford University Press.

Seabra, Sofia G., F. Pina-Martins, E. Marabuto, et al. 2010. "Molecular Phylogeny and DNA Barcoding in the Meadow-Spittlebug *Philaenus spumarius* (Hemiptera, Cercopidae) and Its Related Species." *Molecular Phylogenetics and Evolution* 56 (1): 462–67. https://doi.org/10.1016/j.ympev.2010.03.023.

Seabra, Sofia G., Ana S. B. Rodrigues, Sara E. Silva, et al. 2021. "Population Structure, Adaptation and Divergence of the Meadow Spittlebug, *Philaenus spumarius* (Hemiptera, Aphrophoridae), Revealed by Genomic and Morphological Data." *PeerJ* 9:e11425.

Serio, Francesca, Giovanni Imbriani, Chiara R. Girelli, Pier P. Miglietta, Marco Scortichini, and Francesco P. Fanizzi. 2024. "A Decade after the Outbreak of *Xylella fastidiosa* subsp. *pauca* in Apulia (Southern Italy): Methodical Literature Analysis of Research Strategies." *Plants* 13 (11): 1433. https://doi.org/10.3390/plants13111433.

Servicio de Sanidad Vegetal. 2024. "Situación de *Xylella fastidiosa* en la Comunitat Valenciana." Generalitat Valenciana, Conselleria de Agricultura, Ganidería y Pesca. 19 April 2024. Accessed 8 December 2024. https://portalagrari.gva.es/documents/366567370/368987753/20240419+Situacion+de+Xf+en+la+C+Valenciana+v4.pdf/c55e5f67-eb63-fd5e-a99b-4b61476a462a?t=1713769519023.

Seshu Reddy, K. V., and E. O. Omolo. 1985. "Sorghum Insect Pest Situation in Eastern Africa." In *Proceedings of the International Sorghum Entomology Workshop, Texas A&M University. College Station, Texas, 15–21 July 1984*, edited by V. Kumble. International Crops Research Institute for the Semi-Arid Tropics.

Sevarika, Milos, Gabriele Rondoni, Sonia Ganassi, et al. 2022a. "Behavioural and Electrophysiological Responses of *Philaenus spumarius* to Odours from Conspecifics." *Scientific Reports* 12 (1): 8402. https://doi.org/10.1038/s41598-022-11885-3.

Sevarika, Milos, Andrea Di Giulio, Gabriele Rondoni, Eric Conti, and Roberto Romani. 2022b. "Morpho-Functional Analysis of the Head Glands in Three Auchenorrhyncha Species and Their Possible Biological Significance." *Microscopy and Microanalysis* 28 (6): 2177–87. https://doi.org/10.1017/S1431927622012478.

Severin, Henry H. P. 1950. "Spittle-Insect Vectors of Pierce's Disease Virus. II. Life History and Virus Transmission." *Hilgardia* 19 (11): 357–80.

Shah, Raffique. 2016. "Sugar Industry Was Doomed to Fail: The Truth behind Caroni's Closure." *Wired868* (blog). 27 July 2016. https://wired868.com/2016/07/26/sugar-industry-was-doomed-to-fail-the-truth-behind-caronis-closure/.

Shcherbakov, Dmitry. 2002. "The 270 Million Year History of Auchenorrhyncha (Homoptera)." *Denisia* 4: 29–35.

Shcherbakov, Dmitry E. 1996. "Origin and Evolution of the Auchenorrhyncha As Shown by the Fossil Record." In *Studies on Hemipteran Phylogeny*, edited by C. W. Schaefer. Entomological Society of America. https://doi.org/10.4182/AMYC5234.1996.31.

Shcherbakov, Dmitry E., and Yuri A. Popov. 2002. "Superorder Cimicidea Laicharting, 1781. Order Hemiptera Linne, 1758. The Bugs, Cicadas, Plantlice, Scale Insects, etc." In *History of Insects*, edited by A. P. Rasnitsyn. And D. L. J. Quicke. Kluwer Academic Publishers.

Sherry, Thomas W. 1984. "Comparative Dietary Ecology of Sympatric, Insectivorous Neotropical Flycatchers (Tyrannidae)." *Ecological Monographs* 54 (3): 313–38. https://doi.org/10.2307/1942500.

Sherry, Thomas W., Cody M. Kent, Natalie V. Sánchez, and Çağan H. Şekercioğlu. 2020. "Insectivorous Birds in the Neotropics: Ecological Radiations, Specialization, and Coexistence in Species-Rich Communities." *The Auk* 137 (4): 1–27. https://doi.org/10.1093/auk/ukaa049.

Shih, H. T., and J. T. Yang. 2005. "Supplement on the Species *Aphrophora horizontalis* Kato (Hemiptera: Cercopoidea: Aphrophoridae)." *Journal of the Plant Protection Society* 47 (2): 77–86.

Shih, H. T., H. C. Tsay, and J. T. Yang. 2005. "Notes on the Geographic Distribution and Host Plants for *Poophilus costalis* (Walker, 1851) (Hemiptera: Cercopoidea: Aphrophoridae)." *Plant Protection Bulletin (Taiwan)* 47: 171–78.

Shortman, S. L., S. K. Braman, R. R. Duncan, W. W. Hanna, and M. C. Engelke. 2002. "Evaluation of Turfgrass Species and Cultivars for Potential Resistance to Twolined Spittlebug (Hemiptera: Cercopidae)." *Journal of Economic Entomology* 95 (2): 478–86. https://doi.org/10.1603/0022-0493-95.2.478.

Silva, Jayara D. da Costa, Solange M. de França, Diego O. P. Lopes, Leila L. Dinardo-Miranda, Francisco de A. Neto, and Paulo R. R. Silva. 2020. "Sugarcane Froghoppers in Northeast Brazil." *Arquivos do Instituto Biológico* 87 (1–4): e1192018. https://doi.org/10.1590/1808-1657001192018.

Silva, L. M. da, V. L. dos S. Almeida, A. V. de Almeida, and E. J. Marques. 2009. "História da introdução e controle biológico da cigarrinha da folha *Mahanarva posticata* (Stal) (Hemiptera: Cercopidae) da cana-de-açúcar no Nordeste brasileiro." In *IX Jornada de Ensino, Pesquisa e Extensão–JEPEX 2009. VI Seman Nacional de Ciência e Technologia, 19–23 October 2009*, unpaginated. Recife, Brazil: Universidade Federal Rural de Pernambuco.

Silva, Sara E., Ana S. B. Rodrigues, Eduardo Marabuto, et al. 2015. "Differential Survival and Reproduction in Colour Forms of *Philaenus spumarius* Give New

Insights to the Study of Its Balanced Polymorphism." *Ecological Entomology* 40 (6): 759–66. https://doi.org/10.1111/een.12252.

Silva, S. E. B., A. M. Auad, J. C. Moraes, R. Alvarenga, S. S. Claudino, and T. T. Resende. 2017. "Biological Performance and Preference of *Mahanarva spectabilis* (Hemiptera: Cercopidae) for Feeding on Different Forage Plants." *Journal of Economic Entomology* 110 (4): 1877–85. https://doi.org/10.1093/jee/tox180.

Silva, Thamires F. R. da, Stefan Schwab, Gabriela C. Alves, Albiane C. Dias, and Veronica M. Reis. 2023. "Growth Response and Bacterial Colonization of *Urochloa* Cultivars Inoculated with Different Species of Diazotrophs." *Plant and Soil* 491 (1): 333–53. https://doi.org/10.1007/s11104-023-06120-5.

Singh, Rudra P., H. M. Singh, Hard Baksh, and Om Prakash. 2016. "Influence of Abiotic Factors on Sucking Insect-Pests of Indian Gooseberry (*Emblica officinalis* Gaertn) in Northern Plains of India." *Indian Journal of Horticulture* 73 (1): 124–27.

Singh, Sanjay. 2016. "Guttation: Mechanism, Momentum and Modulation." *The Botanical Review* 82(2): 149–82.

Siwanowicz, Igor, and Malcolm Burrows. 2017. "Three Dimensional Reconstruction of Energy Stores for Jumping in Planthoppers and Froghoppers from Confocal Laser Scanning Microscopy." *eLife* 6:e23824. https://doi.org/10.7554/eLife.23824.

Skinner, Rachel K., Christopher H. Dietrich, Kimberly K. O. Walden, et al. 2020. "Phylogenomics of Auchenorrhyncha (Insecta: Hemiptera) Using Transcriptomes: Examining Controversial Relationships via Degeneracy Coding and Interrogation of Gene Conflict." *Systematic Entomology* 45 (1): 85–113. https://doi.org/10.1111/syen.12381.

Smith, Harlan I. 1925. "Entomology among the Bellacoola and Carrier Indians." *American Anthropologist* 27 (3): 436–40.

Smith, Nigel J. H., Emmanuel A. S. Serrão, Paulo T. Alvim, and Italo C. Falesi. 1995. *Amazonia: Resiliency and Dynamism of the Land and Its People*. Tokyo: United Nations University Press.

Song, Nan, Miao-Miao Wang, Wei-Chao Huang, Zhi-Yi Wu, Renfu Shao, and Xin-Ming Yin. 2024. "Phylogeny and Evolution of Hemipteran Insects Based on Expanded Genomic and Transcriptomic Data." *BMC Biology* 22 (1): 190. https://doi.org/10.1186/s12915-024-01991-1.

Sorias, Leah. 2019. "Caroni Cost Taxpayers $10b since Closure." Trinidad Express Newspapers. 31 January 2019. https://trinidadexpress.com/business/local/caroni-cost-taxpayers-10b-since-closure/article_cded0336-2514-11e9-9b9b-1f0dbb98dbeb.html.

Sosiak, Christine, Pierre Cockx, Pablo A. Suarez, Ryan McKellar, and Phillip Barden. 2024. "Prolonged Faunal Turnover in Earliest Ants Revealed by North American Cretaceous Amber." *Current Biology* 34 (8): 1755–1761. https://doi.org/10.1016/j.cub.2024.02.058.

Soulier, A. 2024. "COOL (Cercopoidea Lists on The Web): A World Knowledge Base Dedicated to Cercopoidea." https://flow.hemiptera-databases.org/cool/.

Sousa, Westefann dos S., Jorge L. C. Silva, Thiago S. Campos, and João V. de Lima Santos. 2021. "Population Dynamics of the Pasture Leafhopper Complex (Hemiptera: Cercopidae) in Brachiaria and Mombasa grass." *Revista Agrogeoambiental* 13 (2): 252–59. https://doi.org/10.18406/2316-1817v13n220211598.

Spadavecchia, Giada, Zeinab Hamouche, Nicola Tagarelli, et al. 2024. "Identifying a Model Host for *Xylella fastidiosa* Transmission by Spittlebugs." *Entomologia Generalis* 44 (5): 1263–74. https://doi.org/10.1127/entomologia/2024/2710.

Stearns, L. A. 1923. "The Family Cercopidae." In *Guide to the Insects of Connecticut. Part IV. The Hemiptera or Sucking Insects of Connecticut*, by W.E. Britton, 206–38. State Geographical and Natural History Survey Bulletin 34. Hartford: State of Connecticut.

Sternberg, Hilgard O. 1973. "Development and Conservation." *Erdkunde* 27 (4): 253–65.

Stewart, Alan J. A., and David R. Lees. 1988. "Genetic Control of Colour/Pattern Polymorphism in British Populations of the Spittlebug *Philaenus spumarius* (L.) (Homoptera: Aphrophoridae)." *Biological Journal of the Linnean Society* 34 (1): 57–79. https://doi.org/10.1111/j.1095-8312.1988.tb01948.x.

Stewart, Alan J. A., and David R. Lees. 1996. "The Colour/Pattern Polymorphism of *Philaenus spumarius* (L.) (Homoptera: Cercopidae) in England and Wales." *Philosophical Transactions of the Royal Society of London. Series B: Biological Sciences* 351 (1335): 69–89. https://doi.org/10.1098/rstb.1996.0005.

Stimmann, M. W., and C. M. Taliaferro. 1969. "Resistance of Selected Accessions of Bermuda Grass to Phytotoxemia Caused by Adult Two-Lined Spittle Bugs." *Journal of Economic Entomology* 62 (5): 1189–90. https://doi.org/10.1093/jee/62.5.1189.

Stokstad, Erik. 2015. "Italy's Olives under Siege." *Science* 348 (6235): 620–620. https://doi.org/10.1126/science.348.6235.620.

Stringer, I., R. Hitchmough, M.-C. Lariviere, et al. 2012. "The Conservation Status of New Zealand Hemiptera." *New Zealand Entomologist* 35 (2): 110–15. https://doi.org/10.1080/00779962.2012.686314.

Strong Jr., Donald R., Earl D. McCoy, and Jorge R. Rey. 1977. "Time and the Number of Herbivore Species: The Pests of Sugarcane." *Ecology* 58 (1): 167–75. https://doi.org/10.2307/1935118.

Stroock, Abraham D., Vinay V. Pagay, Marciej A. Zwieniecki, and N. M. Holbrook. 2014. "The Physicochemical Hydrodynamics of Vascular Plants." *Annual Review of Fluid Mechanics* 46: 615–42.

Stuckey, Joe, and Sarah Stuckey. 2021. "A Brief History of Dairy Farming in Monteverde." In *Monteverde Family Album*, edited by J. Witt and S. Dowell. Asociación de los Amigos de Monteverde. https://monteverdequakers.org/wp-content/uploads/2022/01/FamilyAlbum2021.pdf.

Subba Rao, B. R. 1973. "Description of Two New Species of *Carabunia* Waterston (Hymenoptera: Encyrtidae)." *Oriental Insects* 7 (4): 485–89.

Subba Rao, B. R. 1971. "New Genera and Species of Encyrtids (Hymenoptera: Encyrtidae)." *Journal of Natural History* 5 (2): 209–24. https://doi.org/10.1080/00222937100770121.

Subba Rao, M. S., C. A. Viraktamath, and V. Muniappa. 1988. "Incidence of Leaf Hoppers, Treehoppers and Froghopper (Homoptera) in Sandal Forests of Karnataka in Relation to Sandal Spike Disease." *Annals of Entomology* 6 (2): 25–34.

Sugiura, Shinji, and Kazuo Yamazaki. 2003. "Gall Formation by a Spittlebug, *Aphelaenus nigripectus* (Aphrophoridae: Auchenorrhyncha; Homoptera)." *Entomological Science* 6 (4): 223–28. https://doi.org/10.1046/j.1343-8786.2003.00026.x.

Sujii, Edison R., Maria A. Garcia, and Eliana M. G. Fontes. 2000. "Movimentos de migração e dispersão de adultos da cigarrinha-das-pastagens." *Pesquisa Agropecuária Brasileira* 35: 471–80. https://doi.org/10.1590/S0100-204X2000000300001.

Sujii, Edison R., Maria A. Garcia, Eliana M. G. Fontes, and Robert J. O'Neil. 2002. "Predation as a Mortality Factor in Populations of the Spittlebug, *Deois flavopicta* Stål (Homoptera: Cercopidae)." *Neotropical Entomology* 31 (4): 581–88. https://doi.org/10.1590/S1519-566X2002000400010.

Surano, Antony, Carmine del Grosso, Biagia Musio, et al. 2024. "Exploring the Xylem-Sap to Unravel Biological Features of *Xylella fastidiosa* subspecies *pauca* ST53 in Immune, Resistant and Susceptible Crop Species through Metabolomics and in Vitro Studies." *Frontiers in Plant Science* 14: 1343876. https://doi.org/10.3389/fpls.2023.1343876.

Sutton, G. P., and M. Burrows. 2010. "The Mechanics of Azimuth Control in Jumping by Froghopper Insects." *Journal of Experimental Biology* 213 (9): 1406–16. https://doi.org/10.1242/jeb.036921.

Svanberg, Ivan. 2016. "Cuckoo Spit in Northern European Folk Biology." *Svenska Linnésällskapets Årsskrift* 2016: 117–121.

Szterk, Arkadiusz, Sylwia Flis, Karol Ofiara, and Bartosz Strus. 2024. "Chemical Composition of the Foam Enfolding Juveniles of *Aphrophora alni* (Hemiptera: Aphrophoridae)." *Journal of Asia-Pacific Entomology* 27 (1): 102185. https://doi.org/10.1016/j.aspen.2023.102185.

Szwedo, Jacek. 2016. "The Unity, Diversity and Conformity of Bugs (Hemiptera) through Time." *Earth and Environmental Science Transactions of The Royal Society of Edinburgh* 107 (2–3): 109–28. https://doi.org/10.1017/S175569101700038X.

Szwedo, Jacek, Jowita Drohojowska, Yuri A. Popov, Ewa Simon, and Piotr Wegierek. 2019. "Aphids, True Hoppers, Jumping Plant-Lice, Scale Insects, True Bugs and Whiteflies (Insecta: Hemiptera) from the Insect Limestone (Latest Eocene) of the Isle of Wight, UK." *Transactions of the Royal Society of Edinburgh. Earth and Environmental Science* 110 (3–4): 331–96.

Szwedo, Jacek, B. Wang, and H. Zhang. 2013. "The Eocene Fushun Amber–Known and Unknown." In *International Amber Researcher Symposium. Amber. Deposits-Collections-The Market, 22–23 March 2013, Gdańsk, Poland*, 33–37. Gdańsk International Fair Co. Amberif 2013.

Tanyeri, Rukiye, Selçuk Yurtsever, and Serdar Tezcan. 2023. "Parallel Colour/Pattern Polymorphism in Three Sympatric Spittlebug Species (*Philaenus spumarius*, *P. signatus* and *Mesoptyelus impictifrons*) (Hemiptera: Aphrophoridae) in Mediterranean Türkiye." *Zoology in the Middle East* 69 (4): 384–93.

Taylor, Philip. 2022. "*Xylella*: The Greatest Threat to Australian Agriculture?" *Microbiology Australia* 43 (4): 165–68. https://doi.org/10.1071/MA22055.

Tedders, W. L. 1995. "Identity of Spittlebug on Pecan and Life History of *Clastoptera achatina* (Homoptera: Cercopidae)." *Journal of Economic Entomology* 88 (6): 1641–49. https://doi.org/10.1093/jee/88.6.1641.

Teelucksingh, Daniel. (1997). Comments from the floor on the "Cane Farmers Incorporation and Cess (Amdt.) Bill, 1997." Official Record of the Parliament of Trinidad and Tobago, 16 December 1997, 147–149. https://www.ttparliament.org/wp-content/uploads/2022/01/hs19971216.pdf.

Térano, H. V., I. M. Ramos, J. F. Warumby, A. M. C. L. Netto, A. D. Araújo, and J. V. Oliveira. 1989. "Distribuição e predominância das espécies de cigarrinhas (Homoptera-Cercopidae) associadas com pastagens no Estado de Pernambuco." *Anais da Sociedade Entomológica do Brasil* 18 (1): 21–31.

Terra, Walter R., Clelia Ferreira, and Carlos P. Silva. 2023. "Molecular View of Digestion and Absorption in the Major Insect Orders." In *Molecular Physiology and Evolution of Insect Digestive Systems*, edited by W. R. Terra, C. Ferreira, and C. P. Silva. Springer International Publishing. https://doi.org/10.1007/978-3-031-39233-7_13.

Thompson, Vinton. 1973. "Spittlebug Polymorphic for Warning Coloration." *Nature* 242 (5393): 126–28. https://doi.org/10.1038/242126a0.

Thompson, Vinton. 1984a. "Polymorphism under Apostatic and Aposematic Selection." *Heredity* 53 (3): 677–86. https://doi.org/10.1038/hdy.1984.126.

Thompson, Vinton. 1984b. "Distributional Evidence for Thermal Melanic Color Forms in *Philaenus spumarius*, the Polymorphic Spittlebug." *American Midland Naturalist* 111 (2): 288–95. https://doi.org/10.2307/2425323.

Thompson, Vinton. 1988. "Parallel Colour Form Distributions in European and North American Populations of the Spittlebug *Philaenus spumarius* (L.)." *Journal of Biogeography*, 15 (3): 507–12. https://doi.org/10.2307/2845279.

Thompson, Vinton. 1994. "Spittlebug Indicators of Nitrogen-Fixing Plants." *Ecological Entomology* 19 (4): 391–98. https://doi.org/10.1111/j.1365-2311.1994.tb00257.x.

Thompson, Vinton. 1997. "Spittlebug nymphs (Homoptera: Cercopidae) in *Heliconia* flowers (Zingiberales: Heliconiaceae): Preadaptation and Evolution of the First Aquatic Homoptera." *Revista de Biología Tropical* 45 (2): 905–12. https://revistas.ucr.ac.cr/index.php/rbt/article/view/21075.

Thompson, Vinton. 1999. "Spittlebugs Associated with Actinorhizal Host Plants." *Canadian Journal of Botany* 77 (9): 1387–90. https://doi.org/10.1139/b99-058.

Thompson, Vinton. 2004. "Associative Nitrogen Fixation, C4 Photosynthesis, and the Evolution of Spittlebugs (Hemiptera: Cercopidae) as Major Pests of Neotropical Sugarcane and Forage Grasses." *Bulletin of Entomological Research* 94 (3): 189–200. https://doi.org/10.1079/BER2004293.

Thompson, Vinton. 2011. "Notes on the Biology of *Clastoptera distincta* Doering, the Dwarf Mistletoe Spittlebug (Hemiptera: Cercopoidea: Clastopteridae)." *Cicadina* 12: 81–87. https://opendata.uni-halle.de/bitstream/1981185920/94175/1/cicadina_volume_12_233.pdf.

Thompson, Vinton. 2013. "Spittlebugs in the Genus *Sphenorhina* (Hemiptera: Cercopidae) Associated with Weedy Composite Host Plants (Asteraceae) that May Represent Unusual Cases of Nitrogen Fixation." *Entomotaxonomia* 35 (2): 101–5.

Thompson, Vinton. 2021. "A New Spittlebug of the Genus *Aphrophora* Germar, 1821 (Hemiptera: Cercopoidea: Aphrophoridae) Abundant on Invasive Iceplant, *Carpobrotus Edulis* (L.) N. E. Brown (Aizoaceae), in Coastal California." *The Pan-Pacific Entomologist* 97 (3): 105–28. https://doi.org/10.3956/2021-97.3.105.

Thompson, Vinton. 2022. "Insect-Plant-Fungus Interactions in Mycorrhizal Associations, with a Focus on Spittlebugs and Ectomycorrhizal Host Plants." *Ecological Entomology* 47 (6): 915–29. https://doi.org/10.1111/een.13192.

Thompson, Vinton. 2023a. "New San Francisco Bay Area Spittlebug of the Genus *Clastoptera* Germar, 1839 (Hemiptera: Cercopoidea: Clastopteridae) Makes Unique Mineral-Crusted Spittles." *The Pan-Pacific Entomologist* 99 (2): 111–27. https://doi.org/10.3956/2022-99.2.111.

Thompson, Vinton. 2023b. "Looking for Patterns in a Biosphere Dominated by Symbioses–Bugs, Plants, and the Microorganisms That Mediate Their Interactions." *International Mycorrhiza Society Quarterly E-Newsletter* 4 (1): 6–9.

Thompson, Vinton, and Gervasio S. Carvalho. 2016. "Abrupt Geographical Transition between Aposematic Color Forms in the Spittlebug *Prosapia ignipectus* (Fitch) (Hemiptera: Cercopidae)." *Psyche* 2016:e3623092. https://doi.org/10.1155/2016/3623092.

Thompson, Vinton, Susan Halbert, and Mark Rothschild. 2020. "A New Species of the Spittlebug Genus *Clastoptera* Germar (Hemiptera: Cercopoidea: Clastopteridae) on Florida Oaks." *Insecta Mundi*, September. https://digitalcommons.unl.edu/insectamundi/1288.

Thompson, Vinton, and Olli Halkka. 1973. "Color Polymorphism in Some North American *Philaenus spumarius* (Homoptera: Aphrophoridae) Populations." *The American Midland Naturalist* 89 (2): 348–59. https://doi.org/10.2307/2424039.

Thompson, Vinton, Claire Harkin, and Alan J. A. Stewart. 2023. "The Most Polyphagous Insect Herbivore? Host Plant Associations of the Meadow Spittlebug, *Philaenus spumarius* (L.)." *PLOS ONE* 18 (10): e0291734. https://doi.org/10.1371/journal.pone.0291734.

Thompson, Vinton, and R. León G. 2005. "La identificación y distribución de los salivazos de la caña de azúcar y los pastos (Homoptera: Cercopidae) en Costa Rica." *Manejo Integrado de Plagas y Agroecología (Costa Rica)* (75): 43–51. https://repositorio.catie.ac.cr/handle/11554/6456.

Thompson, Vinton, and Norhashimah Mohd-Saleh. 1995. "Spittle Maggots: Studies on *Cladochaeta* Fly Larvae Living in Association with *Clastoptera* Spittlebug Nymphs." *The American Midland Naturalist* 134 (2): 215–25. https://doi.org/10.2307/2426292.

Thompson, Vinton, and Ruth M. Moscovitch. 2022. "Decline of Melanic Color Forms in Northern Minnesota Populations of *Philaenus spumarius* (L.) (Hemiptera: Aphrophoridae)." *Entomologica Americana* 128 (1–4): 1–10. https://doi.org/10.1664/NYES-D-22-00001.

Tishechkin, D. Y. 2003. "Vibrational Communication in Cercopoidea and Fulgoroidea (Homoptera: Cicadina) with Notes on Classification of Higher Taxa." *Russian Entomological Journal* 12 (2): 139–81.

Tishechkin, D. Y. 2011. "Calling Signals in Sympatric Species of the Far-Eastern *Aphrophora* (Homoptera: Auchenorrhyncha: Aphrophoridae): Regularities of Communication Channel Segregation." *Russian Entomological Journal* 20 (1): 57–64.

Tolotti, Adriana, Wilson S. Azevedo, Victor H. Valiati, and Gervasio S. Carvalho. 2018. *Cigarrinhas-das-pastagens em gramíneas forrageiras no Brasil.* Editora Evangraff.

Tomer, Maor, Liat Gidron-Heinemann, Elad Chiel, and Rakefet Sharon. 2024. "A New Vector of *Xylella fastidiosa*: The Role of *Mesoptyelus impictifrons* as a Vector in Israel." *Phytopathology* 114 (12): 2546–2550. https://doi.org/10.1094/PHYTO-04-24-0111-SC.

Tomlinson, P. B. 2016. *The Botany of Mangroves.* Cambridge University Press.

Tonelli, Mateus. "The Spittlebug *Mahanarva fimbriolata* (Stål) Produces Foam as a Thermoregulatory and Defensive Strategy." DSc diss., University of São Paulo, 2019. https://doi.org/10.11606/T.11.2019.tde-09082019-144717.

Tonelli, Mateus, Simone R. Cotta, Alessandra Rigotto, Armando C. F. Dias, Fernando D. Andreote, and José M. S. Bento. 2020. "The Composition of the Bacterial Community in the Foam Produced by *Mahanarva fimbriolata* Is Distinct from Those at Gut and Soil." *Brazilian Journal of Microbiology* 51 (3): 1151–57.

Tonelli, Mateus, Guilherme Gomes, Weliton D. Silva, et al. 2018. "Spittlebugs Produce Foam as a Thermoregulatory Adaptation." *Scientific Reports* 8 (1): 4729. https://doi.org/10.1038/s41598-018-23031-z.

Tonelli, Mateus, Maria F. G. V. Peñaflor, Luís G. Leite, Weliton D. Silva, Fernanda Martins, and José M. S. Bento. 2016. "Attraction of Entomopathogenic Nematodes to Sugarcane Root Volatiles under Herbivory by a Sap-Sucking Insect." *Chemoecology* 26 (2): 59–66. https://doi.org/10.1007/s00049-016-0207-z.

Trevisan, Olzeno, and Dionatas U. de O. Meneguetti. 2012. "Predação durante a cópula de *Mahanarva fimbriolata* Stål, 1854 (Hemiptera: Cercopidae) por *Montina confusa* Stål, 1859 (Hemiptera: Reduviidae) em posição de acasalamento." *Agro@mbiente On-line* 6 (2): 184–86. https://doi.org/10.18227/1982-8470ragro.v6i2.792.

Tripathi, G. M. 1986. "Record of *Ptyelus nebulosus* (Fab.) (Hemiptera: Cercopidae) as Pest on Sugarcane from U.P." *Bulletin of Entomology* 27: 193–94.

Tripathy, M. K., and M. Rout. 2018. "Diversity of Insect Pests and Their Natural Enemies Infesting Teak (*Tectona grandis*, Verbenaceae) in Coastal Odisha." *International Journal of Current Microbiology and Applied Sciences* 7 (11): 1421–32.

Trjapitzin, V. A., and S. V. Triapitsyn. 2010. "On the Genus *Neocladia* Perkins, 1906 (Hymenotpera: Encyrtidae), with Description of Two New Species." *Russian Entomological Journal* 19 (2): 123–35.

Trotta, Vincenzo, Pierluigi Forlano, Vittoria Caccavo, Paolo Fanti, and Donatella Battaglia. 2021. "A Survey of Potential Vectors of the Plant Pathogenic Bacterium *Xylella Fastidiosa* in the Basilicata Region, Italy." *Bulletin of Insectology* 74 (2): 273–83.

Tsacas, Léonidas, and Guy Couturier. 1993. "Une nouvelle espece de *Cladochaeta* de l'Équateur inquiline des nymphes de *Cephisus erythrocephalus* [Diptera, Drosophilidae; Homoptera, Aphrophoridae]." *Revue Française d'Entomologie (N.S.)* 15: 85–90.

Turner, J. S. 1994. "Anomalous Water Loss Rates from Spittle Nests of Spittlebugs *Aphrophora saratoga* (Homoptera: Cercopidae)." *Comparative Biochemistry and Physiology. A: Comparative Physiology* 107 (4). https://agris.fao.org/search/en/providers/122535/records/647372ef2c1d629bc983df8d.

Turner, J. S. 2000. *The Extended Organism: The Physiology of Animal-Built Structures.* Harvard University Press.

Tyree, Melvin T., and M. H. Zimmermann. 2013. *Xylem Structure and the Ascent of Sap.* Springer.

Urich, F. W. 1913. *The Sugar-Cane Froghopper and Biological Notes on Some Cercopids of Trinidad.* Port-of-Spain, Triniidad: Board of Agriculture, Trinidad and Tobago.

Valenzuela-G., J., A. Lopez M., and A. Garcia B. 1994. "The Activity Pattern and Foraging Habits of *Pachycondyla villosa* (Hymenoptera, Formicidae) in Cocoa Plantations in Soconusco, Chiapas, Mexico." *Folia Entomologica Mexicana* 91: 9–21.

Valério, José R. 1988. "Spittlebugs: Important Pasture Pest in Brazil." *Tymbal,* (12) 14–15.

Valério, José R. 2009. *Cigarrinhas-das-pastagens.* Campo Grande, MS, Brazil: Embrapa Gado de Corte. http://www.infoteca.cnptia.embrapa.br/handle/doc/853370.

Valério, José R., C. Cardona, D. C. Peck, and G. Sotelo. 2001. "Spittlebugs: Bioecology, Host Plant Resistance and Advances in IPM." In *Proceedings of The XIX International Grassland Congress, 11–21 February 2001,* 217–21. São Pedro, São Paulo, Brazil: Fundacao de Estudos Agrarios Luiz Queiroz. https://cgspace.cgiar.org/bitstream/handle/10568/56032/tema5-2.pdf?sequence=1.

Valério, José R., Stephen L. Lapointe, Segment Kelemu, Celso D. Fernandes, and Francisco J. Morales. 1996. "Pests and Diseases of *Brachiaria* Species." In Brachiaria: *Biology, Agronomy, and Improvement, CIAT Publication No. 259,* edited by J. W. Miles, B. L. Maass, C. B. do Valle, and V. Kumble. Centro International de Agricultura Tropical.

Valério, José R., and Octavio Nakano. 1987. "Taxa de excreção e atividade alimentar do adulto da cigarrinha-das-pastagens *Zulia entreriana* (Berg, 1879) (Homoptera: Cercopidae)." *Anais da Sociedade Entomológica do Brasil* 16 (2): 333–40.

Valério, José R., and Octavio Nakano. 1988. "Danos causados pelo adulto da cigarrinha *Zulia entreriana* na produção e qualidade de *Brachiaria decumbens.*" *Pesquisa Agropecuária Brasileira* 23 (5): 447–53.

Valério, José R., and Marlene C. M. Oliveira. 2005. "Parasitism of Spittlebug Eggs (Homoptera: Cercopidae) by the Microhymenopteran *Anagrus urichi* Pickles (Hymenoptera: Mymaridae) in Campo Grande, MS, Brazil." *Neotropical Entomology* 34 (1): 137–38.

Valério, José R., Fabricia Z. V. Torres, Ewerton C. Lira, Glenda M. Weis, W. V. F. de Paula, and Marlene C. M. Oliveira. 2015. "Seleção de híbridos da gramínea forrageira *Panicum maximum* resistente à cigarrinha-das-pastagens *Notozulia entreriana* (Hemiptera: Cercopidae)." In *52a Reunião Anual da Sociedade Brasileira de Zootecnia, Belo Horizonte, MG, Brazil, 19–23 July 2015.*

Valle, Cacilda B. do, Valeria P. B. Euclides, Denise B. Montagner, et al. 2017. *BRS Ipyporã ('Belo Começo' em Guarani): Híbrido de* Brachiaria *da Embrapa. Communicado Technico 137.* Embrapa. http://www.infoteca.cnptia.embrapa.br/handle/doc/1069732.

Valle Ramírez, Segundo, Miguel I. Cruz, Ana P. Arias, et al. 2015. "Evaluación de dos métodos de monitoreo de *Mahanarva andigena* Jacobi en el cultivo de la caña de azúcar (*Saccharum* spp., Híbrido) en la Provincia de Pastaza, Ecuador." *Revista de Protección Vegetal* 30 (3): 185–92.

Valle-Ramírez, Segundo B., Roldán Torres-Gutiérrez, William O. Caicedo-Quinche, Ricardo V. Abril-Saltos, and Darwin J. Sucoshañay-Villalba. 2022. "Aislamiento y caracterización de *Metarhizium* spp. de cultivos de caña de azúcar y su patogenicidad contra *Mahanarva andigena* (Hemiptera: Cercopidae)." *Ciencia y Tecnología Agropecuaria* 23 (1): e2361. http://www.scielo.org.co/scielo.php?pid=S0122-87062022000100019&script=sci_arttext.

van Klink, Roel, Diana E. Bowler, Konstantin B. Gongalsky, Minghua Shen, Scott R. Swengel, and Jonathan M. Chase. 2024. "Disproportionate Declines of Formerly Abundant Species Underlie Insect Loss." *Nature* 628 (8007): 359–64. https://doi.org/10.1038/s41586-023-06861-4.

Van Dinther, J. B. M. 1960. *Insect Pests of Cultivated Plants in Surinam. Landbouwproefstation in Suriname, Bulletin No. 76.* Wageningen, Netherlands: Venman & Zonen.

Vargas P., O. 1970. "Estudio sobre la baba de culebra, *Prosapia distanti* (Homoptera: Cercopidae) y un ensayo sobre su combate en el pasto kikuyo (*P. clandestinium* Hochst.)." Tesis de grado Ingeniero agrónomo, Universidad de Costa Rica.

Velasco P., H. 1975. "Influencía de la quema del pasto pangola sobre la población de Mosca Pinta y otros insectos." *Agricultura Técnica en México* 2: 357–60.

Verbeek, N. A. M. 1998. "Food of Nestling Northwestern Crows." *The Wilson Bulletin* 110 (4): 483–88.

Veríssimo, Bruno A., Alexander M. Auad, Sandra E. B. Silva, and Giani Batista da Silva. 2018. "Biology and Olfactory Response of *Salpingogaster nigra* Schiner (Diptera: Syrphidae)." *Florida Entomologist* 101 (4): 702–4. https://doi.org/10.1653/024.101.0412.

Verma, G. C., D. Singh, and M. Ramzan. 1977. "Occurrence of Spittle Bug on Sorghum." *Entomologists' Newsletter* 7: 4–5.

Vicente-Díez, Ignacio, Rubén Blanco-Pérez, María del M. González-Trujillo, Alicia Pou, and Raquel Campos-Herrera. 2021. "Insecticidal Effect of Entomopathogenic Nematodes and the Cell-Free Supernatant from Their Symbiotic Bacteria against *Philaenus spumarius* (Hemiptera: Aphrophoridae) Nymphs." *Insects* 12 (5): 448. https://doi.org/10.3390/insects12050448.

Villa, María, Isabel Rodrigues, Paula Baptista, Alberto Fereres, and José A. Pereira. 2020. "Populations and Host/Non-Host Plants of Spittlebugs Nymphs in Olive Orchards from Northeastern Portugal." *Insects* 11 (10): 720. https://doi.org/10.3390/insects11100720.

Vogel, Steven. 2005. "Living in a Physical World II. The Bio-Ballistics of Small Projectiles." *Journal of Biosciences* 30 (2): 167–75.

Vreugdenhil, A. 1981. "Froghopper, *Aeneolamia varia*, on Sugarcane in the West-Central Region of Venezuela." In *Proceedings, Second Inter-American Sugar Cane Seminar: Insect and Rodent Pests, Miami, Florida, 6–7 October 1981,25–35.* Miami, FL: Inter-American Sugar Cane Seminar.

Wagner, David L. 2020. "Insect Declines in the Anthropocene." *Annual Review of Entomology* 65 (1): 457–80. https://doi.org/10.1146/annurev-ento-011019-025151.

Wagner, Wilhelm. 1968. "Eine afrikanische Schaumzikade, *Ptyelus grossus* Fabricius, 1781, fossilis, aus dem Pliozän von Willershausen." *Beihefte zu den Berichten der Naturhistorischen Gesellschaft zu Hannover* 6: 21–24.

Walker, Nancy C., Steven M. White, Dan McKay Fletcher, et al. 2023. "The impact of xylem geometry on olive cultivar resistance to *Xylella fastidiosa*: An image-based study." *Plant Pathology* 72(3): 521–35.

Wang, Bo, Gongle Shi, Chunking Xu, et al. 2021. "The Mid-Miocene Zhangpu Biota Reveals an Outstandingly Rich Rainforest Biome in East Asia." *Science Advances* 7 (18): eabg0625.

Wang, Bo, Jacek Szwedo, and Haichun Zhang. 2012. "New Jurassic Cercopoidea from China and Their Evolutionary Significance (Insecta: Hemiptera)." *Palaeontology* 55 (6): 1223–43. https://doi.org/10.1111/j.1475-4983.2012.01185.x.

Wang, Bo, Jacek Szwedo, Haichun Zhang, and Yan Fang. 2010. "The Major Diversification of Cicadomorpha in the Jurassic (Insecta: Hemiptera)." *Earth Science Frontiers* 17: 186–87.

Wang, Qiqi, Lingyan Bai, Xu Chen, and Aiping Liang. 2021. "Laboratory Rearing of *Abidama liuensis* (Hemiptera: Cercopidae) and Description of Immature Stages." *EJE* 118 (1): 82–89. https://doi.org/10.14411/eje.2021.010.

Wang, Tingting, Liuxing Pan, Yalin Zhang, and Wu Dai. 2015. "Morphology of the Mouthparts of the Spittlebug *Philagra albinotata* Uhler (Hemiptera: Cercopoidea: Aphrophoridae)." *Arthropod Structure & Development* 44 (2): 121–30. https://doi.org/10.1016/j.asd.2014.12.001.

Wappler, Torsten, Friðgeir Grímsson, Bo Wang, et al. 2014. "Before the 'Big Chill': A Preliminary Overview of Arthropods from the Middle Miocene of Iceland (Insecta, Crustacea)." *Palaeogeography, Palaeoclimatology, Palaeoecology* 401: 1–12. https://doi.org/10.1016/j.palaeo.2014.03.006.

Washbon, Edwin C. "The Biology and Control of Two Species of Spittlebugs, Genus *Prosapia*, (Homoptera: Cercopidae) in South Florida and Costa Rica." PhD diss., University of Florida, 1968.

Washington, E. S. 2001. "Cover Stories: New Kings of Coke." *Time*, 24 June 2001. https://content.time.com/time/magazine/article/0,9171,157350,00.html.

Weaver, C. R., and D. R. King. 1954. *Meadow Spittlebug, Philaenus leucophthalmus (L.). Research Bulletin 741*. Wooster, Ohio: Ohio Agricultural Experiment Station.

Webber, R. T. 1927. "Some Important Economic Insects of Central Europe." *Journal of Economic Entomology* 20: 301–10.

Webster, Francis M. 1892. "Insects Affecting the Blackberry and Raspberry." *Bulletin of the Ohio Agricultural Experiment Station* 5 (8): 151–217.

Wells, John M., Boligala C. Raju, Hsueh-Yun Hung, William G. Weisburg, Linda Mandelco-Paul, and Don J. Brenner. 1987. "*Xylella fastidiosa* gen. nov., sp. nov: Gram-Negative, Xylem-Limited, Fastidious Plant Bacteria Related to *Xanthomonas* spp." *International Journal of Systematic and Evolutionary Microbiology* 37 (2): 136–43. https://doi.org/10.1099/00207713-37-2-136.

Wheelwright, N. T., and James D. Rising. 1993. "Savannah Sparrow (*Passerculus sandwichensis*)." In volume 2 of *The Birds of North America*, edited by A. Poole, and F. Gill. Academy of Natural Sciences and American Ornithologists' Union. https://doi.org/10.2173/tbna.45.p.

Whitaker Jr., John O., Chris Maser, Robert M. Storm, and Joseph J. Beatty. 1986. "Food Habits of Clouded Salamanders (*Aneides ferreus*) in Curry County, Oregon (Amphibia: Caudata: Plethodontidae)." *The Great Basin Naturalist* 46 (2): 228–40. https://www.jstor.org/stable/41712217.

Whittaker, John B. 1965. "The Distribution and Population Dynamics of *Neophilaenus lineatus* (L.) and *N. exclamationis* (Thun.) (Homoptera, Cercopidae) on Pennine Moorland." *Journal of Animal Ecology* 34 (2): 277–97. https://doi.org/10.2307/2651.

Whittaker, John B. 1969. "Quantitative and Habitat Studies of the Frog-Hoppers and Leaf-Hoppers (Homoptera, Auchenorhyncha) of Wytham Woods, Berkshire." *Entomologist's Monthly Magazine* 105: 27–37.

Whittaker, John B. 1970. "Cercopid Spittle as a Microhabitat." *Oikos* 21 (1): 59–64. https://doi.org/10.2307/3543839.

Whittaker, John B. 1971a. "The Parasitisation of Colour Polymorphism of *Philaenus spumarius* (L.) and of *Neophilaenus lineatus* (L.) (Homoptera) by a Pipunculid." In *Proceedings of the XIII International Congress of Entomology, Moscow, 2–9 August 1968*, 1: 578–79.

Whittaker, John B. 1971b. "Population Changes in *Neophilaenus lineatus* (L.) (Homoptera: Cercopidae) in Different Parts of Its Range." *Journal of Animal Ecology* 40 (2): 425–43. https://doi.org/10.2307/3253.

Whittaker, John B. 1972. "Polymorphism in *Philaenus spumarius* (Homoptera) in the USSR." *Oikos* 23 (3): 366–69. https://doi.org/10.2307/3543176.

Whittaker, John B. 1973. "Density Regulation in a Population of *Philaenus spumarius* (L.) (Homoptera: Cercopidae)." *Journal of Animal Ecology* 42 (1): 163–72. https://doi.org/10.2307/3410.

Whittaker, John B. 1999. "Impacts and Responses at Population Level of Herbivorous Insects to Elevated CO2." *European Journal of Entomology* 96 (2): 149–56.

Whittaker, John B. 2001. "Presidential Address: Insects and Plants in a Changing Atmosphere." *Journal of Ecology* 89 (4): 507–18. https://www.jstor.org/stable/3072207.

Whittaker, John B., and Nicholas P. Tribe. 1996. "An Altitudinal Transect as an Indicator of Responses of a Spittlebug (Auchenorrhyncha: Cercopidae) to Climate Change." *European Journal of Entomology* 93 (3): 319–24.

Whittaker, John B., and Nicholas P. Tribe. 1998. "Predicting Numbers of an Insect (*Neophilaenus lineatus*: Homoptera) in a Changing Climate." *Journal of Animal Ecology* 67 (6): 987–91.

Wiedijk, F. "Variability in the Occurrence of the Sugar Cane Froghopper, *Aeneolamia flavilatera* (Homoptera: Cercopidae), on Sugar Estates in Guyana and Surinam." PhD diss., Wageningen University, 1982.

Wiegert, Richard G. 1964a. "Population Energetics of Meadow Spittlebugs (*Philaenus spumarius* L.) as Affected by Migration and Habitat." *Ecological Monographs* 34 (3): 218–41. https://doi.org/10.2307/1948501.

Wiegert, Richard G. 1964b. "The Ingestion of Xylem Sap by Meadow Spittlebugs, *Philaenus spumarius* (L.)." *The American Midland Naturalist* 71 (2): 422–28. https://doi.org/10.2307/2423300.

Wigglesworth, Vincent. 1982. "Carrington Bonsor Williams, 7 October 1889–12 July 1981." *Biographical Memoirs of Fellows of the Royal Society* 28: 666–84. https://doi.org/10.1098/rsbm.1982.0025.

Wijeweera, Wijeweera P. S. N., K. A. D. W. Senaratne, Kunjithapatham Dhileepan, and M. P. K. S. K. de Silva. 2022. "Insect Diversity on *Calotropis gigantea* (L.) in Sri Lanka." *Ceylon Journal of Science* 51 (2): 121–28.

William, S. J., and K. S. Ananthasubramanian. 1991. "Host Preference and Feeding Habits of the Spittle Bug, *Clovia quadridens* (Walker) (Homoptera: Cercopidae)." *Journal of Ecobiology* 3 (3): 245–47.

Williams, Andrew H. 2004. "Feeding Records of True Bugs (Hemiptera: Heteroptera) from Wisconsin." *The Great Lakes Entomologist* 37 (1–2). https://doi.org/10.22543/0090-0222.2096.

Williams, C. B. 1917. "Notes on a Froghopper Attacking Sugar-Cane at Marienburg Estate, Surinam." *Bulletin of Entomological Research* 7 (3): 271–72. https://doi.org/10.1017/S0007485300017600.

Williams, C. B. 1918. "A Froghopper on Sugar-Cane in British Guiana." *Bulletin of Entomological Research* 9 (2): 163–73. https://doi.org/10.1017/S0007485300037901.

Williams, C. B. 1921. *Report on the Froghopper-Blight of Sugar Cane in Trinidad.* Trinidad, B.W.I: Government Printing Office. https://doi.org/10.5962/bhl.title.15427.

Williams, C. B. 1922. "A Froghopper Damaging Cacao in Panama." *Bulletin of Entomological Research* 13 (3): 271–74. https://doi.org/10.1017/S0007485300045351.

Williams, C. B. 1931. "Entomology in the Tropics. Some Entomological Experiences in the Tropics." *Tropical Agriculture* 8 (5): 119–23.

Wilson, Curtis M., John K. Stewart, and David H. Vail. 1979. "Full Season Impact of the Alfalfa Weevil, Meadow Spittlebug, and Potato Leafhopper in an Alfalfa Field." *Journal of Economic Entomology* 72 (6): 830–34. https://doi.org/10.1093/jee/72.6.830.

Wilson, H. A., and C. K. Dorsey. 1957. "Studies on the Composition and Microbiology of Insect Spittle." *Annals of the Entomological Society of America* 50 (4): 399–406.

Wilson, Louis F. 1987. *Saratoga Spittlebug: Its Ecology and Management.* Agriculture Handbook No. 657. US Department of Agriculture, Forest Service.

Wilson, Louis F. 1991. *Pine Spittlebug: Its Ecology and Management.* Agriculture Handbook No. 695. US Department of Agriculture, Forest Service.

Wilson, Marisa, and Vishala Parmasad. 2014. "Political Economies of Sugar: Views from a Former Sugar Cane Industry." In *Dietary Sugars and Health: From Biology to Policy*, edited by M. Goran, L. Tappy, K. -A. Le, S. Ulijaszek. Taylor & Francis. https://www.research.ed.ac.uk/en/publications/political-economies-of-sugar-views-from-a-former-sugar-cane-indus.

Wilson, Mark V. H. 1977. "New Records of Insect Families from the Freshwater Middle Eocene of British Columbia." *Canadian Journal of Earth Sciences* 14 (5): 1139–55. https://doi.org/10.1139/e77-104.

Wilson, P. N., D. W. Fawkes, and M. G. Emsley. 1962. "Note on a Heavy Infestation of Pangola Grass (*Digitaria decumbens* Stent) by the Sugar-Cane Froghopper (*Aeneolamia varia saccharina* Distant)." *Tropical Agriculture* 39 (1): 49–51. https://journals.sta.uwi.edu/ojs/index.php/ta/article/view/4041.

Wilson, Shannon, Mark S. Thorne, Melissa A. Johnson, Daniel C. Peck, and Mark G. Wright. 2024. "*Prosapia bicincta* (Hemiptera: Cercopidae) Abundance, Plant Associations, and Impacts on Groundcover in Hawai'i Island Rangelands." *Environmental Entomology*, 53 (5): 870–880.

Wilson, Shannon, Mark S. Thorne, Mark G. Wright, Daniel C. Peck, Jennifer Mack, Glen K. Fukumoto, and R. T. Curtiss. 2023. "The Twolined Spittlebug (Hemiptera: Cercopidae) Invades Hawai'i: Establishment, Biology, and Management of a Destructive Forage Grass Pest." *Journal of Integrated Pest Management* 14 (1): pmad023.

Wingate, David B. 1965. "Terrestrial Herpetofauna of Bermuda." *Herpetologica* 21 (3): 202–18.

Wise, Michael J. 2010. "*Limenitis arthemis astyanax* (Fabricius) (Nymphalidae) Sipping Cercopid-Spittle on Tall Goldenrod, *Solidago altissima* L. (Asteraceae)." *Journal of the Lepidopterists' Society* 64 (4): 217–18. https://doi.org/10.18473/lepi.v64i4.a8.

Wise, Michael J., Dwight L. Kieffer, and Warren G. Abrahamson. 2006. "Costs and Benefits of Gregarious Feeding in the Meadow Spittlebug, *Philaenus spumarius.*" *Ecological Entomology* 31 (5): 548–55. https://doi.org/10.1111/j.1365-2311.2006.00814.x.

Withycombe, C. L. 1926. "Studies on the Aetiology of Sugar-Cane Froghopper Blight in Trinidad." *Annals of Applied Biology* 13 (1): 64–108. https://doi.org/10.1111/j.1744-7348.1926.tb04253.x.

Wolcott, George N. 1913. "Report on a Trip to Demerara, Trinidad and Barbados during the Winter of 1913." *Journal of Economic Entomology* 6 (6): 443–57.

Wolcott, George N. 1948. "The Insects of Puerto Rico." *Journal of Agriculture of the University of Puerto Rico* 32 (1): 1–224.

Xie, Y. 1994. "Studies on the Biology of *Hindoloides sparsuta* Jacobi (Homoptera: Machaerotidae)." *Scientia Silvae Sinicae* 30 (5): 445–450.

Xu, R. M. 1986. "Occurrence and Control of the Red-Spotted Black Spittlebug." *Entomological Knowledge* 23 (3): 102–4.

Xu, Xiao-Ting, Jack Szwedo, Di-Ying Huang, et al. 2022. "A New Genus of Spittlebugs (Hemiptera, Cercopidae) from the Eocene of Central Tibetan Plateau." *Insects* 13 (9): 770. https://doi.org/10.3390/insects13090770.

Yaghmaee, F. 2008. "Assessment of Behaviourial Characteristics of *Poophilus nebulosus* Leth Spittlebug on *Alhagi persarum* Boiss and Buhse Camel Thorn Plant in Torbate Jam Region, Khorasan Razavi Province." *Research Journal of Biological Sciences* 3 (8): 807–11.

Yamazaki, Kazuo. 2017. "White Plant Shoots, Wax-Producing Insects and Other White Structures Made by Arthropods: A Mimicry Complex?" *European Journal of Entomology* 114: 342–49.

Yan, H. X., J. W. Zheng, X. F. Wang, J. Z. Cuyi, and H. P. Li. 2019. "Exploring the Potential of a Strain of Entomopathogenic *Pestalotiopsis* spp. in Controlling *Aphrophora flavipes*." *International Journal of Agriculture and Biology* 21 (1): 115–19.

Yang, Han, Chung-Ping Lin, and Ai-Ping Liang. 2016. "Phylogeography of the Rice Spittle Bug (*Callitettix versicolor*) Implies Two Long-Term Mountain Barriers in South China." *Zoological Science* 33: 592–602. https://doi.org/10.2108/zs160042.

Yang, Jinlong, Joseph M. Michaud, Steven Jansen, H. Jochen Schenk, and Yi Y. Zuo. 2020. "Dynamic Surface Tension of Xylem Sap Lipids." *Tree Physiology* 40 (4): 433–44. https://doi.org/10.1093/treephys/tpaa006.

Yoon, Sunhee, and Wang-Hee Lee. 2023. "Assessing Potential European Areas of Pierce's Disease Mediated by Insect Vectors by Using Spatial Ensemble Model." *Frontiers in Plant Science* 14: 1209694. https://doi.org/10.3389/fpls.2023.1209694.

Yousef-Yousef, Meelad, Marina Morente, Natalia González-Mas, Alberto Fereres, Enrique Quesada-Moraga, and Aranzazu Moreno. 2023. "Direct and Indirect Effects of Two Endophytic Entomopathogenic Fungi on Survival and Feeding Behaviour of Meadow Spittlebug *Philaenus spumarius*." *Biological Control* 186: 105348.

Yurtsever, Selçuk. 2000. "On the Polymorphic Meadow Spittlebug, *Philaenus spumarius* (L.) (Homoptera: Cercopidae)." *Turkish Journal of Zoology* 24 (4): 447–60.

Zajac, M. A., and F. R. Hall. 1986. "Development of an Action Threshold for Meadow Spittlebug on Strawberries." In *Fruit Crops 1986: A Summary of Research, Research Circular 290*, 42–44. Wooster, OH: The Ohio State University, Ohio Agricultural Research and Development Center.

Zecharia, Noa, Helena Krasnov, Miri Vanunu, et al. 2022. "*Xylella fastidiosa* Outbreak in Israel: Population Genetics, Host Range, and Temporal and Spatial Distribution Analysis." *Phytopathology* 112 (11): 2296–2309. https://doi.org/10.1094/PHYTO-03-22-0105-R.

Zhang, Jisen, Qing Zhang, Leiting Li, et al. 2019. "Recent Polyploidization Events in Three *Saccharum* Founding Species." *Plant Biotechnology Journal* 17 (1): 264–74. https://doi.org/10.1111/pbi.12962.

Zhang, Liyu, Meiling Zhang, Shuyu Huang, et al. 2022. "A Highly Conserved Core Bacterial Microbiota with Nitrogen-Fixation Capacity Inhabits the Xylem Sap in Maize Plants." *Nature Communications* 13 (1): 3361. https://doi.org/10.1038/s41467-022-31113-w.

Zhang, Qian-Qi, Jun Chen, and Hai-Chun Zhang. 2022. "The First Procercopidae (Insecta: Hemiptera) from the Jurassic Sangonghe Formation of Northwestern China." *Palaeontographica Abteilung A* 322 (5–6): 117–24. https://doi.org/10.1127/pala/2022/0126.

Zhang, Xiao, Hong Chen, Xu Chen, and Aiping Liang. 2024. "Genomic and Transcriptomic Insights into the Genetic Basis of Foam Secretion in Rice Spittlebug *Callitettix versicolor.*" *International Journal of Molecular Sciences* 25 (4): 2012.

Zhang, X. M., Z. L. Han, and Z. K. Yang. 1996. "Biological Characteristics and Prevention of an Aphrophorid Spittlebug." *Entomological Knowledge* 33 (1): 31–33.

Zhong, Haiying, Yalin Zhang, and Cong Wei. 2013. "Anatomy and Fine Structure of the Alimentary Canal of the Spittlebug *Lepyronia coleoptrata* (L.) (Hemiptera: Cercopoidea)." *Arthropod Structure & Development* 42 (6): 521–30. https://doi.org/10.1016/j.asd.2013.04.005.

Zhong, Haiying, Yalin Zhang, and Cong Wei. 2015. "Morphology of the Alimentary Canal of the Leafhopper *Cicadella viridis* (Hemiptera: Cicadellidae)." *Annals of the Entomological Society of America* 108 (1): 57–69. https://doi.org/10.1093/aesa/sau006.

Zhou, T. Y., and G. Q. Zhong. 1986. "A Preliminary Study on *Taihorina* sp." *Entomological Knowledge* 23 (1): 16–18.

Zhu, Qing, Nan Wu, Jolanta Brożek, and Wu Dai. 2019. "Antennal Morphology and Sexual Dimorphism of Antennal Sensilla in *Callitettix versicolor* (Fabricius) (Hemiptera: Cercopidae)." *Insects* 10 (2): 56. https://doi.org/10.3390/insects10020056.

Zimmermann, Ulrich, Heike Schneider, Lars H. Wegner, and Axel Haase. 2004. "Water Ascent in Tall Trees: Does Evolution of Land Plants Rely on a Highly Metastable State?" *New Phytologist* 162 (3): 575–615.

Zolda, Pamela, and Werner E. Holzinger. 2002. "Prey Use in Two Sympatric Species of the Digger Wasp Genus *Bembecinus* (Hymenoptera: Apoidea: Sphecidae)." *Entomologia Generalis* 26 (2): 101–6.

Index

Note: Page numbers preceded by p refer to plates. Page numbers followed by *t* and *f* refer to tables and figures, respectively.